Molecular Genetics and Evolution of Pesticide Resistance

ACS SYMPOSIUM SERIES 645

Molecular Genetics and Evolution of Pesticide Resistance

Thomas M. Brown, EDITOR
Clemson University

Developed from a symposium sponsored
by the ACS Division of Agrochemicals
at The Big Sky Conference VI

American Chemical Society, Washington, DC

Library of Congress Cataloging-in-Publication Data

Molecular genetics and evolution of pesticide resistance / Thomas M. Brown, editor.

p. cm.—(ACS symposium series, ISSN 0097–6156; 645)

"Developed from a symposium sponsored by the Division of Agrochemicals at the Big Sky Conference VI, held from June 18–23, 1995, at Yellowstone Conference Center and Huntley Lodge in Big Sky, Montana.

Includes bibliographical references and indexes.

ISBN 0–8412–3453–1

1. Pesticide resistance—Congresses. 2. Plants—Disease and pest resistance—Genetic aspects—Congresses.

I. Brown, Thomas M. (Thomas Miller) II. American Chemical Society. Division of Agrochemicals. III. Series

SB957.M64 1996
632′.95042—dc20 96–29234
CIP

This book is printed on acid-free, recycled paper.

PRINTED IN THE UNITED STATES OF AMERICA

Advisory Board

ACS Symposium Series

Foreword

THE ACS SYMPOSIUM SERIES was first published in 1974 to provide a mechanism for publishing symposia quickly in book form. The purpose of this series is to publish comprehensive books developed from symposia, which are usually "snapshots in time" of the current research being done on a topic, plus some review material on the topic. For this reason, it is necessary that the papers be published as quickly as possible.

Before a symposium-based book is put under contract, the proposed table of contents is reviewed for appropriateness to the topic and for comprehensiveness of the collection. Some papers are excluded at this point, and others are added to round out the scope of the volume. In addition, a draft of each paper is peer-reviewed prior to final acceptance or rejection. This anonymous review process is supervised by the organizer(s) of the symposium, who become the editor(s) of the book. The authors then revise their papers according to the recommendations of both the reviewers and the editors, prepare camera-ready copy, and submit the final papers to the editors, who check that all necessary revisions have been made.

As a rule, only original research papers and original review papers are included in the volumes. Verbatim reproductions of previously published papers are not accepted.

ACS BOOKS DEPARTMENT

Contents

Resistance Management Strategies

Ecology, Transgenesis, and Regulatory Issues

Indexes

Preface

PESTICIDE RESISTANCE is a simple phenomenon in which Darwinian evolution, catalyzed by our sometimes overzealous attempts to control pests with pesticides, results in pest populations that are no longer controlled by the doses originally applied. I was once asked, "How can resistance to a new pesticide be prevented?" The simple answer is not to use the new pesticide. Pesticides kill, and as such are potent selecting agents that leave the most resistant phenotype.

It may be possible to manage resistance to obtain the optimum benefits from the new pesticide. This was probably done to the best practical extent in Australia with the introduction of pyrethroid insecticides against *Helicoverpa armigera*. Although a reduction of selection succeeded in extending the lives of these safe and potent insecticides and produced great benefits to the growers, resistance continues to evolve slowly.

Improvement is likely if we apply the new tools of molecular biology. These tools may allow much greater sensitivity and accuracy in detecting resistance genes and may yield much more information on the early stages of the evolution of resistance. Such data are very scarce at present and are critically needed to formulate better models of the phenomenon. Of course, this enhanced capability of detecting resistance must be coupled with practical trials in managing resistance and verification of more detailed models.

The problem of resistance provides a fascinating topic for research in agriculture and public health. The most recent challenges are the threats posed against valuable new products such as the extraordinarily effective sulfonylurea herbicides and the new transgenic crops (called plant/pesticides) armed with *Bacillus thuringiensis* toxin to control major insect pests. The point should not be missed that these new technologies have been thrust onto the front lines by resistance to the preceding pesticides.

In recent years, the resistance phenomenon has been turned to an advantage by exploiting resistance genes in herbicide-resistant transgenic crops. This is similar to the older technology of insecticide-resistant beneficial arthropods, such as predatory mites, selected for acaride resistance and employed to control orchard pests.

As resistance to pesticides, antibiotics, and other drugs continued to pose a serious threat to mankind, the American Chemical Society convened a meeting of experts in the fields of bactericides, fungicides, herbicides, and insecticides. This ACS Agrochemicals Division Special Conference VI, "Molecular Genetics and Ecology of Pesticide Resistance", was held on June 18–23, 1995, in Big Sky, Montana.

Papers presented and ideas generated at Special Conference VI serve as the nucleus for this book. Because of the current explosion of information in molecular biology, it was impossible to include all the topics of interest to several disciplines within the scope of the meeting and the book. Instead of attempting comprehensive coverage, we provide a wide sampling of the most important discoveries in each area of pests and pesticides, including a comparison of antibiotic and pesticide resistance. As this approach stimulated lively discussion at the meeting, the goal of the book is to present interesting topics across several disciplines, all challenged by resistance.

Acknowledgements

We thank Thomas J. McCoy of Montana State University for providing a welcome to our international group at the meeting and William Dyer and Bruce Maxwell, who assisted with local arrangements.

Planning the meeting was a task tackled by a Program Committee of experts in the control of pest bacteria, fungi, weeds, and insects. As Chairman of that committee, I thank them for the countless hours they spent responding to questionnaires, listing possible speakers, and discussing which topics to include or exclude. They are as follows:

Carol Bender	Oklahoma State University
Henri Darmency	Institut National de la Recherche Agronomique
Richard H. ffrench-Constant	University of Wisconsin
Linda Field	Rothamsted Experiment Station
Makato Fujimura	Sumitomo Chemical Company, Ltd.
Jonathan Gressel	Weizmann Institute of Science
Robert M. Hollingworth	Michigan State University
Hideo Ishii	Fruit Tree Research Station, MAFF
Wolfram Koller	New York State Agricultural Experiment Station
Leonard L. Saari	DuPont Company

Once the speakers had been chosen and invited, we sought the financial support necessary to assemble an international group of experts. We gratefully acknowledge the following organizations for support of Special Conference VI:

- Cotton Incorporated
- the Fungicide Resistance Action Committee (FRAC)
- the Herbicide Resistance Action Committee (HRAC)
- the Insecticide Resistance Action Committee (IRAC)

Special thanks are due to Nancy Ragsdale, Don Baker, and Barry Cross, who, as successive chairpersons of the Division of Agrochemicals, were very supportive of this project, and to Paul Hedin who advised us in the planning stages. Many staff at ACS Meetings, including Dianne Ruddy, Nicole Rodgers, and Nancy Todd, and at ACS Books, including Michelle Althuis, Anne Wilson, and Vanessa Johnson-Evans, were very helpful. The graphics used on the cover were specifically designed for this volume by Don Gregory and his associates at Molecular Simulations, Inc. The book would not have been produced without the willingness of T. E. Skelton of Clemson University to dedicate secretarial assistance for my editing chores. This was provided most capably by Angie Justice, whom I also thank.

THOMAS M. BROWN
Department of Entomology
Clemson University
Clemson, SC 29634–0365

June 14, 1996

Chapter 1

Applications of Molecular Genetics in Combatting Pesticide Resistance

An Overview

Thomas M. Brown

Department of Entomology, Clemson University, Clemson, SC 29634–0365

Resistance to pesticides continues as a major threat to agriculture and public health despite many years of intensive research. Initially observed against insecticides, recently bactericides, fungicides, and herbicides are becoming challenged increasingly. In addition, many pest species have developed multiple mechanisms of resistance. Molecular genetics is bringing new understanding of the genetic basis of resistance and the ways in which resistance evolves and spreads. Recent molecular characterization of several resistance genes has provided the means to study the evolution, population genetics and ecology of resistance at the genotypic level. This paper will review recent discoveries of resistance mechanisms, contributions of genetic technology to monitoring for resistance, and will consider the role of molecular genetics in assessing the risks of resistance to new pesticides and pest control strategies, including transgenic methods.

Molecular genetics has led to a better understanding of many cases of pesticide resistance. It will lead to improved strategies in managing this serious problem of agriculture and public health. Also, molecular transgenesis has provided insecticide-bearing crops and herbicide-resistant crops with future potential for crops resistant to phytopathogens, and insecticide resistant beneficial insects such as silkworm, honey bee and pest controlling parasitoids. Scientists from several disciplines employ molecular genetics in coping with resistance to bactericides (*1*), fungicides (*2*), herbicides (*3*) and insecticides (*4*). It may be useful to compare the progress among disciplines to determine whether similar phenomena are involved in the development of resistance in molecular targets, in organisms, or in populations of organisms.

An overview of pesticide resistance mechanisms will be presented including a review of recent developments in applying molecular genetics to examining pest populations, to countering resistance, and to exploiting resistance in beneficial transgenic applications.

Crop Protection and Pesticide Resistance

In the next century it is estimated that world population will double to

0097–6156/96/0645–0001$15.00/0

approximatley 10-12 million people; therefore, agricultural production must be expanded or made more efficient to meet nutritional needs. Crop yields are increasing; however, the percentage of yield lost to pests is increasing as well in most cases. The situation of pest control viewed over recent decades is one in which both pesticide use and pest damage are increasing. When eight major crops were surveyed (*5*), it was estimated that 42% of attainable value of production was lost to pests. Rice sustained over 50% loss, while 30-40% of other crops were lost. The authors noted that pesticide sales are increasing, especially herbicides but also other pesticides. There are several factors involved in increased loss which include pushing high-yielding varieties into new areas of production, but resistance to pesticides is a likely factor in some cases. The authors have warned that the spread of pesticide resistance must be watched. There are approximately 100 species resistant to fungicides, 100 species resistant to herbicides and 600 species resistant to insecticides. Many of these are species of practical impact and some are very troublesome due to having evolved resistance to most of the pesticides registered for use against them.

An emerging theme for the future is sustainable agriculture in which the goal is to preserve agricultural production using environmentally compatible farming practices and to eliminate practices which destructively exploit natural resources for short-lived benefits. An important component of sustainable agriculture is integrated pest management, the use of ecologically sound methods, including cultural methods, biological control, and appropriate chemical and genetically engineered pesticides to limit losses to pests. In practice, pesticides and pest-resistant crops are major tools of integrated pest management along with crop rotations, biological control agents, mechanical control of pests, *etc.*

Pesticide resistance management is a developing concept which must include strategies to delay resistance to all types of pesticides, including natural products and crops possessing genetically engineered insecticides and fungicides. sometimes called plant-pesticides. One general management strategy is to limit genetic selection pressure by avoiding the exclusive use of one chemical class of pesticide repeatly against the same pest population. This implies, and requires, that there be a choice of effective pesticides with different modes of action available.

As for insecticides, there are few major classes available. Market share of synthetic pyrethroids has increased while organochlorines and cyclodienes have declined due to environmental problems and resistance. Note that the cholinesterase inhibiting organophosphorus and carbamate classes remain very important while crops genetically engineered to produce the *Bacillus thuringiensis* insecticidal toxin may reduce the need for conventional insecticides. Also, share may accrue to insect development inhibitors, and to acetylcholine mimicking nitromethylene heterocycles, and other unconventional compounds having relatively narrow spectra of activity, but fitting very importantly in some situations; e. g. the cotton IPM project in Israel, has made effective use of these compounds against whiteflies (*6*).

There is a greater variety of chemistry among available fungicides and herbicides with several classes of very potent new pesticides gaining market share rapidly in the last decade. However, resistance has been observed against these new low-rate fungicides and herbicides and, in some cases, it has developed more rapidly than against the older, and in some cases, less specifically acting fungicides and herbicides. Therefore, sustainable agriculture will require careful management of the pesticides available to provide a wide variety of actions and to fit into integrated pest management strategies in the long term.

Mechanisms of Resistance

Molecular genetics has enhanced the understanding of resistance mechanisms which is the basis for developing resistance management strategies. The study of resistance mechanisms has advanced from the level of the protein in the form of an enzyme or ion channel, to the level of the gene.

Pharmacodynamic Mechanisms. For many categories of pesticides, there are examples of resistance due to selection for a target protein which is less sensitive to a pesticide. The original example of this knowledge was the discovery by Hirschberg and McIntosh of the point mutation in chloroplast DNA altering a thylakoid protein to produce resistance to triazine herbicides (*7*). In recent years it has become clear for this photosystem II target that the original mutation found in *Amaranthus* is the necessary path to resistance in many species of weeds, apparently due to the tight structural requirements for this highly evolved protein (A. Trebst, this volume). Similar findings are emerging with acetolactate synthase, the target for the much newer and more potent sulfonylurea herbicides, in which there are many possible mutations giving resistance in the laboratory, but only a few mutations found in weeds in the field (M. J. Guttieri, C. V. Eberlein, C. A. Mallory-Smith and D. C. Thill, this volume). A recent report describes amplification of the acetolactate synthase gene in cell lines of carrot, *Daucus carota,* selected for chlorsulfuron resistance (*8*).

Recently, entomologists as well have discovered point mutations in genes for targets such codon 302 of *Rdl* which controls the GABA receptor target of cyclodienes (R. ffrench-Constant,N. M. Anthony, D. Andreev and K. Aronstein, this volume); mutation at this codon is the mode of pharmacodynamic resistance in many diverse pests. Molecular genetic analysis of house fly gene for the resistant sodium ion channel, the target of DDT as well as the synthetic pyrethroids, revealed a single mutation conferring *kdr* which was supplemented by a second mutation to produce *super-kdr* (M. Williamson, D. Martinez-Torres, C. A. Hick, N. Castells and A. L. Devonshire, this volume).

One lesson learned is that target mutations may give negatively correlated cross-resistance in which the mutation confers greater susceptibility to another pesticide. Benzimidazole fungicides were affected by a mutation in the gene for the target, ß-tubulin which was more susceptible to *N*-phenylcarbamates (*9*). Negative correlation within the organophosphorus insecticide class was observed in resistant acetylcholinesterase of the house fly and *Heliothis virescens* (reviewed in T. M. Brown, P. K. Bryson, F. Arnette, M. Roof, J. L. B. Mallett, J. B. Graves and S. J. Nemec, this volume). In *Drosophila melanogaster*, this insecticide target appears to have several mutable sites for resistance (D. Fournier, S. Berrada and V. Bongibault, this volume).

Pharmacokinetic Mechanisms. Pharmacokinetic resistance mechanisms include uptake, translocation, penetration and elimination of pesticides. Several common genes for detoxication occur in multigene families, can be found in gene clusters, and are sometimes amplified so that many copies of the same gene are present in the genome. In the case of P450, it is now clear from genetic linkage analysis that the Rutgers strain of house fly possesses a regulatory type of resistance which increases the amount of P450 gene transcription (*10*).

Target-site resistance, when combined with detoxication, can present a synergistic increase of resistance. A recent example is an intensely resistant Argentinian strain of aphid in which an insensitive acetylcholinesterase (*11*), previously unknown in this species, was combined with the widespread mechanism of gene-amplified carboxylesterase (L. M. Field, C. A. Hick, A. L. Devonshire, N. Javed, J. M. Spence and R. L. Blackman, this volume). The pharmacokinetic mechanism resulting from the amplified carboxylesterase of aphids and mosquitoes is considered sequestration, because there is high affinity binding, but only very slow catalysis of insecticide hydrolysis. Recently, the term sequestration was applied as a mechanism of herbicide resistance in paraquat-resistant hairy fleabane in which the intact paraquat was sequestered to trichomes in the resistant biotype (*12*).

Detoxication also plays a role in the evolution of bactericide resistance. Multicopper oxidases have been selected by agricultural applications of copper (D. A. Cooksey, this volume). Rapid experimental evolution of resistant ß-lactamase was demonstrated using selection by a bactericide between rounds of artificial genetic recombination in a process known as sexual PCR (*13*).

Evolution and Population Genetics

Evolution of resistance is a process in which the frequency of genes for resistance increases in a population of a pest so that an increased proportion of that population survives when the pesticide is applied at the originally efficacious dose. When a pesticide has failed, it is important to determine whether the failure was due to a change in susceptibility of the pest or due to some other factor, such as application technique. Susceptibility tests are performed with samples of the pest from that population and compared to baseline results, populations from fields without failure, or laboratory strains to determine the degree of resistance.

While documentation of the evolution of resistance by susceptibility testing is common, there are few examples in which the dynamics of specific genes has been observed as resistance evolved. Observing resistance on a genotypic level will be critical for the understanding of resistance evolution and for the accurate formulation of resistance management strategies. Counter strategies will be most effective if aimed at populations early in the transition to resistance and will require techniques to measure resistance at very low levels (*14*).

If resistance were always due to only one mechanism, then a simple susceptibility test with doses to discriminate between homozygous and heterozygous survivors would provide enough information to observe the evolution of resistance. This situation may be the case for many weeds resistant to herbicides; however, multiple mechanisms are known for herbicide resistance in *Lolium rigidum* (C. Preston, F. J. Tardif and S. B. Powles, this volume) and it is quite common for pestiferous insects to possess an insensitive target along with a detoxicative mechanism (*15*). To study the evolution of resistance involving two or more genetic loci, susceptibility testing must be augmented by determinations of the mechanisms present in the population.

Mechanism-based monitoring and analysis of resistance in field populations of pests is becoming more common. An example is our program of surveillance for methyl parathion resistance due to insensitive acetylcholinesterase in the tobacco budworm, *H. virescens*, the key pest of cotton in the southeastern USA (T. M. Brown, P. K. Bryson, F. Arnette, M. Roof, J. L. B. Mallett, J. B. Graves and S. J. Nemec, this volume). The genotype *AceIn* RR, RS or SS was determined in individual heads of pheromone-trapped adults by microtiter plate analysis. Frequency of the R allele was 90% in one population sampled in 1983, but has declined to about 14%. The fact that heterozygotes were detected 10 years after the general replacement of methyl parathion by pyrethroids, suggests that a return to methyl parathion would rapidly select high resistance again. Several additional mechanisms of resistance to methyl parathion have evolved in *H. virescens* (*16*) and one collection from North Carolina was highly resistance by pharmacokinetic mechanisms only (*17*).

Pyrethroid resistance in *H. virescens* has steadily increased in Louisiana and has been a localized problem in other mid-south states (*18*). East of Mississippi, there were no practical problems with pyrethroid resistance until failures were reported in Alabama in 1995. So it would appear that there was a greater risk of resistance in the mid-south than in the southeast, perhaps due to pest genetics and the lack of resistance genes in the eastern population. Current interest is focussed on the first commercial plantings of transgenic cotton which expresses insecticidal toxin from *B. thuringiensis;* a sensitive assay has been developed to monitor the susceptibility of *H. virescens* (S. R. Sims, J. T. Greenplate, T. B.Stone, M. A. Caprio and F. L. Gould, this volume).

One application of molecular genetics is the molecular monitoring for specific resistance alleles using polymerase chain reaction techniques; this versatile technique has been applied to monitoring of fungicide resistance (*19*) and the variations of the technique have been reviewed (*20*). Amplification of specific alleles by discriminating primers is the most simple technique. Allele-specific endonuclease digestion is perhaps the most common in which all alleles are amplified and the R and S subsequently discriminated based upon the presence or absence of a restriction endonuclease cutting site giving a diagnostic restriction fragment length polymorphism. This technique has been criticized as liable to the false negative in which the endonuclease failure to cut may be interpreted as absence of the cutting site; therefore, the more complicated method of "PCR double-RFLP" was developed with restriction sites introduced in a nested PCR step to mark each allele for cutting by a different endonuclease (*21*). The result is that wild-type or mutant homozygote is cut by its diagnostic endonuclease, while the heterozygote is cut by both enzymes. Disadvantages are that introduced restriction sites must be designed with five primers prepared, and each sample must be divided for two nested PCRs and digestions.

Two techniques are available for screening for any unknown polymorphism in a particular sequence. Single-stranded conformational polymorphism analysis is relatively simple and commonly used, while a new, commercialized kit from Ambion (no endorsement implied) employs several steps leading to RNase cleavage of only mismatched wild-type plus mutant di-RNA hybrids to detect mutations (22).

Are resistant strains handicapped by a fitness deficit? Recent evidence to the contrary includes highly fit fungicide-resistant Spanish strains of *Botrytis cinerea* (*23*) and sulfonylurea-resistant *Kochia scoparia* possessing enhanced germination at low temperature (*24*). These findings recall the experience with insecticide resistance which was thought to confer a handicap in the early cases during the 1940's and 1950's, but is generally considered to evolve without a fitness deficit in many populations in more recent years.

Can a resistance gene spread around the world in one pest species? Evidence for this phenomenon in the northern house mosquito, *Culex pipiens,* was based on the molecular analysis of a gene for a detoxicative carboxylester hydrolase (M. Raymond and N. Pasteur, this volume). Regarding the spread of resistance, it often occurs dramatically with whiteflies and aphids.

Resistance Management

Agriculture will benefit from resistance management for sustained productivity, but first we must answer the basic questions and make informed decisions toward resistance management strategies. The potential application of molecular genetics to pesticide resistance will depend on positive interactions among growers, manufacturers, research and extension personnel and the regulatory agencies.

Manufacturers are now promoting research on resistance and formulating management strategies for their new products. This effort is underway for insecticides (G. D. Thompson and P. K. Leonard, this volume), fungicides (*25*) and herbicides (*26*).

There are general prophylactic strategies which are based simply on applying the principles of integrated pest management; e. g., to apply pesticides when needed as based on scouting of the crop and knowledge of other control factors such as biological and cultural control. To delay resistance in the cotton bollworm, *Helicoverpa armigera*, pyrethroid insecticide applications were limited to a window of time corresponding to the most crucial period for control of this pest during the growing season in Australia (N. W. Forrester and L. J. Bird, this volume). Other propylactic strategies include the used of high doses, alternating dose levels, alternating chemistries, or mixtures to prevent resistance; however, most of these operations are based on assumptions regarding the genetic basis of resistance and not upon active monitoring for the genes actually involved. Because the actual evolution of resistance has never been observed at the in the field by measuring resistance gene frequencies at specific genetic loci as they increase in a population, we must test these assumptions before refining such management strategies.

Practical application of the molecular genetic information should include studies on the evolution of resistance. Simple questions remain unanswered; e. g. resistance to sulfonylureas herbicides conferred by target insensitivity is a dominant trait, were initially populations heterozygous and how long did it take for homozygous resistant populations to evolve? Much more complicated questions can be asked about weed species which are capable of both target insensitivity and detoxicative mechanisms.

Ecology, Transgenesis and Regulation

The future holds much promise for engineered resistant crops and beneficial insects; however, there are ecological considerations to be addressed with regulatory implications. Can resistance genes move across species? This ecological interaction appears to have been the source of resistance to agricultural bactericides in some plant pathogens (G. W. Sundin and C. A. Bender, this volume). Herbicide-resistant crops are an example of a beneficial application of resistance (*27*), but there is concern about movement of resistance from transgenic crops into weeds (Darmency, this volume) and experimental introgression of transgenic glufonsinate resistance has been demonstrated recently within species of the genus *Brassica* (*28,29*).

Resistance to organophophorus insecticides has been expressed in *D. melanogaster* transformed with the bacterial *opd* gene in a P-element (*30*). Transgenesis in insects had been confined to such experimental studies (reviewed by K. J. Hughes, S. K. Narang, R. A. Leopold, O. A. Johnson and J. K. DeVault, this volume); however, this technology has advanced with the recent successful transformations of a more important dipteran pest, *Ceratitis capitata,* the Mediterranean fruit fly (*31*) and, very recently, the first lepidopteran transformation was achieved in the corn earworm, *Helicoverpa zea* (*32*). Insecticide resistance will be involved in many aspects of transgenesis in insects from the use of resistance genes as selectable markers for preparing vectors to the application of transgenic resistance in protecting beneficial insects from insecticide sprays (*33*). As this work advances, we must be vigilant to avoid providing insects with resistance genes not already present in their genome, and to assess the risk of unintended gene transfer across insect species.

In the next decade we may witness, or participate in, new applications of resistance through transgenesis. Regulatory decisions will play a major role in these applications. The Environmental Protection Agency has formed a working group to study various aspects of resistance as it may affect regulatory policy in the future (S. R. Matten, Paul I. Lewis, Gail Tomimatsu, Douglas W. S. Sutherland, Neil Anderson, and L. Colvin-Snyder, this volume).

Conclusions

It is very clear that molecular genetics will bring a better understanding of the resistance phenomenon and it is revealing many similarities in the evolution of resistance among various pests organisms. From the basic understanding of the mutations involved, we can determine the roles of specific genes in the evolution of resistance and from that knowledge, produce better strategies for delaying resistance in the future.

Resistance has its positive applications and molecular genetic techniques have enabled the use of herbicide resistance in transgenic crops, while the engineering of insecticide resistance into beneficial insects may be on the horizon. Pest resistance in transgenic crops (plant/pesticides) must be managed to delay resistance from evolving in the pest just as resistance has evolved to conventional pesticides.

Literature Cited

1. Sundin, G. W.; Bender, C. L. *App. Environ. Microbiol.* **1993**, *59*, 1018-1024.
2. Köller, W. In *Modern Fungicides and Antifungal Compounds*; H. Lyr, P. E. Russel and H. D. Sisler, Ed.; Intercept Ltd.: Andover, 1996; pp 301-311.
3. Powles, S.; Holtum, J. *Herbicide Resistance in Plants: Biology and Biochemistry*; Lewis Publishers: Chelsea, Michigan, 1994.
4. Mullin, C. A.; Scott, J. G. *Am. Chem. Soc.* **1992**, *505*, 174-190.
5. Oerke, E. C.; Dehne, H.-W.; Schönbeck, F.; Weber, A. Crop Production and Crop Protection. Estimated Losses in Major Food and Cash Crops; Elsevier Science B. V.: Amsterdam, 1994, 808 pp.
6. Horowitz, A. R.; Forer, G.; Ishaaya, I. *Pestic. Sci.* **1994**, *42*, 113-122.
7. Hirschberg, J.; McIntosh, L. *Science* **1983**, *222*, 1346-49.
8. Caretto, S.; Giardina, M. C.; Nicolodi, C.; Mariotti, D. *Theor. Appl. Genet.* **1994**, *88*, 520-524.
9. Fujimura, M.; Oeda, K.; Inoue, H.; Kato, T. In *Managing Resistance to Agrochemicals*; M. B. Green, H. M. LeBaron and W. K. Moberg, Ed.; American Chemical Society: Washington, D. C., 1990; Vol. 421; pp 224-236.
10. Cariño, F. A.; Koener, J. F.; F. W. Plapp, J.; Feyereisen, R. *Insect Biochem. Mol. Biol.* **1994**, 24, 411-418.
11. Moores, G. D.; Devine, G. J.; Devonshire, A. L. *Brighton Crop Protection Conference Pests and Diseases - 1994* **1994**, *1*, 413-418.
12. Norman, M. A.; Smeda, R. J.; Vaughn, K. C.; Fuerst, E. P. *Pestic. Biochem. Physiol.* **1994**, *50*, 31-42.
13. Stemmer, W. P. C. *Bio/technology* **1995**, *13*, 549-553.
14. Brown, T. M.; Brogdon, W. G. *Ann. Rev. Entomol.* **1987**, *32*, 145-62.
15. Brattsten, L. B.; Holyoke, C. W.; Leeper, J. R.; Raffa, K. *Science* **1986**, *231*, 1255-1260.
16. Brown, T. M. In *Managing Resistance to Agrochemicals*; M. B. Green, H. M. LeBaron and W. K. Moberg, Ed.; American Chemical Society: Washington, D. C., 1990; Vol. 421; pp 61-76.
17. Konno, T.; Hodgson, E.; Dauterman, W. C. *Pestic. Biochem. Physiol.* **1989**, *33*, 189-99.
18. Graves, J. B.; Leonard, B. R.; Micinski, S.; Burris, E.; Martin, S. H.; White, C. A.; Baldwin, J. L. In *Proc. Beltwide Cotton Res. Conf.*; 1993; pp 788-794.
19. Wheeler, I. E.; Kendall, S. J.; Butters, J.; Hollomon, D. W. *Pestic. Sci.* **1995**, *43*, 201-209.

20. ffrench-Constant, R.; Aronstein, K.; Anthony, N.; Coustau, C. *Pestic. Sci.* **1995**, *43*, 195-200.
21. Grau, O.; Griffais, R. *Nucleic Acids Res.* **1994**, *22*, 5773-5774.
22. Ambion "Faster Mutation Analysis," Ambion, 1995.
23. Raposo, R.; Delcan, J.; Melgarejo, P.; Gomez, V. *Brighton Crop Protection Conference Pests and Diseases - 1994* **1994**, *1*, 493-498.
24. Dyer, W. E.; Chee, P. W.; Fay, P. K. *Weed Sci.* **1993**, *41*, 18-22.
25. Russell, P. E. In *Pesticide Outlook*; 1995; p. 17.
26. Jutsum, A. R.; Graham, J. C. In *Pesticide Outlook*; 1995; p. 3.
27. Dekker, J.; Duke, S. O. *Adv. Agron.* **1995**, *54*, 69-116.
28. Mikkelsen, T. R.; Anersen, B.; Jørgensen, R. B. *Nature* **1996**, *380*, 31.
29. Abbott, A. *Nature* **1996**, *380*, 94.
30. Benedict, M. Q.; Scott, J. A.; Cockburn, A. F. *Insect Molec. Biol.* **1994**, *3*, 247-252.
31. Loukeris, T. G.; Livadaras, I.; Arca, B.; Zabalou, S.; Savakis, C. *Science* **1995**, *270*, 2002-2005.
32. DeVault, J. D.; Hughes, K. J.; Leopold, R. A.; Johnson, O. A.; Narang, S. K. *Genome Res.* **1996**, in press.
33. Hoy, M. A. *Insect Molecular Genetics: an Introduction to Principles and Applications;* Academic Press, Inc.: New York, 1994, 546 pp.

Molecular Genetics and Resistance Mechanisms

Chapter 2

Molecular Genetics of Target-Site Resistance to Acetolactate Synthase Inhibiting Herbicides

Mary J. Guttieri[1], Charlotte V. Eberlein[1], Carol A. Mallory-Smith[2], and Donn C. Thill[3]

[1]Department of Plant, Soil, and Entomological Sciences, Aberdeen Research and Extension Center, University of Idaho, Aberdeen, ID 83210
[2]Crop and Soil Science Department, Oregon State University, Corvallis, OR 97331
[3]Department of Plant, Soil, and Entomological Sciences, University of Idaho, Moscow, ID 83844

Weed biotypes resistant to herbicides that inhibit acetolactate synthase (acetohydroxyacid synthase, E.C. 4.1.3.18) are now widespread in North America. The molecular genetics of resistance in crop species and *Arabidopsis* are well characterized. Several point mutations confer resistance to the sulfonylurea, imidazolinone, triazolopyrimidine, or pyrimidinyl oxybenzoate herbicides. Some of these point mutations also confer herbicide resistance in field-selected herbicide resistant weed biotypes. Specific cross-resistance patterns appear to be associated with specific point mutations.

Acetolactate synthase (acetohydroxyacid synthase, E.C. 4.1.3.18) is a key enzyme in the synthesis of the branched chain amino acids valine, leucine, and isoleucine. Acetolactate synthase (ALS) catalyzes two parallel reactions: condensation of two moles of pyruvate to form one mole of acetolactate, and condensation of one mole of pyruvate with one mole of 2-oxobutyrate to form one mole of acetohydroxybutyrate. Acetolactate is a precursor to valine and leucine; acetohydroxybutyrate is a precursor to isoleucine. The molecular biology and physiology of ALS-inhibitor resistance in weeds and crops has been reviewed recently by Saari et al. (*1*), Shaner (*2*), and Devine and Eberlein (*3*).

ALS is encoded in the nucleus and is active in the chloroplast. The *in vivo* oligomeric structure of ALS in higher plants has not been definitively established. An N-terminal chloroplast transit peptide of approximately 59 to 92 amino acids, depending on species, is presumed to direct localization in the chloroplast (*4,5*). Most diploid plant species have a single ALS locus, with corn (*Zea mays*) being a notable exception with two loci (*6*); tetraploid tobacco (*Nicotiana tabacum*) has two loci (*7*); and *Brassica* species have numerous loci (*8*). The mature protein is approximately 575 amino acids, depending on species. The mature ALS amino acid sequence is highly conserved across species. The

0097–6156/96/0645–0010$15.00/0

consensus sequence derived from published ALS amino acid sequences of *Arabidopsis thaliana* (*csr1* allele), *Brassica napus* (*ALS1*, *ALS2*, *ALS3* alleles), *Zea mays* (two alleles), *Xanthium strumarium* (one allele), and *N. tabacum* (*SuRA* and *SuRB* alleles) is shown in Figure 1. Beginning from the first fully conserved domain, these nine sequences have 63% homology. Throughout this manuscript, amino acids will be designated using the numbering in the original reference and, in parentheses, the corresponding amino acid in the *A. thaliana* sequence reported by Mazur et al. (*4*).

ALS-Inhibiting Herbicides

Four families of herbicides inhibit ALS activity in plants: the sulfonylurea, imidazolinone, triazolopyrimidine, and pyrimidinyl oxybenzoate herbicides. The sulfonylurea herbicides were the first ALS inhibitors introduced, and have been marketed in North America since the introduction of chlorsulfuron (Glean®) in 1982. Sulfonylurea herbicides have been used widely due to their high efficacy, low use rates, and environmental safety. Imidazolinone herbicides, such as imazethapyr (Pursuit®) and imazaquin (Scepter®) increasingly are used for weed control. The triazolopyrimidine and pyrimidinyl oxybenzoate herbicides are relatively new herbicides; flumetsulam (Broadstrike®) is a triazolopyrimidine herbicide recently registered for weed management in corn and soybeans.

Herbicides that inhibit ALS activity are generally considered to be environmentally friendly due to their exceptionally low mammalian toxicity and low use rates. Some ALS inhibiting herbicides are remarkably selective. For example, rimsulfuron (Matrix®), recently registered for weed control in potatoes, *Solanum tuberosum*, is a highly effective herbicide for control of the related species, hairy nightshade, *Solanum sarrachoides* (*9*).

Laboratory Selection for Herbicide Resistance

In an effort to expand the range of herbicides available for weed control, a number of laboratories have selected for ALS inhibitor resistance in crop species and *Arabidopsis thaliana* (cloned resistance genes from *A. thaliana* could be used to generate resistant crops through transformation). In general, selection for ALS inhibitor resistance has led to isolation of mutants with resistance as a consequence of herbicide-insensitive ALS. Exceptions include primisulfuron-tolerant corn, which was tolerant due to increased metabolism (*10*), and chlorsulfuron-tolerant soybean, in which a recessive resistance allele conferred non-target site resistance (*11*).

The first report of selection for ALS inhibitor resistance in higher plants was the selection of chlorsulfuron and metsulfuron methyl resistant tobacco cell cultures (*12*). Two resistant regenerants were identified, the C3 and S4 mutants. Further selection of S4 cell cultures led to the identification of the S4-Hra mutant, which was significantly more resistant than either the C3 or original S4 mutant (*13*). DNA sequence analysis of the C3 mutant ALS genes identified a point mutation in the $Pro_{196(197)}$ (Figure 1, Pro_{197}) codon encoding a Gln substitution (*14*). DNA sequence analysis of the S4-Hra mutant ALS genes identified a point mutation in the codon for $Pro_{196(197)}$ (Figure 1, Pro_{197}), encoding an Ala substitution, and a second point mutation in the codon for $Trp_{573(591)}$ (Figure 1, Trp_{591}), encoding a Leu substitution (*14*). There also is a report of a point mutation encoding an $Ala_{199(205)}$ (Figure

1, Ala_{205}) to Asp substitution conferring resistance to sulfonylurea herbicides in tobacco cell cultures (*15*).

Selection of mutagenized *A. thaliana* seed with ALS inhibitors has led to identification of a series of mutations conferring resistance to specific families of ALS inhibitors. Selection of mutagenized seed with chlorsulfuron led to identification of the *Csr1-1* allele for resistance (*16*). This dominant allele conferred high levels of resistance to chlorsulfuron (a sulfonylurea herbicide) and triazolopyrimidine sulfonamide, and conferred low levels of resistance to imazapyr (an imidazolinone) and pyrimidinyl oxybenzoate (*17*). A point mutation in the codon for Pro_{197} (Figure 1, Pro_{197}), encoding a Ser substitution, was associated with resistance in the *Csr1-1* mutant (*18*).

Selection of mutagenized *A. thaliana* seed with imazapyr led to identification of the *Csr1-2* allele for resistance (*19*), also known as the *imr* allele. In contrast to the *Csr1-1* allele, the dominant *Csr1-2* allele conferred high levels of resistance to imazapyr (an imidazolinone) and pyrimidinyl oxybenzoate, and conferred low levels of resistance to chlorsulfuron and triazolopyrimidine sulfonamide (*17*). The basis of resistance conferred by the *Csr1-2* allele was determined to be a point mutation in the codon for $Ser_{653(670)}$ (Figure 1, Ser_{670}), near the 3' end of the coding region, that encoded an Asn substitution (*20*).

Selection of mutagenized *A. thaliana* seed with triazolopyrimidine sulfonamide led to identification of the *Csr1-3* allele, which, like the *Csr1-1* and *Csr1-2* alleles, is a dominant allele (*21*). The molecular basis of resistance in *Csr1-3* has not been reported. However, it is interesting to note that *Csr1-3*, like the *Csr1-1* allele, conferred high levels of resistance to both chlorsulfuron and triazolopyrimidine sulfonamide, and conferred low levels of resistance to imazapyr or pyrimidinyl oxybenzoate (*17*).

Selection of corn cultures with the imidazolinone herbicides imazethapyr and imazaquin has led to development of imidazolinone-resistant corn genotypes (*22*). ALS gene sequences from two herbicide resistant corn genotypes have been reported. In ICI 8532 IT, resistance is a consequence of a point mutation in the codon for $Ala_{57(155)}$ (Figure 1, Ala_{155}), encoding a Thr substitution (*23*). ALS isolated from ICI 8532 IT was resistant to imidazolinone herbicides and pyrimidinyl oxybenzoate, but was not resistant to chlorsulfuron and flumetsulam (a triazolopyrimidine) (*24*). In Pioneer 3180 IR, resistance was the result of a point mutation in the codon for $Trp_{542(591)}$ (Figure 1, Trp591), encoding a Leu substitution (*24*). ALS isolated from Pioneer 3180 IR was resistant to representatives of all four families of ALS inhibitors (*24*).

Selection of *B. napus* cell cultures led to development of a highly resistant cell line. The DNA sequences of the ALS genes from this cell line were evaluated (*25*). One of the ALS alleles, *ALS3*, was found to have a point mutation in the codon for $Trp_{557(591)}$ (Figure 1, Trp_{591}), encoding a Leu substitution, as was observed in Pioneer 3180 IR corn and the tobacco S4-Hra mutant. This mutant *B. napus ALS3* gene was used to transform tobacco. Transgenes with a range of herbicide sensitivity were recovered. Three of these transgenes were characterized. Transgene 22 had approximately 10-fold resistance to chlorsulfuron and no cross-resistance to imazethapyr or triazolopyrimidine sulfonamide. Transgene 35 had approximately 4-fold resistance to chlorsulfuron, 10-fold resistance to imazethapyr, and 1000-fold resistance to triazolopyrimidine sulfonamide. Transgene 43 had >100-fold resistance to chlorsulfuron and >1000-fold resistance to both imazethapyr and tri-

Figure 1. *Aligned wild-type ALS amino acid sequences of B. napus, N. tabacum, Z. mays, X. strumarium, and A. thaliana, beginning at the first conserved domain. Conserved amino acids are shown, and dashes represent nonconserved amino acids. Amino acid positions implicated in resistance are shown in reverse type and are referred to in the text by the code letter placed immediately above. Numbering is according to Mazur et al. (4) for A. thaliana.*

azolopyrimidine sulfonamide. This wide variation in resistance was attributed by the authors to differences in transgene expression (*25*). This work clearly illustrates the inherent difficulty in evaluating gene action via transformation.

Herbicide-Resistant Weed Biotypes

The first ALS inhibitor resistant weed biotype, chlorsulfuron resistant prickly lettuce (*Lactuca serriola*), was identified in northern Idaho in 1987, five years after the introduction of chlorsulfuron (Glean®) (*26*). The basis of resistance in this biotype was determined to be modified ALS (*27*). The resistant lettuce biotype was cross-resistant, at the whole-plant level, to other sulfonylurea herbicides and imazethapyr and imazapyr, but not imazaquin (*26*). DNA sequence analysis of a region encompassing $Pro_{173\,(197)}$ (Figure 1, Pro_{197}) identified a point mutation in the codon for $Pro_{173\,(197)}$ encoding a His substitution in the resistant biotype (*28*).

Since 1987, ALS inhibitor resistance has been documented in 27 species worldwide (J. C. Cotterman, personal communication). The majority of the reported incidences of sulfonylurea herbicide resistance are in kochia (*Kochia scoparia*). The mechanism of resistance in all resistant kochia biotypes reported to date is modified ALS (*1,29-31*). Evaluation of the DNA sequence in the region encompassing $Pro_{173(197)}$ (Figure 1, Pro_{197}) in ten chlorsulfuron-resistant kochia biotypes from across North America indicated that seven of the ten biotypes had mutations in the codon for $Pro_{173(197)}$ (*31*). All six possible single point mutations conferring amino acid substitutions were identified in these seven resistant biotypes. The three biotypes that did not have mutation in the codon for $Pro_{173(197)}$ were resistant due to chlorsulfuron-insensitive ALS, and ALS isolated from these resistant kochia biotypes was 3 to 11-fold less sensitive to inhibition by imazethapyr than susceptible biotype ALS (*31*).

Cross-resistance of 298 chlorsulfuron-resistant kochia biotypes (resistance defined as a two-fold increase in I_{50}) to three sulfonylurea herbicides (metsulfuron methyl, sulfometuron methyl, and triasulfuron) and imazapyr has been evaluated (*1*). Of 179 chlorsulfuron-resistant biotypes tested, all were resistant to metsulfuron methyl. Of 298 chlorsulfuron-resistant biotypes tested, all were resistant to metsulfuron methyl, and all but one were resistant also to triasulfuron. Of 158 chlorsulfuron-resistant biotypes tested, 145 were resistant to imazapyr. However, R/S ratios of I_{50} values for imazapyr were generally in the range of 5 to 7, while R/S ratios for the sulfonylurea herbicides were generally in the range of 30 to 60.

Russian thistle (*Salsola iberica*) biotypes resistant to chlorsulfuron and metsulfuron methyl have been identified throughout Eastern Washington. The DNA sequence of the ALS gene from one biotype was determined in a region encompassing $Pro_{173(197)}$ (Figure 1, Pro_{197}). This biotype was found to have a point mutation encoding a Pro to Leu substitution in one of its copies of the ALS gene (Guttieri, M. J., Eberlein, C. V., University of Idaho, unpublished data).

Recently, common cocklebur (*Xanthium strumarium*) biotypes resistant to imazaquin have been reported. One of these biotypes, isolated in Mississippi (*32*), arose from three years of banded applications of imazaquin. ALS isolated from this biotype was not resistant to flumetsulam (a triazolopyrimidine), or chlorimuron (a sulfonylurea). The basis of resistance in this biotype was determined to be a point mutation in the codon for $Ala_{133(155)}$ (Figure 1, Ala_{155}) encoding a Thr substitution (*24*). This Ala to Thr substitution is analogous to the Ala to Thr substitution in ICI 8532 IT corn. Like ICI 8532 IT corn, ALS isolated from this cocklebur biotype was cross-resistant to pyrimidinyl oxybenzoate, but not to chlorsulfuron or flumetsulam (*24*).

A second imazaquin-resistant common cocklebur biotype was isolated from a field in Missouri that had received multiple applications of imazaquin over four years (*33*). Three unique point mutations encoding amino acid substitutions were identified in this biotype. These amino acid substitutions included: Gln_{269} to His, Asn_{522} to Ser, and $Trp_{552(591)}$ (Figure 1, Trp_{591}) to Leu . ALS isolated from the Missouri cocklebur biotype was resistant not only to imazaquin and pyrimidinyl oxybenzoate, but also was resistant to chlorsulfuron and flumetsulam (*24*). Because of the similar pattern of cross-resistance observed in Pioneer 3180 IR corn, the Trp to Leu substitution was identified as causal to resistance.

Bernasconi and coworkers also have fused the common cocklebur ALS coding sequence to glutathione-S-transferase and evaluated the expressed fusion product in *Escherichia coli*. All possible single and double point mutations encoding substitutions at $Trp_{552(591)}$ (Figure 1, Trp_{591}), excluding those that would have resulted in a termination codon, were introduced by site-directed mutagenesis. Of the seven modified ALS sequences introduced, only the Leu substitution yielded active enzyme (*24*).

Summary

Substitutions at Pro_{197}, as observed in *Arabidopsis Csr1-1*, tobacco C3, prickly lettuce, and kochia (see Figure 1) can confer high levels of resistance to sulfonylurea herbicides. In *Arabidopsis*, substitution at Pro_{197} also conferred high levels of resistance to a triazolopyrimidine herbicide. Asp substitution for Ala_{155} resulted in sulfonylurea resistant tobacco cell cultures; there are no reported cross-resistance patterns for this mutation. The $Ser_{653(670)}$ to Asn substitution observed in *Arabidopsis Csr1-2* (see Figure 1) conferred high levels of imidazolinone and pyrimidinyl oxybenzoate resistance, and very little resistance to sulfonylurea and triazolopyrimidine herbicides. The $Ala_{57(155)}$ to Thr substitution observed in ICI 8532 IT corn and the imazaquin resistant cocklebur biotype from Mississippi (see Figure 1) conferred resistance to imidazolinone and pyrimidinyl oxybenzoate herbicides. The $Trp_{542(591)}$ to Leu substitution observed in Pioneer 3180 IR corn, the Missouri cocklebur biotype, the *AHAS3* allele from highly resistant *B. napus* cell cultures, and tobacco S4-Hra (see Figure 1) appears to confer high levels of resistance to all four families of ALS inhibitors. The cross-resistance patterns of laboratory-derived mutants correlate very well with cross-resistance patterns of field-selected weed biotypes with the same point mutations (*24*).

Literature Cited

1. Saari, L. L.; Cotterman, J. C.; Thill, D. C. In *Herbicide Resistance in Plants: Biology and Chemistry*; Powles, S. B.; Holtum, J. C., Eds.; Lewis Publishers: Boca Raton, FL, **1994**, pp 83-140.
2. Shaner, D. L.; Sing, B. K. In *Reviews in Pesticide Toxicology, Vol. IV*; Roe, M; Kuhr, R.; Burton, J., Eds; NCS: Raleigh, NC, **1995**.
3. Devine, M. D.; Eberlein, C. V. In *Reviews in Pesiticide Toxicology, Vol. IV*; Roe, M; Kuhr, R.; Burton, J., Eds; NCS: Raleigh, NC, **1995**.

4. Mazur, B. J.; Chui, C. F.; Smith, J. K. *Plant Physiol.* **1987**, *85*, pp 1110-1117.
5. Wiersma, P. A.; Hachey, J. E..; Crosby, W. L.; Moloney, M. M. *Mol. Gen. Genet.* **1990**, *224*, pp 155-159.
6. Fang, L. Y.; Gross, P. R.; Chen, C.; Lillis, M. *Plant Molecular Biol.* **1992**, *18*, pp 1185-1187.
7. Chaleff, R. S.; Bascomb, N. F. *Mol. Gen. Genet.* **1987**, *210*, pp 33-38.
8. Rutledge, R.; Ouellet, T.; Hattori, J.; Miki, B. L. *Mol. Gen. Genet.* **1991**, *229*, pp 31-40.
9. Eberlein, C. V.; Whitmore, J. C.; Stanger, C. E.; Guttieri, M. J. *Weed Technol.* **1994**, *8*, pp 428-435.
10. Harms, C. T.; Montoya, A. L.; Privalle, L. S.; Briggs, R. W. *Theor. Appl. Genet.* **1990**, *80*, pp 353-358.
11. Sebastian, S. A.; Chaleff, R. S. *Crop Sci.* **1987**, *27*, pp 948-952.
12. Chaleff, R. S.; Ray, T. B. *Science* **1984**, *223*, pp 1148-1151.
13. Creason, G. L.; Chaleff, R. S. *Theor. Appl. Genet.* **1988**, *76*, pp 177-182.
14. Lee, K. Y.; Townsend, J.; Tepperman, J.; Black, M.; Chui, C. F.; Mazur, B.; Dunsmuir, P.; Bedbrook, J. *EMBO Journal* **1988**, *7(5)*, pp 1241-1248.
15. Hartnett, M. E.; Chui, C. F.; Mauvais, C. J.; McDevitt, R. E.; Knowlton, S.; Smith, J. K.; Falco, S. C.; Mazur, B. J. *ACS Symp. Ser. Am. Chem. Soc.* **1990**, *421*, pp 459-473.
16. Haughn, G. W.; Somerville, C. *Mol. Gen. Genet.* **1986**, *204*, pp 430-434.
17. Mourad, G.; King, J. *Planta* **1992**, *188*, pp 491-497.
18. Haughn, G. W.; Smith, J.; Mazur, B.; Somerville, C. *Mol. Gen. Genet.* **1988**, *211*, pp 266-271.
19. Haughn, G. W.; Somerville, C. R. *Plant Physiol.* **1990**, *92*, pp 1081-1085.
20. Sathasivan, K.; Haughn, G. W.; Murai, N. *Plant Physiol.* **1991**, *97*, pp 1044-1050.
21. Mourad, G.; Pandey, B.; King, J. *J. Hered.* **1993**, *84*, pp 91-96.
22. Anderson, P. A.; Georgeson, M. *Genome* **1989**, *31*, pp 994-999.
23. ICI Patent WO92/08794.
24. Bernasconi, P.; Woodworth, A. R.; Rosen, B. A.; Subramanian, M. V. *J. Biol. Chem.* **1995**, (in press).
25. Hattori, J.; Brown, D.; Mourad, G.; Labbe, H.; Ouellet, T.; Sunohara, G.; Rutledge, R.; King, J.; Miki, B. *Mol. Gen. Genet.* **1995**, *246*, pp 419-425.
26. Mallory-Smith, C. A.; Thill, D. C.; Dial, M. J. *Weed Technology* **1990**, *4*, pp 163-168.
27. Mallory-Smith, C. A. **1990,** Identification and inheritance of sulfonylurea herbicide-resistant prickly lettuce (*Lactuca serriola*). *Ph.D. Dissertation*, Univ. of Idaho, Moscow, ID.
28. Guttieri, M. J.; Eberlein, C. V.; Mallory-Smith, C. A.; Thill, D. C.; Hoffman, D. L. *Weed Sci.* **1992**, *40*, pp 670-678.
29. Saari, L. L.; Cotterman, J. C.; Primiani, M. M. *Plant Physiol.* **1990**, *93*, pp 55-61.
30. Sivakumaran, K.; Mulugeta, D.; Fay, P. K.; Dyer, W. E. *Weed Sci.* **1993**, *43*, pp 159-165.
31. Guttieri, M. J.; Eberlein, C. V.; Thill, D. C. *Weed Sci.* **1995**, *43*, pp 175-182.
32. Schmitzer, P. R.; Eilers, R. J.; Cseke, C. *Plant Physiol.* **1993**, *103*, pp 281-283.
33. Kendig, J. A.; DeFelice, M. S. *WSSA Abstracts* **1994**, *34*, pp 12.

Chapter 3

Molecular Genetics of Acetylcholinesterase in Insecticide-Resistant *Drosophila melanogaster*

D. Fournier, S. Berrada, and V. Bongibault

Laboratoire d'Entomologie Appliquée, Université Paul Sabatier, 118 Route de Narbonne, 31062 Toulouse, France

Most of insect cholinesterase (EC 3.1.1.7) is found in the central nervous system where it is a key component of cholinergic synapses. Its function is to rapidly terminate neurotransmission by hydrolyzing acetylcholine. Thus inhibition of cholinesterase is lethal and this action of inhibitors is the basis of their use as insecticides. Organophosphorus and carbamate pesticides react as acetylcholine analogues but irreversibly complex the enzyme. Extensive use of these compounds resulted in several resistant species which possess altered cholinesterase less sensitive to inhibition by the insecticides.

The Gene Coding for Cholinesterase and Expression.

Only one gene has been found coding for *Drosophila melanogaster* cholinesterase. To clone this gene, several mutants had been isolated which permitted genetic localization the *Ace* locus (*1-3*). A chromosome walk in the region was performed (*4*). Detection of transcripts within this region, isolation of corresponding cDNA clones and sequencing have been achieved (*5*). The transcription unit is 34 kb long and encompasses ten exons (*6*). The *Drosophila* coding sequence is more split than its vertebrate counterparts, the presence of numerous introns favors recombination between the different exons and hence resistance to insecticides occurs when two points mutations are present in adjacent exons (see below). In insects, cholinesterase is mainly expressed in the central nervous system (*7*), total activity shows a transient peak at the first larval stage and is maximal in adults (*8*). The gene has been expressed in deficient flies. A minigene was constructed; it lacked the intronic regions and was flanked with 1.5 kb genomic sequence upstream from the start of transcription thought to contain all major promoter elements of the gene. Once the minigene was injected in *Drosophila* via P-mediated transformation, it was possible to rescue *Ace* lethal mutants and to obtain a tissue-specific expression (*9*). Expression has been achieved *in vitro* using two systems: Transient expression was obtained in *Xenopus* oocytes (*10*). The protein produced in this system is active, glycosylated, but not processed for the glycophosphatydyl-inositol anchor (see below). *Drosophila Ace* cDNA was expressed in a *Baculovirus*-lepidoteran cells expression system. The

0097–6156/96/0645–0017$15.00/0

protein produced in this system revealed molecular forms analogous to those found for cholinesterase purified from *Drosophila* heads (*11*). Conversely, we failed in expressing an active protein in *E. coli* or in yeast.

Structure of the Enzyme.

Drosophila cholinesterase was easily purified to homogeneity by affinity chromatography as the enzyme from housefly (*12-13*). It has been characterized as an amphiphilic dimer linked to the membrane via a glycolipid anchor (Figure 1). The two active subunits are each composed of two polypeptides, 16 and 55 kDa. The 55 kDa polypeptide bears the serine of the active site, the glycolipid anchor and the cysteine responsible for dimerization. Several isozymes can be detected; they originate from the main form by protease and lipase digestion (*14-15*).

Dimeric Protein. In all developmental stages, the major molecular form is an amphiphilic membrane-bound dimer, but a significant proportion of amphiphilic monomer is also observed in larvae and young pupae which could represent the precursor of the dimeric form (*8*). Partial reduction of purified dimeric forms by 2-mercaptoethanol or dithiothreitol gave rise to an amphiphilic monomeric active form (*16*) suggesting that one or several cysteines are involved in the dimeric structure. There are 9 cysteine residues in the *Drosophila* cholinesterase sequence. The amino acid sequence deduced from *Drosophila* cholinesterase gene was compared with *Torpedo* cholinesterase (*5*) and afterwards with human butyrylcholinesterase (*17-18*). *Drosophila* cholinesterase exhibits approximatively 30% overall residue identity with vertebrate cholinesterases. Important structural and functional features such as cysteine residues involved in intrachain disulfide bonds or active site regions are conserved among these proteins. In peptidic sequence alignment, six cysteine residues were found to be conserved in *Drosophila* cholinesterase (*5*). It is likely that all these proteins share an identical folding pattern. We deduced putative positions of internal disulfide bonds in *Drosophila* cholinesterase; Cys_{700} is localized in the peptide precursor which is exchanged with the glycolipid anchor (see below); however two cysteines, at position 328 and 615 remained available for interchain linkages. In order to assign inter-disulfide bridges, we independently mutagenized the two free cysteines. We expressed the genes and analyzed the protein in gradient gels: the intersubunit bond involves Cys_{615}. The monomer is active and excreted, suggesting that dimerization is not essential for the enzyme function. The monomer is more sensitive to protease degradation than the dimer suggesting that the role of dimerization is to protect the enzyme (*19*).

Subunit Composition. The apparent molecular weight of native cholinesterase dimer from insects has been estimated at 150 kDa by gel filtration, Ferguson plot or pore limit electrophoresis in the presence of anionic detergents (*12-14*). The putative protein encoded by the *Ace* cDNA was 70 kDa (*5*) which accounted for the monomer. Analysis on SDS-PAGE of purified *Drosophila* cholinesterase revealed two major polypeptides of 55 and 16 kDa. We hypothesized that the protein is translated as a precursor of 70 kDa which generates two polypeptides by proteolytic cleavage. In two-dimensional denaturing electrophoresis, the first dimension performed in non-reducing conditions and the second one in the presence of 2-mercaptoethanol, a 110 kDa polypeptide was reduced into the 55 kDa polypeptide: therefore, the 55 kDa subunit possesses the sulfhydryl group involved in the inter-subunit linkage and corresponds to the C-terminal end. The 16 kDa polypeptide corresponds to the N-terminal end of the 70 kDa precursor . This result was obtained by sequencing the N-terminal end of the 16 kDa polypeptide (*20*) and by using specific polyclonal antibodies raised against fusion proteins containing either N-terminus or C-terminus portions of the cDNA-deduced protein sequence (*21*). The processing of a single cholinesterase precursor into two polypeptides was not observed in vertebrates. In contrast to vertebrate sequences, *Drosophila* cholinesterase includes an insertion of 33 amino acids in position 147-180 (Figure 2). This additional peptide is hydrophilic and was supposed to contain the cleavage site(s) of the precursor into 16 and 55 kDa subunits since its location is consistent with apparent molecular weights of the two subunits and no disulfide bond passes through this region. In order to obtain some insight on the role of this hydrophilic portion of the *Drosophila* protein, a cDNA deleted for this region, was constructed and used in a *Xenopus* oocytes expression system. We obtained an active protein composed of one polypeptide with an apparent molecular weight of 65 kDa. This result indicates that the supplementary hydrophilic polypeptide is responsible for the proteolytic cleavage found in purified extracts. Deletions of half the coding sequence of the hydrophilic peptide resulted in a maturated protein composed of two peptides indicating that the maturation consisted of a cut inside the hydrophilic peptide. Furthermore, the cut was not site specific and may have arisen either in one half or in the other of the hydrophilic peptide. We did not find any role for the hydrophilic peptide, but the cut is related to the secretion of the protein outside the cell since we found uncut protein inside the oocyte and a cut protein outside (*10*). A similar hydrophilic peptide, although not conserved, was found at the same location in other known dipteran cholinesterase sequences (*22*, Williamson and Devonshire, pers. comm.), but not in cholinesterases from other insects so far known.

Signal Peptide. The cDNA-deduced protein sequence exhibits an NH_2-terminal peptide sufficiently hydrophobic to play the role of signal peptide as found in membrane-associated and exported protein precursors (*5*). It is removed from the mature protein. The N-terminal amino acid of the mature protein was determined by microsequencing the 16 kDa subunit at position Val39 (*20*).

Glycophosphatidyl-inositol Anchor. *D. melanogaster* cholinesterase was shown to be anchored to the membrane via a glycophosphatidyl-inositol (G-PI) anchor as previously described in human and bovine erythrocyte cholinesterases (*23-25*) and *Torpedo* cholinesterase (*26*). The amphiphilic dimer was shown to be converted into hydrophilic dimer and monomer on autolysis of the extract. The binding of insect cholinesterase to detergent was first shown by Arpagaus and Toutant (*27*). Then, Gnagey *et al.* (*13*) found ethanolamine and glucosamine in purified cholinesterase from *Drosophila* suggesting the presence of a G-PI anchor. This anchor was then shown to be sensitive to phosphatidylinositol-phospholipase C (PI-PLC) from *Bacillus cereus* or *Trypanosoma brucei* by Triton X-114 partitioning and electrophoresis in non denaturing gels (*15*). The PI-PLC sensitivity of *Drosophila* cholinesterase was confirmed by labeling experiment of the protein. [^{125}I] TID, a photoactivatable affinity probe specific of the lipid moiety of G-PI anchored protein was selectively removed by PI-PLC (*20*). In addition, the PI-PLC digestion of *Drosophila* cholinesterase was shown to uncover a complex carbohydrate, the cross-reacting determinant antigen (CRD) using CRD-specific antibodies. The CRD antigen was originally described on the soluble form of the variant surface glycoprotein (VSG) of *Trypanosoma brucei* (*28*). Anti-CRD antibodies recognized the hydrophilic dimeric form present in head *Drosophila* extracts suggesting the existence of an endogenous phospholipase (*15*).

The protein sequence deduced from cDNA exhibits a C-terminal hydrophobic polypeptide of 30 amino acids. Such a C-terminal hydrophobic extension was shown to occur in all G-PI anchored protein precursors (see for reviews *29-31*). It is thought to be used as a temporary anchorage to the membrane before its rapid exchange to a G-PI anchor. By analogy with *Torpedo* cholinesterase, the C-terminal amino acid of the mature protein which is amide-linked to the ethanolamine is supposed to be the Cys615 in *Drosophila* (*20*). The proof of C-terminal hydrophobic peptide cleavage was found using polyclonal antibodies raised against a fusion protein containing the hydrophobic peptide sequence. These antibodies failed to recognize the protein

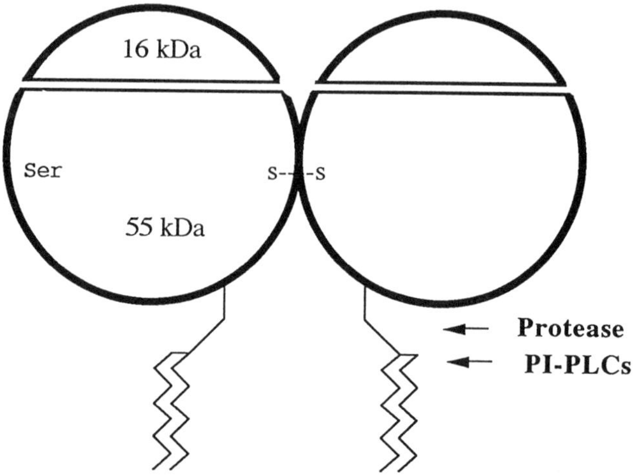

Figure 1. A model for *Drosophila* cholinesterase structure. The enzyme has been characterized as a dimeric protein composed of two active units covalently associated. each is composed of two polypeptides (55 and 16 kDa) noncovalently associated. Cholinesterase is an amphiphilic protein linked to the membrane via a glycolipid anchor located at the C-terminal end of the 55 kDa polypeptide.

```
MAISCRQSRV LPMSLPLPLT IPLPLVLVLS LHLSGVCGVI DRLVVQTSSG  50
PVRGRSVTVQ GREVHVYTGI PYAKPPVEDL RFRKPVPAEP WHGVLDATRL 100
SATCVQERYE YFPGFSGEEI WNPNTNVSED CLYINVWAPA KARLRHGRGA 150
NGGEHPNGKQ ADTDHLIHNG NPQNTTNGLP ILIWIYGGGF MTGSATLDIY 200
NADIMAAVGN VIVASFQYRV GAFGFLHLAP EMPSEFAEEA PGNVGLWDQA 250
LAIRWLKDNA HAFGGNPEWM TLFGESAGSS SVNAQLMSPV TRGLVKRGMM 300
QSGTMNAPWS HMTSEKAVEI GKALINDCNC NASMLKTNPA HVMSCMRSVD 350
AKTISVQQWN SYSGILSFPS APTIDGAFLP ADPMTLMKTA DLKDYDILMG 400
NVRDEGTYFL LYDLIDYFDK DDATALPRDK YLEIMNNIFG KATQAEREAI 450
IFQYTSWEGN PGYQNQQQIG RAVGDHFFTC PTNEYAQALA ERGASVHYYY 500
FTHRTSTSLW GEWMGVLHGD EIEYFFGQPL NNSLQYRPVE RELGKRMLSA 550
VIEFAKTGNP AQDGEEWPNF SKEDPVYYIF STDDKIEKLA RGPLAARCSF 600
WNDYLPKVRS WAGTCDGDSG SASISPRLQL LGIAALIYIC AALRTKRVF  649
```

Figure 2. Primary structure of *Drosophila* cholinesterase. The arrows indicate the N-terminal end and the presumed C-terminal amino-acid of the mature protein. The 33 amino-acids of the hydrophilic peptide which is absent in vertebrate cholinesterases are underlined. The cystein involved in the intersubunit disulfide linkage is encircled. The star marks the serine of the active site and amino-acids involved in resistance are boxed.

purified from *Drosophila* head or expressed in *Baculovirus*. By contrast, *Xenopus* oocytes were unable to perform this exchange with the insect precursor. As a consequence, the enzyme was no longer externalized and the proteolytic cutting did not occur (*10*). Expression of a cDNA, lacking the C-terminal hydrophobic peptide extension that signals for the G-PI anchor attachment, led to the secretion of the protein into the medium. This confirmed the role of this sequence as signal in the anchor attachment (*11*).

Glycoprotein. Cholinesterase purified from *Drosophila* heads is a glycoprotein that binds to lectins such as Concanavaline A-sepharose. This property has been used in purification schemes (*12*). The deduced sequence of the protein presents five potential sites of asparagine-linked glycosylation at positions 126, 174, 331, 531 and 569. Among them, four are effective sites of asparagine-linked glycosylation. Glycosylation significantly affects neither the activity of the enzyme nor its targeting. In contrast, glycosylation protects the protein against proteolytic digestion and hence is partly responsible for the high stability of the protein (*19*).

Folding of Cholinesterase.

We found mutations involved in the folding of the enzyme in two different experiments; first, by mutagenizing some amino-acids highly conserved in cholinesterases and related sequences (*32*), and second, by analyzing thermosensitive mutant flies. In order to identify amino acids important for the enzyme function, we first mutated the Asx_{248} (aspartate) to an asparagine; this position is highly conserved in cholinesterases as well as in cholinesterase-like proteins devoid of esterase activity. This mutation had a drastic effect on the activity and on the secretion of the protein. *Xenopus* oocytes injected with a gene bearing this mutation produced an inactive protein which remained sequestered inside the cell, misfolded and aggregated with other proteins. Ace^{J40} is a heat sensitive mutant isolated by Greenspan *et al.* (*33*). In homozygous conditions, this mutation is lethal when flies are raised above 25°C. At permissive temperature, flies have 30% of wild type activity, but the mutation does not disturb the overall structure of the enzyme (*34*). We amplified by PCR the exons of the gene encoding cholinesterase from DNA extracted from homozygous flies and we found one mutation, the Pro_{75} changed to a leucine. We expressed the mutated protein. The secreted enzyme was not modified in its catalytic or structural properties, but secretion was temperature dependent, higher at 20°C than at 25°C. Specific activity of the enzyme found inside the oocyte, *en route* to the external medium was lower, suggesting that a part of the protein was misfolded and remained sequestered in the secretory pathway (*35*). We tested this hypothesis by comparing patterns obtained in gels run in denaturing and native conditions. We observed a difference between the two gels suggesting that thermosensitive mutation results in the misfolding of a part of the protein, which is inactive and not secreted. Proline 75 is highly conserved in cholinesterases and related proteins and might be important for the folding due to its cyclic structure. Identically, we studied a cold sensitive mutant, Ace^{J29}. This mutant is lethal when flies are raised under 23°C (*33*). No enzyme alteration could explain the lethality of the flies at restrictive temperature. We found one mutation: the Ser_{314} is changed to a phenylalanine. We expressed the mutated protein. Excretion was temperature dependent (higher at 25°C than at 20°C) due to the misfolding of a part of the protein.

Kinetics.

Substrate specificity of insects cholinesterase is intermediate between those of mammalian cholinesterase and butyrylcholinesterase (EC 3.1.1.8) (*36*). Knowledge of structure-activity relationship of wild type cholinesterase is a prerequisite for studying the effect of mutations found in resistant insects. We first studied the catalytic behavior of the enzyme with respect to substrate hydrolysis. Cholinesterase catalysis occurs via an acyl-enzyme mechanism with two enzyme-substrate intermediates: first, an addition (Michaelis) complex (ES) and, then, an acetylated enzyme (EA). However, most cholinesterases do not follow the Michaelis-Menten profile. Vertebrate acetylcholinesterase displays an inhibition by excess of substrate and vertebrate butyrylcholinesterase displays an activation at intermediate substrate concentration (*37*). In insects, hydrolysis of choline and thiocholine esters also deviates from the simple Michaelis-Menten model: kinetics of *Drosophila* cholinesterase is triphasic, displaying complexities of both butyrylcholinesterase and cholinesterase. At 25°C and pH7, the K_m for acetylthiocholine was 4 μM. At intermediate substrate concentration (0.03 mM < [ASCh] < 1 mM), there was an activation (Km_2 = 40 μM) which was followed by inhibition (K_{ss} = 38 mM) at higher substrate concentration ([ASCh] > 1 mM) (Figure 3).

In vertebrate cholinesterase, the inhibition is due to the binding of a second substrate molecule at the mouth of the active site gorge which involves several aromatic residues (*37-38*). Only one of them (Tyr_{408}) is conserved among cholinesterases according to the alignment of Krejci *et al.* (*32*) suggesting that other amino acids may be responsible for substrate inhibition in *Drosophila*.

The activation binding site seems to be different from the peripheral binding site since the same protein exhibits both activation and inhibition behavior depending on substrate concentration range. In order to test this hypothesis, inhibition studies were performed with substrate analogs; *i.e.* compounds bearing a quaternary nitrogen atom such as propidium, edrophonium, tetramethylammonium and choline. At a fixed concentration, these compounds are strong inhibitors of the activation but had no effect on substrate inhibition suggesting that concentration-dependent activation of inhibition by acetylthiocholine is due to binding at two different sites which have different affinities for the substrate as well as for substrate analogs.

Activation may result either from the binding of a substrate molecule on a specific site or on the choline binding site (Trp_{121}) of the acyl enzyme intermediate as hypothesized by Ericksson and Augustisson (*39*) for horse serum butyrylcholinesterase. At low substrate concentration, the Arrhenius plot was linear, and nucleophile competition using methanol showed that deacylation was the rate limiting step ($k2>k3$). On the contrary, at substrate concentration leading to activation (up to 30 μM), acylation became partly limiting at 25°C ($k2=k3$) and was the rate limiting step at 15°C. Thus, activation results in a temperature induced increase of deacylation rate which favor the second hypothesis

Resistance to Insecticides.

Mutations Found in Resistant Strains. Since the early 1950s, organophosphates and carbamates have been widely used to control insect pests around the world. These insecticides are hemisubstrates that inactivate cholinesterase by phosphorylating or carbamylating the active serine (*40*). A mechanism of insect resistance to these insecticides consists in the alteration of cholinesterase which becomes less sensitive to their inhibition. In 1964, Smissaert described for the first time a resistant acarine carrying a modified cholinesterase(*41*). Altered cholinesterases were detected in several resistant insect species such as aphids, Colorado potato beetles or mosquitoes (review in *42*). Since this review, we found another resistant cholinesterase in the pear psylla (*43*) suggesting that this list is far from exhaustive. Resistance is very variable, it varies from 2 to 200,000 fold depending on species or strains. To understand this variability, twenty-two *Drosophila* field strains were selected for parathion resistance. Most of them harbored an altered cholinesterase resistant to paraoxon. The level of resistance was variable among the strains suggesting that several mutations may be responsible for resistance of the enzyme. The first method to test this hypothesis was to investigate cross-resistance towards several inhibitors. Resistance patterns obtained for each cholinesterase allowed four types of proteins to be distinguished, confirming that resistance originates from several mutations (*44*). Analysis of the *Ace* gene sequence from these strains resulted in the identification of five point mutations (*45*).

Three dimensional structure of the *Torpedo californica* acetylcholinesterase has recently been resolved (*46*). This structure allowed then to localize the mutations which all occur around the active site gorge. Interestingly, some of the mutations were found to be identical in other insects showing that *Drosophila* may represent a model insect to study insecticide resistance mechanisms in agricultural and medical pests. On the other hand, in some insects such as the house fly, other mutations different from those of *Drosophila* were detected (Williamson and Devonshire, pers. comm.). This indicates that there are a lot of modifications which can modify the active site conformation and/or reactivity leading to a resistant enzyme while preserving the acetylcholine hydrolase activity. Each mutation provides a weak level of resistance which is not sufficient to confer alone a selective advantage to the insect when exposed to an insecticide. But a weak mechanism may be selected when it is in association with other resistance mechanisms such as either increased degradation of insecticides by oxidases, esterases or glutathione transferases or decreased penetration of the insecticide through the insect cuticle (*47*).

In most *Drosophila* resistant strains, several mutations were found in the same protein. Combinations of several mutations in the same protein gave highly resistant enzymes (Figure 4). This result suggests that resistance originates from recombination between single mutated alleles preexisting in natural populations besides recombination between different genes (*45*). This mechanism explains the high diversity of mutated proteins found in natural populations and would allow insects to rapidly adapt for new selective pressures. As recombination implies mixing of differents populations, outbreeding would favor occuring of resistance. Thus, resistance by this mechanism does not appear as a ramdom mechanism but depends on the biology of species.

Putative Mutations. Insecticides have been designed to be more effective against insects than against vertebrates. One component of this specificity originates from the cholinesterase active site. *Drosophila* cholinesterase is more susceptible to insecticides than vertebrate enzymes. The two active sites are conserved except for some residues. For example, Ile199 in the *Drosophila* sequence corresponds to a valine in vertebrates. In some resistant *Drosophila* strains, we found a valine at position 199 instead of isoleucine, showing that mutations which convert the insect enzyme form to a vertebrate enzyme form will result in resistance. From this observation we may hypothesize that several of the differences in the active site between the insect and the vertebrate enzyme may be responsible for resistance. One of them is Tyr_{109} in *Drosophila* cholinesterasewhich corresponds to an aspartate in vertebrate sequences. Mutation of this amino acid to glycine in the human butyrylcholinesterase gives rise to an "atypic" phenotype characterized by a reduced activity for charged compounds (*48*). We investigated the importance of Tyr_{109} by substituting a glycine, an aspartate, or a lysine using *in vitro* mutagenesis. We then expressed the mutant proteins in *Xenopus* oocytes. These mutations affected some catalytic properties of the enzyme and its sensitivity to insecticides. The mutated enzymes were different from the wild type, either more susceptible or more resistant (*49*).

Biochemical Markers Related to Resistance. Resistance is due to modifications of the active site; these modifications change the catalytic activity of the enzyme towards insecticides and probably towards substrates or inhibitors. These compounds may be used as biochemical markers to identify a mutation in field populations. As an example, we studied Phe_{368} which is mutated to tyrosine in some resistant strains. We mutated this residue in Tyr, Trp, Ser and Gly and expressed the protein in

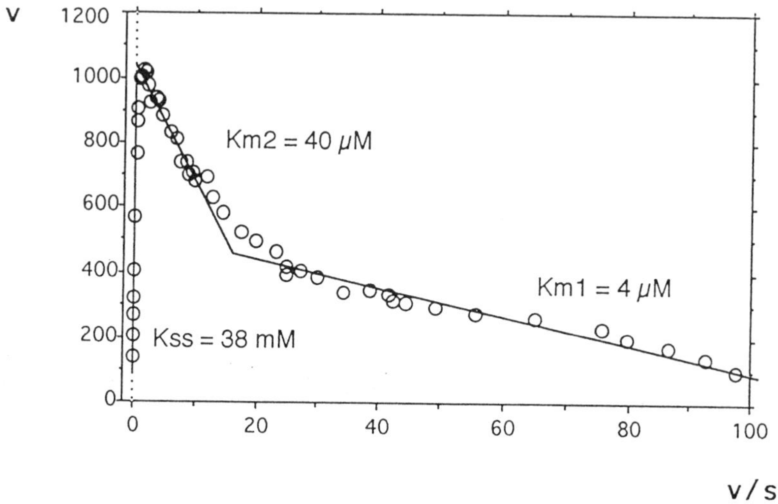

Figure 3. Eadie-Hoffstee substrate activity curves for acetylthiocholine showing the activation (Km2) and inhibition by excess of sbstrate (Kss)

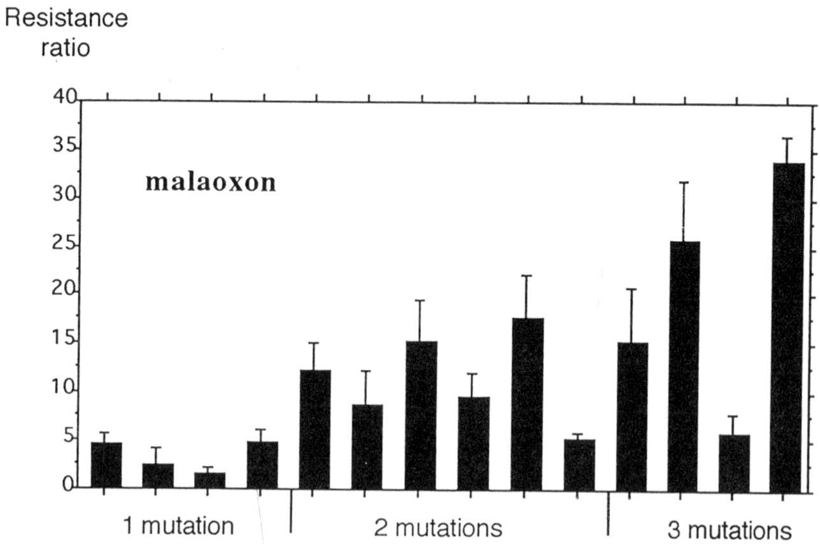

Figure 4. Effects of single and combined mutations on resistance of acetylcholinesterase to malaoxon.

Baculovirus infected cells. The amino acid at this position is important for affinity of the enzyme to organophosphates and hence for resistance. Harel *et al.* (*50*) and Vellom *et al.* (*51*) showed that the corresponding phenylalanine in the Mouse and Torpedo enzyme is located at the bottom of the gorge and may account for the specificity of substrate of the enzyme in vertebrates. As most insect cholinesterase metabolize butyrylcholine, although less efficiently than acetylcholine (*36*), we hypothesized that depending on the amino acid found at that position, we have either an increase or a decrease in affinity for butyrylcholine. Preliminary results seem to confirm the importance of mutations on Phe_{368} in insect for substrate specificity (Table 1).

Table 1. Substrate specificity and mutations at position 368

Phe (wild type)	ASCh (1) > PrSCh (0.7) > BuSCh (0.57)
Tyr	BuSCh (1.2) > PrSCh (1.2) > ASCh (1)
Trp	ASCh (1) > PrSCh (0.79) > BuSCh (0.04)
Gly	PrSCh (1.23) > ASCh (1) > BuSCh (0.77)
Ser	PrSCh (1.2) > ASCh (1) = BuSCh (1)

ASCh: acetylthiocholine, BuSCh: butyrylthiocholine, PrSCh: propionylthiocholine

If other mutations do not produce this effect, we will be able to predict the mutations involved in resistance from biochemical data in insects from which the cholinesterase gene has not been cloned. Changes in butyrylthiocholine hydrolysis and affinity were observed along with resistance. In some cases, the resistant enzymes were more active or had higher affinity for butyrylcholine (*52-53*), while in other cases the affinity for butyrylcholine decreased (*54*). Phe_{368} appears as a good candidate to be involved in resistance in these resistant strains.

Sensitivity to Insecticides and Cholinesterase Content. In *Drosophila*, cholinesterase is encoded by one gene. Amount of enzyme produced by the two copies of this gene is easily determined by scoring its activity with a spectrophotometric assay by using acetylthiocholine as substrate. P-element transformation with a minigene allowed to construct flies with different cholinesterase content in the central nervous system.Transformation of wild type flies with the *Ace* minigene gives rise to an extra gene. These flies have about 120-130% of wild type activity. Although the deficiency of the *Ace* gene is lethal, null flies can be rescued with a *Ace* minigene. These flies display 20-25% of the wild type activity (*9*). Furthermore, several *Ace* mutations have been isolated (*1-33*). Strains bearing a lethal *Ace* mutation can be maintained in heterozygous state, and such flies display half activity of wild type which corresponds to the inactivation of one allele not compensated by the active one. Thus, we obtained flies with 20-25%, 50%, 100% and 120% cholinesterase activity in the central nervous system. Toxicological analysis of these strains using malathion revealed a good correlation between the cholinesterase amounts and resistance to this insecticide (*55*). Does this mechanism exist in field populations? El-abidin Salam and Pinsker (*56*) reported an increase in cholinesterase activity by selecting *D. melanogaster* for resistance to parathion and fenthion. Similarly, other authors reported an increase in Vmax of cholinesterase in other insects, but this activity increase may reflect either an overproduction of the enzyme or a qualitative change leading to a modification of catalytic parameters.

Thus, we do not know yet if this resistance mechanism is significantly participating in insecticide resistance of field populations. Nevertheless, this dose effect explains the genetics of qualitative changes, why resistance is controlled by an incompletely dominant factor. Heterozygous insects bear only 50% of resistant AChE and hence are more susceptible to high doses of insecticides than homozygous insects which bear 100% of resistant AChE (*42*).

References

1. Hall, J.C.; Kankel, D.R. *Genetics* **1976**,*83*, 517-535.
2. Gausz, J.; Hall, L.M.C.; Spierer, A.; Spierer, P. *Genetics* , **1986**, *112*, 65-78.
3. Nagoshi, R.N.; Gelbart, W.M. *Genetics* **1987**, *117*, 487-502.
4. Bender, W.; Spierer, P.; Hogness, D.S. *J. Mol. Biol.*, **1983**, *168*, 17-33.
5. Hall, L.M.C.; Spierer, P. *EMBO J.* **1986**, *5*, 2949-2954.
6. Fournier, D.; Karch, F.; Bride, J.M.; Hall, L.M.C.; Berge, J.B.; Spierer, P. *J. Mol. Biol.* **1989**, *210*, 15-22.
7. Cohloun, E.H. *J. Agric. Food Chem.* **1960**, *8*, 252.
8. Arpagaus, M.; Fournier, D.; Toutant, J.P. *Insect. Biochem.* **1988**, *6*, 539-549.
9. Hoffmann, F.; Fournier, D.; Spierer, P. *J. Mol. Biol.* **1992**, *223*,17-22.
10. Fournier, D.; Mutero, A.; Rungger, D. *Eur. J. Biochem.* **1992a**, *203*, 513-519.
11. Chaabihi, H.; Fournier, D.; Fedon, Y.; Bossy, J.P.; Ravalec, M.; Devauchelle, G.; Cérutti, M. *Biochem. Biophys. Res. Comm.* **1994**, *203*, 734-742.
12. Steele, R.W.; Smallman, B.N. *Bioch. Biophys. Acta.* **1976**, *445*, 147-157.
13. Gnagey, A.L.; Forte, M.; Rosenberry, T.L. *J. Biol. Chem.* **1987**, *262*, 1140-1145.
14. Fournier, D.; Cuany A.; Bride J.M.; Bergé J.B. *J. Neurochem.* **1987**, *49*, 1455-1461.
15. Fournier, D.; Bergé, J.B.; Cardoso de Almeida, M.L.; Bordier C. *J. Neurochem.* **1988a**, *50*, 1158-1163.
16. Toutant, J.P.; Arpagaus, M.; Fournier, D. *J. Neurochem.* **1988**, *50*, 209-218.
17. Prody, C.A.; Zevin-Sonkin, D.; Gnatt, A.; Goldberg, O.; Soreq, H. *Proc. Natl. Acad. Sci. USA* , **1987**, *84*, 3555-3559.
18. Lockridge, O.; Bartels, C.F.; Vaughan, T.A.; Wong, C.K.; Norton, S.E.; Johnson, L.L. *J. Biol. Chem.*, **1987**, *262*, 549-557.
19. Mutero, A.; Fournier, D. *J. Biol. Chem.* **1992**, *267*, 1695-1700.
20. Haas, R.; Marshall, T.C.; Rosenberry, T.L. *Biochemistry* **1988**, *27*, 6453-6457.
21. Fournier, D.; Bride, J.M.; Karch, F.; Bergé, J.B. *FEBS Lett.* **1988b**, *238*, 333-337.
22. Malcolm, C.A.; Hall, L.M.C. 1990, In *Molecular Insect Science*, Hagedorn, H.H. *et al.* Eds, Plenum Press, New York, pp 57-64.
23. Haas, R.; Brandt, P.T.; Knight, J.; Rosenberry, T.L. *Biochemistry* **1986**, *25*, 3098-3105.
24. Roberts, W.L.; Rosenberry, T.L. *Biochemistry* **1986**, *25*, 3091-3098.
25. Roberts, W.L.; Kim, B.H.; Rosenberry, T.L. *Proc. Natl. Acad. Sci. USA* **1987**, *84*, 7817-7821.
26. Futerman, A.H.; Low, M.G.; Michaelson, D.M.; Silman, I. *J. Neurochem.* **1985**, *45*, 1487-1494.
27. Arpagaus, M.; Toutant, J.P., *Neurochem. Int.*, **1985**, *7*, 793-804.
28. Cardoso de Almeida, M.L.; Turner, M.J.; *Nature*, **1983**, *302*, 349-352.
29. Cross, G.A.M., *Cell* , **1987**, *48*, 179-181.
30. Low, M.G. *Biochem. J.*, **1987**, *244*, 1-13.
31. Ferguson, M.A.J.; Williams, A.F. *Ann. Rev. Biochem.*, **1988**, *57*, 285-320.

32. Krejci, E.; Duval, N.; Chatonnet, A.; Vincens, P.; Massoulié, J. *Proc. Natl. Acad. sci. USA* **1991**, *88*, 6647-6651.
33. Greenspan, R.J.; Finns J.A.; Hall J. *J. Comp. Neurol.* **1980**, 189, 741-774.
34. Toutant, J.P.; Arpagaus, M. **1989**, In *Insect Neurochemistry and Neurophysiology*, Borkovec, A.B. Ed., Humana press zz, 239-242.
35. Mutero, A.; Bride, J.M.; Pralavorio, M.; Fournier, D. *Mol. Gen. Genet.* **1994**, *243*, 699-705.
36. Toutant, J.P. *Progress Neurobiol.* **1989**, *32*, 423-446.
37. Radic, Z.; Pickering, N.A.; Vellom, D.; Camp, S.; Taylor, P. *Biochemistry* **1993**, *32*, 12074-12084.
38. Schafferman, A.; Velan, B.; Ordentlich, A.; Kronman, C.; Grosfeld, H.; Leitner, M.; Flashner, Y.; Cohen, S.; Barak, D.; Ariel, N. *EMBO J.* 1992, **11**, 3561-3568.
39. Ericksson, H.; Augustinsson, K.B. *Biochem. Biophys. Acta* **1979**, *567*, 161-173.
40. Aldridge, W.N. (1950) *Biochem. J.* 1950, **46**, 451-460.
41. Smissaert, H.R. *Science* **1964**,*143*, 129-131.
42. Fournier. D.; Mutero, A. *Comp. Biochem. Physiol.* **1994**, *108C*, 19-31.
43. Berrada, S.; Fournier, D.; Cuany, A.; Nguyen, T.X. *Pest. Biochem. Physiol.* **1994**, 48, 41-47.
44. Pralavorio, M.; Fournier, D. *Biochem. Genet.* **1992**, *30*, 77-83.
45. Mutero A.; Pralavorio M.; Bride J.M.; Fournier D. *Proc. Natl. Acad. Sci. USA* 1994,**91**, 5922-5926.
46. Sussman, J.L.; Harel, M.; Frolow, F.; Oefner, C.; Goldman, A.; Toker, L.; Silman, I. *Science*, **1991**, *253*, 872-878.
47. Raymond, M.; Heckel, D.G.; Scott, J.G. *Genetics*, **1989**, *123*, 543-551.
48. Lockridge, O.; LaDu, B.N. *J. Biol. Chem.*, **1978**, *253*, 361-366.
49. Mutero, A.; Pralavorio, M.; Simeon, V.; Fournier, D. *Neuroreport* **1992**,*3*, 39-42.
50. Harel, M.; Sussman, J.L.; Krejci, E.; Bon, S.; Chanal, P.; Massoulié, J.; Silman, I. *Proc. Natl. Acad. Sci. U.S.A* **1992**, *89*, 10827-10831.
51. Vellom, D.C.; Radic, Z.; Li, Y.; Pickering, N.A.; Camp, S.; Taylor, P. *Biochemistry* **1993**, *32*,12-17.
52. Hama H.; Iwata T.; Miyata T.; Saito T. *Appl. Ent. Zool.* **1980**,*15*, 249-261.
53. Brown, T.M.; Bryson, P.K. *Pestic. Biochem. Physiol.* **1992**, *44*, 155-164.
54. O'Brien, R.D.; Tripathi, R.K; Howell, L.L. *Biochim. Biophys. Acta* **1978**, *526*, 129-134.
55. Fournier, D.; Bride, J.M.; Hoffmann, F.; Karch, F. *J. Biol. Chem.* **1992b**, *267*, 14270-14274.
56. El-Abidin Salam, A.Z.; Pinsker,W. *Genetica* **1981**, 5, 11-14.

Chapter 4

An Overview of Auxinic Herbicide Resistance: Wild Mustard (*Sinapis arvensis* L.) as a Case Study

J. Christopher Hall, Steven R. Webb, and Satish Deshpande

Department of Environmental Biology, Ontario Agricultural College, University of Guelph, Guelph, Ontario N1G 2WI, Canada

This paper provides an overview of the mode of action of auxinic herbicides and the physiological and biochemical basis for resistance to these herbicides using the wild mustard (*Sinapis arvensis* L.) biotypes as example. The resistant (R) wild mustard, found in Western Canada, was resistant to the auxinic herbicides picloram, dicamba, 2,4-D and MCPA while the susceptible (S) biotype became severely epinastic with increases in ACC synthase activity, and ACC, MACC and ethylene levels upon herbicide treatment. Morphologically, the R biotype was shorter and more branched than the S biotype and had more numerous leaves with a higher chlorophyll and cytokinin content. The auxin binding characteristics of the two biotypes were also found to be different. Light scattering studies showed that picloram affected the calcium transport processes in the S but not the R biotype. A study of the wild mustard at the molecular, cellular and whole plant level promises to answer many questions about the mechanism(s) of action and resistance to this important class of herbicides.

The auxinic herbicides were the first selective organic herbicides developed and have been used in agriculture for over fifty years primarily for the selective control of broadleaf weeds in grass crops (1). Auxinic herbicides include members of the phenoxyalkanoic acid (eg. 2,4-D, MCPA, MCPP), benzoic acid (eg. dicamba, chloramben), and pyridine (eg. picloram, clopyralid, triclopyr) classes of herbicides. These herbicides are thought to be mimics of the natural plant hormone IAA (indole-3-acetic acid, auxin). When applied at the proper doses to plants, a number of physiological and morphological changes including cell elongation, epinasty, hypertrophy, root initiation, and ethylene biosynthesis occur which result in the death of susceptible species (1,2). These physiological and morphological changes can be attributed to several biochemical responses resulting from auxinic herbicide (or auxin) exposure, namely a rapid Ca^{2+} influx, activation of plasma membrane ATPase, and increases in both nucleic acid and protein biosynthesis. The biochemical changes are believed to be elicited by the interaction of the auxinic-herbicide ligand with a putative

0097–6156/96/0645–0028$15.00/0

auxin binding protein (ABP) that results in a signal that initiates the observed cascade of biochemical, physiological and morphological events.

Only a few cases of resistance to the auxinic herbicides have been reported despite their prolonged and wide spread use (3,4,5). However, auxinic-herbicide resistant biotypes of several weed species have been identified, including *Sphenoclea zeylanica* Gaertn. (gooseweed) (6), *Stellaria media* L.(chickweed) (3) and *Ranunculus acris* L. subsp. acris (buttercup or meadow buttercup) and *Carduus nutans* (nodding thistle) (7,8). Recently, populations of wild mustard (*Sinapis arvensis* L.) with cross resistance to dicamba, dichloroprop, 2,4-D, mecoprop, MCPA and picloram have been described (5,9,10). The purpose of this paper will be to: i) briefly review the history and mode of action with specific emphasis on selectivity, physiological and biochemical effects, and the mechanism of action of the auxinic herbicides, ii) discuss the cases of plant species resistant to the auxinic herbicides, and iii) specifically review the basis of resistance of wild mustard to the auxinic herbicides in the context of biochemical and molecular genetic mechanisms and discuss how these studies have provided new insights into the mode of action of the auxinic herbicides.

History of the Auxinic Herbicides

The history of the discovery and development of the auxinic herbicides has been reviewed (11). Initially, Darwin reported in the 1880's on phototropism prompting other scientists to search for its cause. Phototropism was linked to a growth substance which was later found to be IAA. Zimmermann and Hitchcock working at the Boyce Thompson Institute synthesized many compounds belonging to the phenoxyacetic acid and benzoic acid families of plant growth regulators with auxin-like effects on plants (cited in 11). In 1942, they reported that the growth effects of 2,4-D were considerably more powerful than IAA. Subsequently, Mitchell and Hamner suggested that 2,4-D could be used as a herbicide. In August of 1944, Mitchell and Marth using 2,4-D selectively eliminated dandelions infesting a lawn within 3 weeks.

Selectivity of the Auxinic Herbicides

Selectivity between dicotyledonous and monocotyledonous species has made the auxinic herbicides one of the most widely used and important herbicide groups (1,2,12,13). This selectivity is primarily attributed to differences between their morphology and/or the rate of translocation and metabolism of the auxinic herbicides. Destruction of the phloem of dicotyledonous species as the result of abnormal tissue proliferation is avoided in tolerant monocotyledonous species where the phloem is scattered in bundles each surrounded by protective sclerenchyma tissue. In addition, monocotyledonous species, unlike dicotyledonous species, have vascular bundles that do not have a cambium and pericycle both of which are sensitive to these herbicides. Translocation of the auxinic herbicides is restricted in monocotyledonous species mainly because of the presence of the intercalary meristem in the stems and leaves (1). Furthermore, rapid metabolism of auxinic herbicides and differences in metabolic products have been suggested as possible mechanisms for the observed selectivity between the two classes (14,15). It is suggested that the herbicide is more likely to conjugate with a carbohydrate moiety in monocotyledonous species whereas in dicotyledonous species, conjugation to amino acids is more common. Amino acid conjugates in dicotyledonous species appear to be less metabolically stable than their corresponding carbohydrate-auxin conjugates. Therefore, even though auxinic herbicides can be conjugated in susceptible dicotyledonous species, the herbicide can easily be mobilized by hydrolysis and continue to exert its toxic effects.

Similar mechanisms as described above have been proposed to explain selectivity differences within dicotyledonous species. For instance, differences in 2,4-D susceptibility between red and black currants were attributed to differences in metabolism (16). Within 24 h of foliar application tolerant red currant degraded approximately 50% of the 2,4-D acetic acid side chain, compared with 2% for susceptible black currant. As a result of increased metabolism, red currant is less susceptible to 2,4-D than black currant. Another factor attributed to selectivity differences within dicotyledonous species is differences in sensitivity at the target site(s) as indicated by studies involving rapeseed (17,18). Differences in rapeseed response to clopyralid and picloram provide indirect evidence of differential target site sensitivity. Rapeseed is tolerant to clopyralid but susceptible to picloram, even though both are structurally similar and members of the pyridine class of auxinic herbicides. There were no differences in the uptake, translocation or metabolism of these two herbicides in rapeseed that could account for differences in sensitivity. Additional evidence supporting differential sensitivity at the target site(s) was provided by demonstrating that pretreatment of rapeseed with clopyralid protected plants from damage by picloram (19). These studies suggest clopyralid and picloram act at a common site, but the binding of the two herbicides at this site is different.

Physiological and Biochemical Effects. Auxinic herbicides generally share several characteristics with IAA which include: i) dose-response patterns where stimulatory effect on plant cell growth is seen at low doses while phytotoxic effects are seen at high doses, ii) replacement of IAA as the hormone supplement in tissue culture, iii) differential sensitivity to auxinic herbicides among different tissue types (roots, buds, active meristem and callus) as well as between tissues at different physiological stages of growth, and iv) induction of cell growth by cell elongation, as opposed to cell division; the cell elongation response is composed of an initial rapid growth response occurring within 7 to 10 minutes and a long-lasting growth response occurring within 30 to 45 minutes (12,13,20). In addition, because IAA and 2,4-D appear to regulate plant growth by similar modes of action (21), many studies characterizing IAA effects or mode of action, replace IAA with 2,4-D or other synthetic auxins (22,23). The following summarizes numerous effects of auxinic herbicides.

Physiological effects include:

1. Epinasty (twisting or downward curling of leaves, stems, or petioles) in susceptible plants resulting from differential rates of cell growth in these organs. The upper tissue side growing faster than the lower side, resulting in the characteristic downward curling.
2. Stimulation of ethylene production. The relationship between auxin and ethylene is important because ethylene is also an endogenous plant hormone.

3. Abnormal or amplified stem tissue growth which may result in senescence and desiccation of tissues. Abnormal tissue enlargement and excessive growth is termed hypertrophy.
4. The production of adventitious roots following auxin treatment.
5. Changes in cell wall integrity following auxinic herbicide treatment which involves a loosening or plasticization of the cell wall structure. This is followed by wall extension due to cell turgor pressure. Several studies have shown auxinic herbicide induced cell expansion and decreased cell wall integrity are prevented by treatment with inhibitors of RNA and/or protein synthesis (24). These studies indicate continued cell elongation depends on stimulation of mRNA and protein synthesis which likely code for cell wall modifying enzymes. The physiological response of decreased cell wall integrity which results in cell elongation, appears to be linked to auxin induced increases in mRNA and protein synthesis, and increased H^+ extrusion through the plasma membrane to the cell wall.

Biochemical effects include:

1. Auxinic herbicide stimulation of nucleic acid and protein biosynthesis which results in the physiological changes associated with stimulation of auxin induced cell growth (e.g., stem tissue proliferation, root initiation, ethylene evolution, etc.).
2. Hydrogen ion extrusion through the plasma membrane which occurs quickly (7 to 8 minutes) following auxin or auxinic herbicide treatment. The increased proton pumping is generally attributed to the activation of a plasma membrane ATPase. The mechanism of ATPase activation is unclear, but it is thought to be triggered by a rapid influx of calcium ions resulting from the binding of auxinic herbicides to specific receptors on the plasma membrane. This induces a cascade of biochemical events possibly involving calcium as a secondary messenger (25,26). The involvement of calcium as the secondary messenger suggests auxin causes a rapid and transient calcium increase in the cytoplasm opening of calcium channels resulting in a rapid depolarization of the plasma membrane and a subsequent hyperpolarization all within 15 minutes of auxin treatment.

Mode of Action of the Auxinic Herbicides

The phytotoxic action of auxinic herbicides is likely associated with the eventual disruption of the hormone balance in plant cells. Under normal physiological conditions plants closely regulate the concentration of IAA in their tissues. Auxin levels are maintained by controlling the rates of IAA synthesis, import, export, and degradation as well as reversible and irreversible conjugation of IAA in tissues. The major reason an auxinic herbicide is phytotoxic is probably due to the inability of the plant to regulate the internal concentration of the auxinic herbicide. Therefore, concentration of auxinic herbicides in tissues becomes high, and as a result, auxin interaction with other plant hormones (cytokinin etc.) required for regulating plant metabolism and growth is disrupted.

The critical event believed to initiate the cascade of biochemical effects leading to hormonal imbalance is the interaction of the herbicide with a specific receptor or auxin-binding protein. Several research groups have identified several auxin-binding proteins (ABPs) that may have receptor function (27). Auxin-binding proteins have been localized to the plasma membrane, the endoplasmic reticulum and the nucleus. Numerous locations of auxin-binding proteins within the cell have made the task of assigning a particular role difficult.

Although ABPs have been demonstrated to bind auxin and auxinic herbicides, direct experimental evidence for the auxin-binding protein transmitting the binding event to elicit an appropriate intracellular response is absent. There is indirect evidence these proteins have receptor function. For instance, antibodies specific for ABPs block some auxin induced biochemical and physiological changes (28). The addition of purified ABP to normal protoplasts was found to increase sensitivity to auxin treatment. Furthermore, evidence for the existence of receptor function was provided by the discovery that clopyralid is an antagonist of picloram in rapeseed (18,19).

The close relationship between ABP and the mode of auxinic herbicide action will be examined using the example of wild mustard in a later section.

Resistance to the Auxinic Herbicides

Because the auxinic herbicides are believed to have multiple modes of action and do not persist in the soil, they are placed in the low risk category in terms of the development of herbicide resistance. Resistance to these herbicides was not anticipated because they had been used for over 40 years with only a few reports of resistance. However, repeated applications of herbicides under field conditions has selected weeds with resistance to auxinic herbicides. Populations of *Kochia scoparia* L. with differential sensitivity to 2,4-D and dicamba, were collected in four states across the United States after repeated applications (29). Repeated use of 2,4-D was cited as the factor responsible for strains of *Erechtites hieracifolia* (L.) Raf. ex DC. and *Commelina diffusa* Burm. f. with increased resistance to 2,4-D in sugar cane fields in Hawaii (30,31). Populations of *Tripleurospermum inodorum* (L.) Schultz Bip. were found in Britain with differing levels of resistance to MCPA which could be correlated to previous exposure to the herbicide (30). 2,4-D-resistant *Carduus nutans* L. in New Zealand pastures was correlated to historical exposure to MCPA (7,8). Yellow starthistle (*Centaurea solstitialis* L.) resistant to picloram was detected in a non-arable pasture in Washington, USA that had been frequently treated with picloram over the preceding ten years (32). Also, resistance to picloram was selected in tobacco (*Nicotiana tabacum* L.) cell culture (33). Resistance to other auxinic herbicides in field-grown weed species has been reviewed in detail (2,34,35). Mecoprop-resistant *Stellaria media* L. in the United Kingdom appears to be the only case thus far in which resistance is not due to selection pressure from many years of repeated herbicide use (36,37). Generally, little is known about mechanisms of auxinic herbicide resistance in resistant weeds. Mecoprop resistance in chickweed (*Stellaria media* L.) (38) and picloram resistance in yellow starthistle (39) were not due to altered absorption, translocation or metabolism. It was determined that mecoprop resistance was not mediated by changes in H^+-ATPase activity (38).

Comparison of Resistant and Susceptible *Sinapis arvensis* L.

Populations of *S. arvensis* L. (wild mustard) resistant to the auxinic herbicides were first reported by Heap and Morrison (9). The resistant populations were found in a field in west-central Manitoba (Canada) that had been treated with a mixture of dicamba/MCPA/mecoprop for over 10 years. In growth room experiments, Heap and Morrison found that these plants were resistant to 2,4-D and MCPA and highly resistant to dicamba. In subsequent studies conducted in cooperation with Morrison, Peniuk et al. (5) found that the R biotype described by Heap and Morrison was more resistant to picloram than dicamba. Although the extent of resistance of wild mustard to the auxinic herbicides had been well documented and characterized (9), there was little information available at that time on the basis of the mechanism of resistance. Consequently, physiological and biochemical studies were conducted to elucidate the mechanism of resistance of *S. arvensis* to some of the auxinic herbicides in our laboratory. This on-going research is described in the following section.

Absorption, Translocation, Metabolism and Ethylene Biosynthesis Studies. Initial studies showed that absorption, translocation and metabolism of ^{14}C-radiolabelled 2,4-D, dicamba, and picloram were similar in the R and S biotypes of *S. arvensis* but the S biotype produced more ethylene in response to the herbicide treatment than the R biotype (5). Since it has been well documented that auxinic herbicide-induced ethylene plays a critical role in eliciting the observed morphological and physiological changes in susceptible plants after treatment with the auxinic herbicides (19,40,41,42,43), Peniuk et al. (5) investigated the differences in ethylene evolution following application of 2,4-D to the R and S biotypes. There was little or no difference between untreated R and S control plants and R plants treated with 100 g a/ha, whereas there was a six-fold increase in ethylene levels emanating from the S biotype (Figure 1). In more detailed studies, severe epinasty was observed within 24 h after picloram was applied to the S biotype with concomitant increases in ACC (1-aminocyclopropane-1-carboxylic acid), ACC synthase, MACC (1-malonylaminocyclopropane-1-carboxylic acid) and ethylene (44). No epinasty occurred in the R biotype, nor was there an increase above basal levels of ACC synthase, ACC, MACC and ethylene in this biotype. Both biotypes became epinastic when fumigated with 120 ml L^{-1} of ethylene. When the leaf discs from both biotypes were treated with 1 mM ACC after pretreatment with aminooxyacetic acid (1 mM), an inhibitor of ACC synthase, both biotypes produced ethylene indicating that ethylene forming enzyme was not impaired in the resistant biotype. On this basis, it was suggested that picloram-induced ethylene biosynthesis in the S biotype probably results from *de novo* synthesis of ACC synthase; this, however, was not the case in the R biotype (44).

Effects of auxinic herbicides on auxin binding and seedling growth. Based on these results, Webb and Hall (45) hypothesized that sensitivity differences between the R and S biotypes may be due to a different interaction of the herbicide with an auxin-binding protein(s) (ABP). ABP preparations from both biotypes have similar substrate (^{3}H-IAA binding) and time-course profiles. Using ABP preparations derived from the S biotype, Scatchard analysis revealed the presence of two populations of [^{3}H]-indole-3-acetic acid

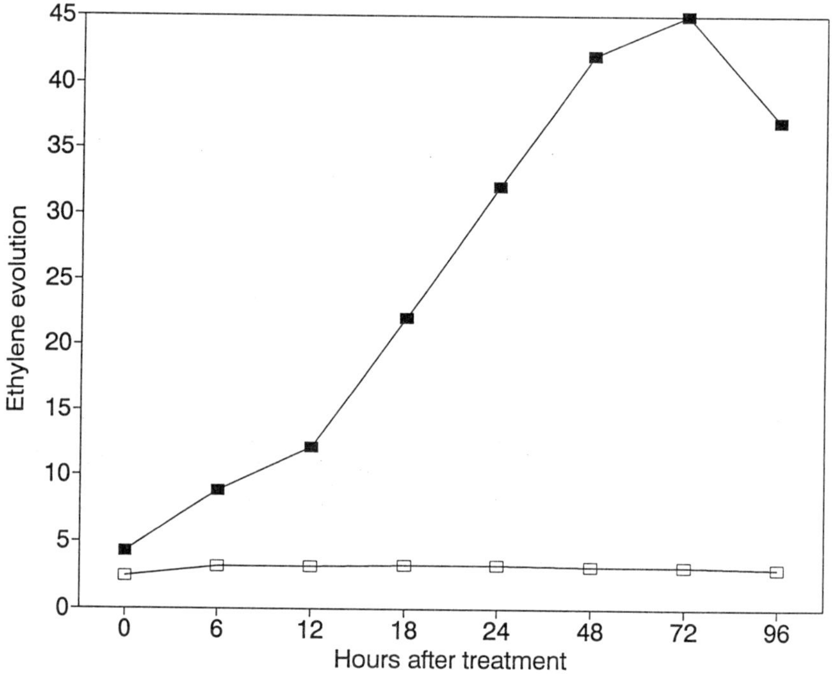

Figure 1: Ethylene evolution measured in nl hr^{-1} (g fresh weight)$^{-1}$ from the S (Filled squares) and R (Open squares) biotypes of wild mustard plants following foliar application of picloram (100 g/ha). Ethylene evolution from untreated plants from both biotypes was less than 3 nl hr^{-1} (g fresh weight)$^{-1}$

(IAA) binding sites (Figure 2, filled squares). The approximate K_d values and number of binding sites of the high- and low-affinity populations were 0.46 nM with 350 fmol/mg protein and 8.32 nM with 2110 fmol/mg protein, respectively. In contrast, Scatchard analysis of the R biotype resulted in a single population of binding sites with an estimated K_d and number of sites comparable to the S biotype's low-affinity population (Figure 2, open squares). Several auxinic herbicides were examined for their effect on [^{3}H]IAA binding activity from extracts of both biotypes. Mecoprop was the most potent inhibitor of [^{3}H]IAA binding to S biotype ABP preparations, while picloram and dicamba were substantially less potent inhibitors. The same pattern of inhibition was also observed when the effect of these herbicides on seedling growth was examined in the susceptible biotype. When picloram and dicamba were used to inhibit ^{3}H-IAA binding and seedling growth, the susceptible biotype was significantly more sensitive to inhibition than the resistant biotype ABP preparations; however, no differences between the two biotypes were observed following mecoprop treatment. It was therefore suggested that there is a relationship between ABP binding activity and sensitivity to auxinic herbicides and this relationship may eventually provide clues to explain the mechanism of auxinic herbicide resistance in this biotype (45).

Morphological and Germination Differences. There were morphological differences between the two biotypes of *S. arvensis* (46). For example, the R biotype was shorter, more branched, and had a smaller root system than the S biotype. Leaves of the R biotype were smaller, darker green, and had a higher chlorophyll content than those of the S biotype. On a physiological basis the seed germination percentage in the dark at 5, 10 and 15°C was the same for both biotypes; but at 30°C, germination was found to be 20% greater in the R biotype. Although there was no difference in percentage germination between the two biotypes at 5°C, the R biotype seedlings were approximately two times longer than the S biotype. Regardless of the biotype, the percentage germination at 24°C was greater in the light. As in the dark, more R (89%) than S (72%) seed germinated under light.

Differential Effects of Cytokinins. When compared to controls, the cytokinin benzyladenine, inhibited seedling growth of the susceptible biotype by 34, 52, and 57%, respectively, at 0.1, 1 and 10 mg/L whereas no concentration was inhibitory to the R biotype. Furthermore, the synthetic cytokinin thidiazuron (TDZ) inhibited growth of susceptible seedlings by 49, 57, and 84% at 0.1, 1 and 10 mg/L, whereas only the highest concentration of TDZ inhibited the R biotype (55%). Senescence occurred more rapidly in leaf discs of the S biotype. Furthermore, the cytokinin levels, as determined by a cucumber cotyledon greening bioassay (47), were higher in the R (1200 ng/10 g root fresh weight; 0.533 μM) than the S (320 ng/10 g root fresh weight; 0.142 μM) biotype.

High cytokinin levels are known to promote chloroplast development, chlorophyll synthesis, and lateral bud formation; they also reduce shoot and root growth (48,49). The plant hormones cytokinin and IAA are known to interact and in many cases cytokinins are auxin antagonists (48,49). This may explain why the R biotype is darker green with higher chlorophyll levels, shorter, and more branched than the S biotype (46). Furthermore, seed germination can be enhanced by the addition of kinetin or benzyladenine (50) which may explain the higher percentage germination of the R than the S biotype observed at 30°C and higher temperatures.

S. arvensis and *Arabidopsis*: Molecular Genetic Comparisons

Estelle and Sommerville (51) also found that a single gene mutation in *Arabadopsis thaliana* resulted in several auxin resistant lines which display pleiotropic morphological effects. These lines were 50 and 8 times more resistant to 2,4-D and IAA, respectively. The resistance was due to a recessive mutation at the *axr*1 locus. Resistance was not due to differences in uptake and/or metabolism of 2,4-D as was the case with *Sinapis arvensis* resistance to 2,4-D, dicamba, and picloram (5). Furthermore, the *A. thaliana* auxin resistant mutants had one distinct morphological phenotype which resulted in plants that were bushy and short with small, thin roots, and small leaves (51). These researchers hypothesized that the *axr*1 gene may encode for a different auxin receptor and resistance may be due to an alteration that has a greater effect on the affinity of this receptor for 2,4-D and IAA. In subsequent research on 20 *axr*1 mutant lines of *A. thaliana*, at least five different *axr*1 alleles were described, but all the mutants had similar phenotypes (52). The extent of auxin resistance of each mutant line could be directly correlated with severity of the morphological alterations. The distinct morphological phenotypes of the mutant *A. thaliana* described by these authors are similar to some of those described for the R biotype of *S. arvensis*. It has been shown that dicamba resistance in *S. arvensis* was determined by a single, completely dominant nuclear allele (10). Therefore, the mutation of *S. arvensis* may cause the pleiotropic morphological effects seen in the R biotype and also may be responsible for a general reduction in the sensitivity of this biotype to exogenously applied auxinic herbicides. Furthermore, our laboratory has shown that there is a difference in the ABP of the R biotype of *S. arvensis* which reduces their affinity for the auxinic herbicides dicamba, picloram, and 2,4-D (45) which is in agreement with the hypothesis of Estelle and Sommerville (51). However, it should be noted that other factors associated with resistance such as enhanced cytokinin levels or altered auxin binding proteins such as have not yet been investigated in *A. thaliana*.

To this end, our laboratory has found that endogenous levels of calcium are 2 to 3 times higher in the R than the S biotype (unpublished results). An interaction between calcium and cytokinin that results in delaying senescence has been shown (53,54). These authors found that without calcium, cytokinin was ineffective in delaying senescence. However, the effect of the cytokinin was restored in the presence of calcium. It was also shown that bud formation by cytokinin was dependent upon the presence of calcium (55). These results suggest that in the presence of high levels of calcium and cytokinin, such as those found in the R biotype as a result of the mutation, the classical cytokinin responses observed, such as delayed senescence and morphological alterations, may be involved in the resistance mechanism of the R biotype. Further research must be conducted to unequivocally link the interaction between calcium and cytokinin with the resistance mechanism of *S. arvensis* to the auxinic herbicides.

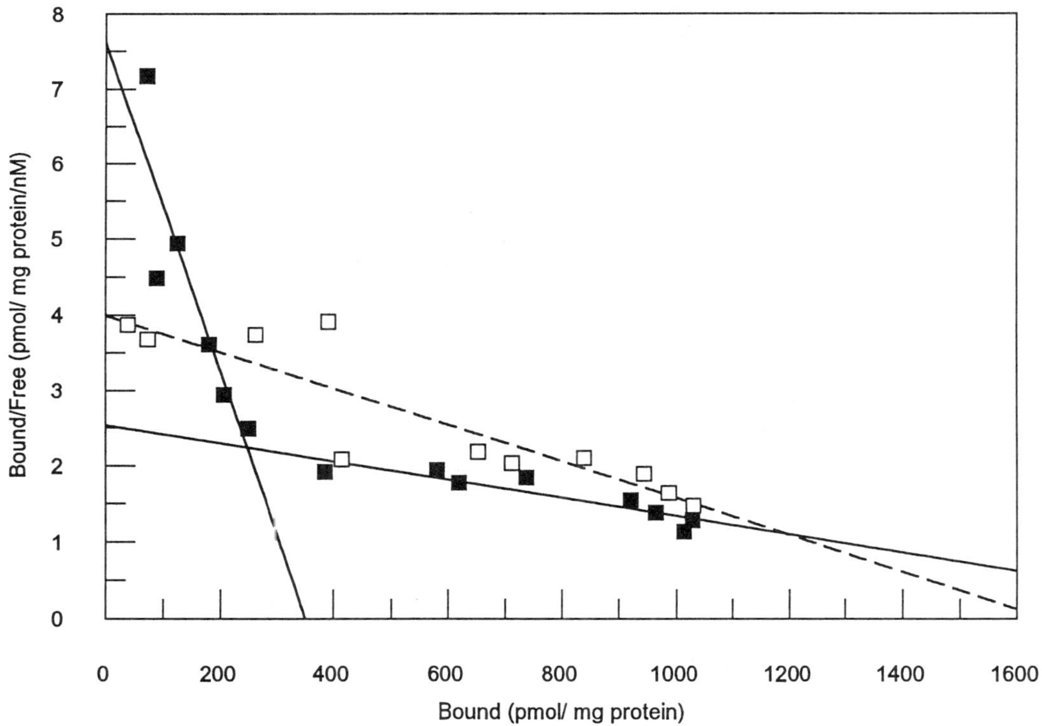

Figure 2: Scatchard plot of [^{3}H]IAA binding to auxin-binding protein preparations from the S (Filled squares) and R (Empty squares) biotypes of the wild mustard.

Light Scattering Spectroscopy

The dynamics of herbicide resistance at the cellular level were explored with light scattering spectroscopy. Light scattering spectroscopy is a non-invasive probe that allows for the detection of changes in cell morphology in real-time (56,57). Changes in cell morphology result from physiological processes and can be used to elucidate cell response as well as study biochemical events initiated by auxinic herbicides. Flash-initiated nucleotide-independent and Mg-ATP dependent transients were studied at forward scattering angles (58).

Nucleotide-Independent Flash-Induced Signals. Nucleotide-independent flash signals were acquired by suspending protoplasts in an optical cuvette containing the incubating media. The change in scattering intensity due to photo-bleaching of the protoplast suspension at a selected angle was recorded as a function of time. Figure 3 shows the signals acquired from R (Trace A) and S (Trace B) protoplasts at 5° scattering angle. The general profile of the signal was comparable to a step function. The signal from the R protoplasts was larger in amplitude than that obtained from the S protoplasts. Picloram reduced the amplitude of this signal in the S protoplasts in proportion to its concentration (Figure 4, Traces C (50 μM) and D (100 μM); Trace B is the control signal) while little change was seen in the R protoplasts (Figure 4, Trace A (100 μM picloram).

ATP-Dependent Flash-Induced Signals. When these signals were obtained after the introduction of 0.2 mM ATP in the incubation medium, ATP-dependent activity manifested as a conformational change was found to be different between the biotypes (Figure 5). There are qualitative and quantitative differences between the signals obtained in the presence of ATP. Most significantly, the signal obtained in the presence of ATP was characterized by a slower rising component in addition to the initial rapid phase. Since the slower rising component could be abolished by the ATPase inhibitor vanadate, it most likely resulted from ATPase activity (58). In the S biotype, this signal was inhibited by picloram in direct proportion to its concentration. The signal amplitude from the **S** protoplasts decreased 25% in 10 μM picloram and 60% in 25 μM picloram (Figure 5, Traces C and D; Trace B is the control signal). The signal from the R biotype was not affected by similar levels of picloram (Figure 5, Trace A). Incubation of the **S** protoplasts with the calcium ionophore A23187 in the presence of 0.5 mM calcium, which increased the cytosolic calcium level, reduced the inhibitory effect of picloram (Figure 6). In Figure 6, trace A is the control signal while trace B was obtained after incubation with the ionophore A23187 (5 μM). Trace C was obtained with 25 μM picloram after incubation with the ionophore to increase intracellular calcium. Trace D, obtained with 25 μM picloram only, is shown for comparison. As can be seen, the signal (Trace C) obtained with picloram after incubation in A23187 to increase intracellular calcium levels was not affected to the same extent as that obtained with picloram alone (Trace D). Based on these results, we hypothesize that picloram affects the calcium and hydrogen dynamics in the **S** biotype. In essence, the scattering characteristics of **S** protoplasts incubated with picloram can be nearly reproduced in the **R** protoplasts by simultaneous incubation with picloram and verapamil. This implicates the calcium ion with a significant role in modulating auxinic herbicide resistance.

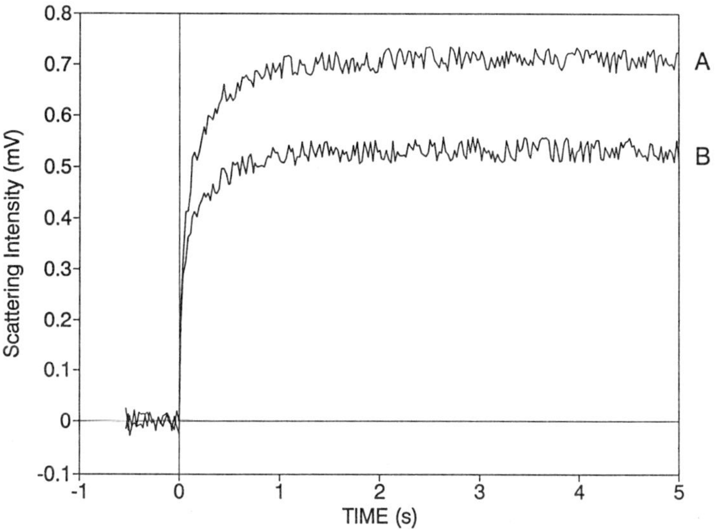

Figure 3: Nucleotide-independent flash-induced signals from the R (Trace A) and S (Trace B) protoplast suspensions of wild mustard. The media used for these signals was 0.4 M sorbitol, 5 mM MOPS and 0.5 mM $CaCl_2$ at a pH of 6.5. The flash was activated at time 0.

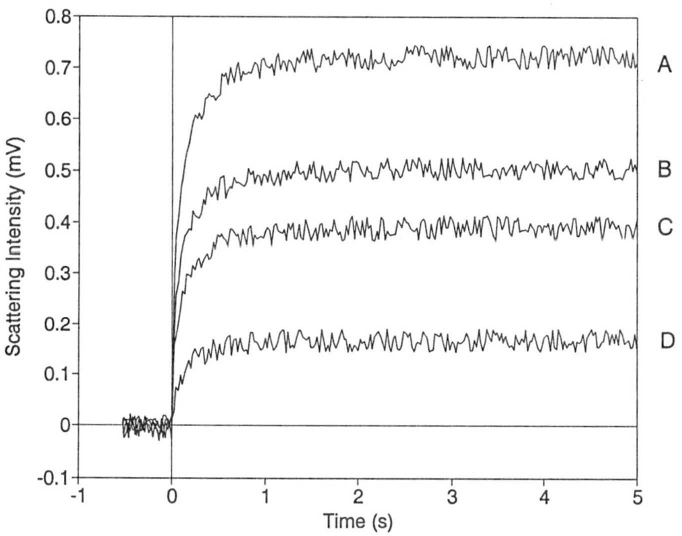

Figure 4: Nucleotide-independent flash-induced signals from the R and S protoplast suspensions of wild mustard in the presence of picloram. Trace A: R protoplasts with 100 µM picloram; Trace B: Control signal with S protoplasts; Traces C and D: S protoplasts with 50 and 100 µM picloram, respectively.

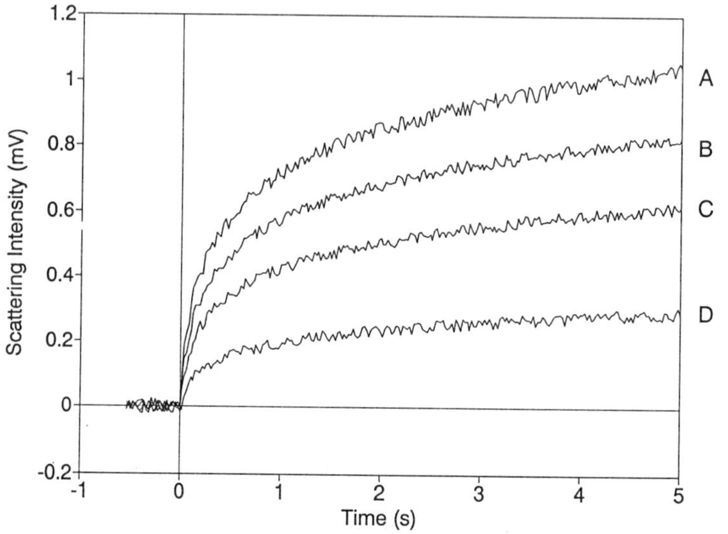

Figure 5: ATP-dependent flash-induced signals from R and S biotypes in the presence of picloram. Trace A: R protoplasts with 25 μM picloram; Trace B: Control signal from S protoplasts; Trace C and D: S protoplasts with 10 and 25 μM picloram, respectively.

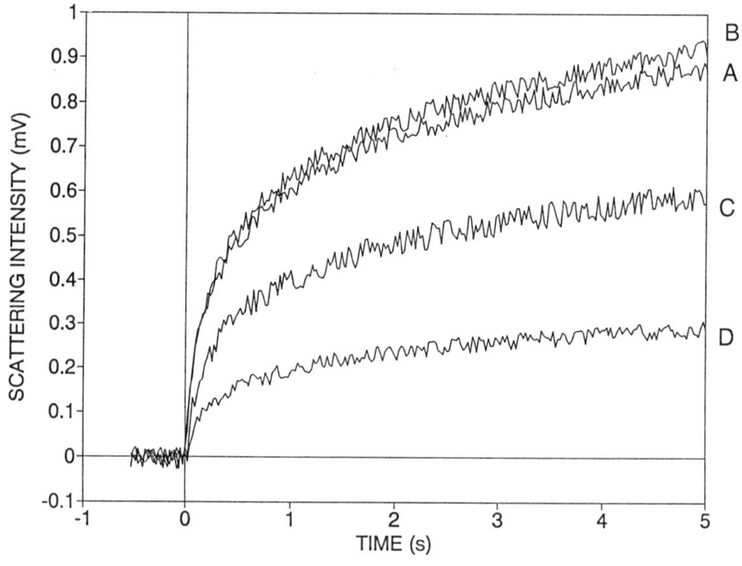

Figure 6: ATP-dependent signals obtained after incubation with the calcium ionophore A23187 to increase intracellular calcium. Traces A and B are the signals without and with A23187, respectively. Trace C is obtained in the presence of 25 μM picloram after treatment with A23187. Trace D is obtained with 25 μM picloram without A23187 treatment.

Conclusions. The resistant *S. arvensis* presents an intriguing model of auxinic herbicide resistance. Furthermore, the elucidation of the resistance mechanism may also hold significant clues about the mode of auxin action in general. A comparison of the responses of the R and S biotypes at the plant, cellular, and molecular levels upon treatment with auxinic herbicides promises valuable answers to many questions about auxins and auxinic herbicides. Our laboratory continues to probe these biotypes using a variety of biochemical and physiological techniques.

Acknowledgements. This research was supported by grants to J.C.H. from the Natural Sciences and Engineering Research Council of Canada, Sandoz Agro Canada Inc., and Cargill Limited. We thank Drs. I.M. Heap and I.N. Morrison, Department of Plant Science, University of Manitoba, Winnipeg, Manitoba, Canada for providing seeds of the wild mustard biotypes.

Literature Cited

1. Devine, M.D.; Duke, S.O.; Fedtke, C. In *Physiology of Herbicide Action*; Prentice Hall Inc., Englewood Cliffs, NJ, 1993; pp. 295-309.
2. Coupland, D. Resistance to the auxin analog herbicides, In *Herbicide Resistance in Plants*; Powles, S.B.; Holtum, J.A.M. Eds.; Lewis Publishers, Boca Raton, Fla. 1994; pp 171-214.
3. Barnwell, P.; Cobb, *Weed Res.* ***1989,*** *29*,136-140.
4. LeBaron, H.M. ; Gressel, J. *Herbicide Resistance in Plants*; Wiley Publishers: New York, NY, 1992; 401pp.
5. Peniuk, M.G.; Romano, M.L.; Hall, J.C. *Weed Res.* **1993,** *33*, 431-440.
6. Migo, T.R.; Mercado, B.L. ; De Datta, S.K. *Philippine J. Weed Sci.* **1986,** *13*, 28-38.
7. Bourdot, G.W.; Harrington, K.C.; Popay, A.I. *Proc. Brit. Crop Prot. Conf. - Weeds.* ***1989,*** *1*,309-316.
8. Bourdot, G.W.; Hurrell, G.; Saville, D.J. *Weed Res.* **1990,** *38*,449-457.
9. Heap, I.; Morrison, I.N. *Weed Sci. Soc. Amer. Abstr.* **1992,** *32*, 164.
10. Jasieniuk, M.; Morrison, I.N.; Brule-Babel, A. *Weed Sci.* **1995,** *13*, 192-195.
11. Peterson, G.E. *Agric. History* **1967,** *41*, 244-253.
12. Ashton, F.M.; Crafts, A.S. *Mode of Action of Herbicides*, John Wiley and Sons Inc.: Toronto, ON, 1981; pp 525.
13. Loos, M.A. In *Herbicides: Chemistry, Degradation, and Mode of Action*; Kearney, P.C.; Kaufman, D.D., Eds.; Marcel Dekker Inc.: New York, NY, 1975; pp 1-128.
14. Hatzios, K.K. In *Environmental Chemistry of Herbicides;* Grover, R.; Cessna, A.J. Eds.; CRC Press: Boca Raton, FL, 1991, Vol II; pp 141-185.
15. Hatzios, K.K.; Penner, D. *Metabolism of Herbicides in Higher Plants;* Burgess Publishing Co.: Minneapolis, MN, 1982; 142 pp.
16. Luckwill, L.C.; Lloyd-Jones, C.P. *Annuals Applied Biol.* **1960,** *48*, 613-675.
17. Hall, J.C.; Vanden Born, W.H. *Weed Sci.* **1988,** *36*, 9-14.
18. Hall, J.C.; Soni, M. *Pestic. Biochem. Physiol.* **1989,** *33*, 1-10.

19. Hall, J.C.; Soni, M. *Z. Naturforsch.* **1991,** *46*, 957-962.
20. Fedtke, C. *Biochemistry and Physiology of Herbicide Action*; Springer-Verlag: New York, NY, 1982; 202 pp.
21. Theologis, A. *Ann. Rev. Plant Physiol.* **1986,** *37*, 407-438.
22. Barbier-Brygoo, H. *Crit. Rev. Plant Sci.* **1995,** *14*,1-25.
23. Guilfoyle, T.J.; Hagen, G.; Li, Y.; Ulmasov, T.N.; Liu, Z.; Strabala, T.; Gee, M. *Aust. J. Plant Physiol.* **1993,** *20*,489-502.
24. van der Zaal, E.J., Mennes, A.M.; Libbenga, K.R. *Planta.* **1987,** *172*, 514-519.
25. Hertl, R. *Z. Pflanzenphysiol.* **1983,** *112*, 53-67.
26. Cross, J.W. *Plant Cell Environ.* **1985,** *8*, 351-359.
27. Jones, A.M. *Ann. Rev. Plant Physiol. Plant Mol. Biol.* **1994,** *45*, 393-420.
28. Barbier-Brygoo, H.; Ephritikihne, G.; Klambt, D.; Ghislain, M.; Guern, J. *PNAS USA.* **1989,** *86*, 891-895.
29. Bell, A.R.; Nalewaja, J.D.; Schooler, A.B. *Weed Sci.* ***1972,*** *20*,458-462.
30. Ellis, M.; Kay, Q.O.N. *Weed Res.* **1975,** *15*, 307-315.
31. Hanson, N.S. *Proc. Internat. Soc. Sugar Cane Technol.* **1959,** *10*, 538-549.
32. Callihan, R.H.; Prather, T.S.; Northam, F.E. *Proc. 1989 Knapweed Symposium*, Montana State University: Montana 1989; *45*, 73-78.
33. Chaleff, R.S. *Science* **1983,** *219*, 676-682.
34. Holt, J.S. ; LeBaron, H.M. *Weed Tech.* **1990,** *4*, 141-149.
35. Holt, J.S.; Poweles, S.B.; Holtum, J.A.M. *Ann. Rev. Plant Physiol. Plant Mol. Biol.* **1993,** *44*, 203-229.
36. Lutman, P.J.W.; Lovegrove, A.W. *Proc. Brit. Crop Prot. Conf. - Weeds* **1985,** *1*, 411-418.
37. Lutman, P.J.W. ; Snow, H.S. *Proc. Brit. Crop Prot. Conf. - Weeds.* **1987,** *3*, 911.
38. Coupland, D.; Cooke, D.T.; James, C.S. *J. Expt. Botany.* **1991,** *42*,1065-1071.
39. Sterling, T.M.; Lownds, N.K.; de Soyza, T.V. *Plant Physiol.* **1993,** *102S*, 64.
40. Abeles, F.B. *Weed Sci.* **1968,** *16*,498-500.
41. Holm, R.E.; Abeles, F.B. *Planta.* **1968,** *78*, 293-297.
42. Morgan, P.W. In *Herbicides;* Audus, L.J. Ed.; Academic Press: New York, NY, 1976, Vol. 1; pp. 256-280.
43. Hall, J.C.; Bassi, P.K.; Spencer, M.S.; Vanden Born, W.H. *Plant Physiol.* **1985,** *39*, 18-23.
44. Hall, J.C.; Alam, S.M.M.; Murr, D.P. *Pestic. Biochem. Physiol.* **1993,** *47*, 36-43.
45. Webb, S. R.; Hall, J.C. *Pestic. Biochem. Physiol.* **1995,** *52*, 137-148.
46. Hall, J.C.; Romano, M.L. *Pestic. Biochem. Physiol.* **1995,** *52*, 149-155.
47. Fletcher, R.A.; Kallidumbill, V.; Steele, P.*Plant Physiol.* **1982,** *69*, 675-677.
48. Bidwell, R.G.S. *Plant Physiology*; Macmillan: New York, NY, 1974; 643 pp.
49. Salisbury, F.B.; Ross, C.W. *Plant Physiology*; Wadsworth Publishing Company: Belmont, CA, 1992, 682 pp.
50. Khan, A.A. *Plant Physiol.* **1968,** *43*, 1463-1467.
51. Estelle, M.A.; Sommerville, C. *Mol. Gen. Genet.* **1987,** *206*, 200-206.
52. Lincoln, C.; Britton, J.H.; Estelle, M. *Plant Cell.* **1990,** *2*, 1071-1077.
53. Poovaiah, B.W.; Leopold, A.C. *Plant Physiol.* **1973,** *52*, 236-239.

54. Leopold, A.C.; Poovaiah, B.W.; Dela Fuente, R.K.; Williams; R.J. *Plant Growth Substances 1973*; Hirokawa Press: Tokyo, Japan, 1973; pp. 780-788.
55. Poovaiah, B.W.; Reddy, A.S.N. *CRC Crit. Rev. Plant Sci.* **1987,** *6*, 47-103.
56. Latimer, P. *Ann. Rev. Biophys. Bioeng.* **1982,** *11*, 129-150.
57. Latimer, P.; Moore; D.M. ; Bryant, F.D. *J. Theoret. Biol.* **1968,** *21*, 348-367.
58. Deshpande, S.; Hall, J.C.*Biochem. Biophys. Acta.* **1995,** *1244*,69-78.

Chapter 5

The Molecular Basis of Plant Resistance to Photosystem II Herbicides

Achim Trebst

Plant Biochemistry, Ruhr-Universität, 44780 Bochum, Germany

Photosystem II herbicides replace a plastoquinone by binding to a polypeptide subunit of photosystem II. The target protein is called the herbicide binding or Q_B binding protein, but also the D1 protein. A three dimensional model of the folding of the aminoacid sequence of the D1 protein and of its herbicide binding niche is available. Amino acid substitutions or deletions in the herbicide binding niche lead to herbicide tolerance. In algae numerous mutants were constructed by site-selected and site-directed mutagenesis. In higher plants acquired herbicide tolerance leads always to a Ser_{264} to Gly change in the D1 protein. Only in cell cultures were a few further amino acid substitutions reported.

Resistance to photosystem II herbicides either acquired in the field or constructed in the laboratory is well established. Resistance or tolerance may be due to faster degradation, overproduction of a binding protein or mutations in the binding site that lead to changing affinity to the primary target protein. This review concentrates on tolerance to herbicides due to the third case, i. e. to amino acid substitutions in the D1 protein of photosystem II, the primary target for PS II herbicides. For this the mode of action of the herbicides and the structure of the herbicide binding receptor will be introduced. Then the response in both positive (diminished affinity) and negative (supersensitivity) cross resistance to mutations in the target protein will be described.

Mode of Action and QSAR of Herbicides and Related PS II Inhibitors

It is known since three decades that well known herbicides like diuron or atrazine inhibit photosynthetic electron flow. This topic has been reviewed previously *(1-4)*. The inhibition site was located on the acceptor side of photosystem II. As this photosystem oxidizes in the light water to oxygen (the donor side of the electron for the photosystem) and reduces plastoquinone (on the acceptor side which accepts the electron from the excited photosystem) it was suspected and then shown *(5)* that the herbicides are analogous to plastoquinone and displace it from its binding site. The protein that binds this plastoquinone and the herbicide was identified as a 32 kDa

0097–6156/96/0645–0044$15.00/0

protein by photoaffinity labelling with a photoreactive azido triazine derivative *(6)*. It was called therefore the herbicide binding protein. It is also called the Q_B binding protein because the herbicide displaces plastoquinone Q_B from this protein. There are two plastoquinone binding sites in photosystem II, Q_A and Q_B, of which only the Q_B site can be occupied by herbicides as well.

The scheme (Figure 1) shows the reaction sequence, i. e. the electron flow from water being oxidized via a manganese cluster by the excited chlorophyll dimer of the reaction center P_{680}, a pheophytin intermediate to the primary plastoquinone Q_A being reduced. Plastoquinone Q_A (that is only reduced to the semiquinone state) then reduces plastoquinone Q_B, which in a second turn of the photocycle gets fully reduced to the hydroquinone. Plastoquinol leaves the protein allowing a new plastoquinone to enter the site, if not in competition with a herbicide present, that because of higher binding affinity blocks the binding site on that protein. Then electron flow comes to a halt.

As of recently the herbicide binding protein is called the D1 protein of PS II. This is because it became clear that the Q_B and herbicide binding protein is identical to a thylakoid membrane protein already known as a chloroplast encoded rapidly turning over D1 protein *(7)*. This identity of the herbicide binding protein and turning over protein was important as the messenger RNA and the DNA of this rapidly turning over D1 protein had already been detected in the plastome and sequenced very early *(8)*. This *psb*A gene encodes the D1 protein in the chloroplasts; its DNA sequence made the amino acid sequence of the herbicide protein known.

In 1986 it came as a surprise that the sequence of the D1 protein is homologous to two protein sequences in purple bacteria that were shown to be the reaction center binding protein of the photosystem of these photosynthetic bacteria *(9)*. With the crystallization and the X-ray structure of this photosystem of the bacteria *(10)* it was possible to clarify that indeed the D1 protein with the Q_B binding site (together with the until then neglected homologous D2 protein of the thylakoid membrane with the Q_A binding site) forms even the reaction center of photosystem II *(2)*. The D1/D2 protein complex also binds the reaction center chlorophyll P_{680}, pheophytin in addition to the quinones. This great step forward in the membrane structure of photosystems established that photosystem II herbicides interact indeed with the reaction center of PS II - at its quinone binding site. From the comparison with the high resolution X-ray structure of the bacterial photosystem *(10)* now also a topology of the photosystem II protein and in particular the dimensions of the quinone binding and herbicide binding site could be deduced *(2, 11)*. Accordingly the herbicide binding niche in the folding of the amino acid sequence of the D1 protein is embedded in a hydrophobic pocket close to the surface of the membrane towards the matrix space. The pocket is formed by the end of two transmembrane helices, a short tilted parallel helix and a stretch of amino acids that connects the parallel and transmembrane helix V *(12)*. The connection between the transmembrane helix IV and the parallel helix is a long hydrophilic sequence that covers up the herbicide binding niche towards the hydrophilic matrix space (see Figure 2). The role of individual amino acids in the niche became clear via the response of herbicide binding affinity to amino acid substitutions. This is discussed below.

The structure activity relationships (QSAR) of herbicide activity of a large numer of inhibitors of photosystem II is studied in great detail since a long time with increasing sophistication over the years (see *1-4, 13*). QSAR compares the inhibitory potency in photosystem II preparations with substitution parameters of the various chemical compounds. QSAR allowed already a prediction of the steric dimensions in the herbicide target before the protein was actually known *(14)*. Computer aided modelling of the herbicides into the folding of the amino acid sequences of the D1 protein correlated the QSAR results of the compounds with the

$$H_2O \rightleftarrows [Mn_4] \rightarrow RC\ [P_{680}] \rightarrow \text{pheophytin} \rightarrow Q_A \rightarrow Q_B \; (2H^+ / QH_2) \quad \text{<-target site of herbicides}$$
O_2

Figure 1: Electron Flow through Photosystem II
Charge separation of the reaction center P_{680} in the light leads to pheophytin reduced and P680 oxidized. Pheophytin reduces plastoquinone Q_B via Q_A. P_{680} is rereduced by a manganese cluster, that in turn oxidizes water. The plastoquinone Q_B binding site is the target site for herbicides.

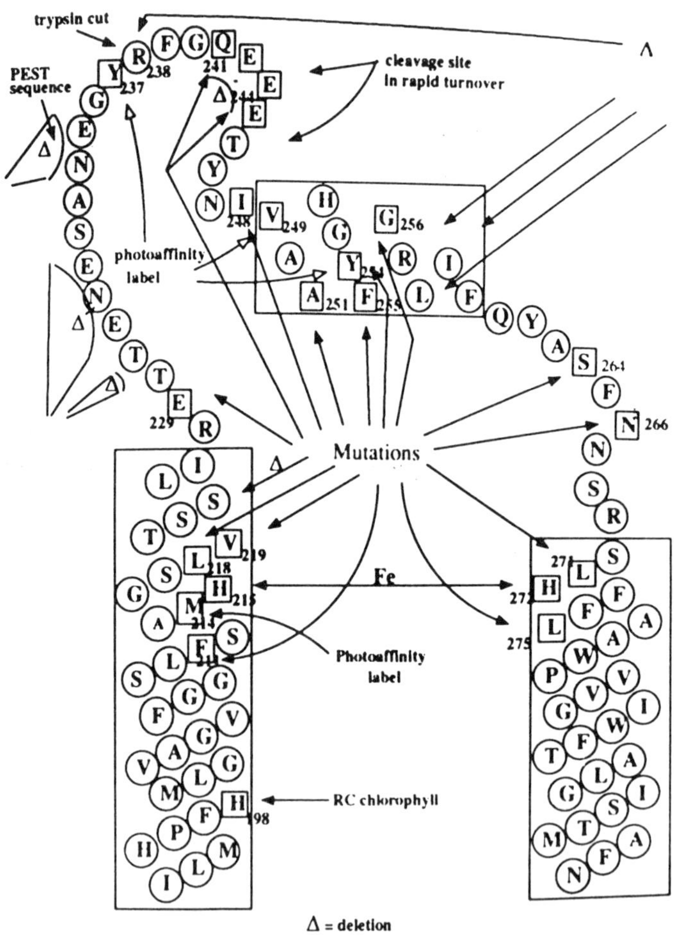

Figure 2: Folding of the Amino Acid Sequence of the Herbicide Binding Niche in the D1 Protein
Arrows indicate amino acid changed in herbicde tolerant mutants
Δ stands for deletions; also indicated are amino acids tagged by photoaffinity labeling by azido derivatives of triazines, ureas and phenols. The cleavage site in rapid turnover is indicated as well as the PEST site that may direct it. The trypsin cut in the loop was informative on the orientation of herbicides in the site.

topological information of the protein *(15)*. 3D-QSAR is the latest improvement in this rapidly developing field *(13)*.

Target Site Mutations in Photosystem II Yielding Herbicide Tolerance. The first atrazine resistant weeds were discovered as early 1976 *(16)*. It was shown that the resistance is maternally inherited *(17)* (i. e. due to a change of a protein that is encoded by the chloroplast genome) and is the consequence of a change in binding affinity in the primary mode of action *(6)*. The protein target at that time was not known. But then these first mutants were instrumental in identifying this herbicide binding protein, as the photoaffinity labelling technique with azido atrazine *(6)* (mentioned above) failed to mark a protein band in the gel electrophoresis of chloroplast polypeptides in the mutant but did in the wildtype. When the *psb*A gene was known *(8)* relatively quickly also the base change in the tolerant *Amaranthus* was identified that leads to a substitution of Ser_{234} to Gly in the D1 protein *(18)*. With the topological information coming available this amino acid turned out to be in the stretched connection between the parallel helix and transmembrane helix V. Ser_{264} is involved in hydrogen bonding to plastoquinone Q_B *(11, 12)*. Many herbicide tolerant weeds are now known (for review see *19-22*) where the change of amino acid Ser_{264} of the D1 protein is responsible for the tolerance. By random mutagenesis herbicide resistant mutants were isolated from green and bluegreen algae (for review see *23-25*). In *Chlamydomonas reinhardtii* 5 amino acid substitutions were identified as of 1988 that led to diuron and/or atrazine tolerance; Val_{219}, Ala_{251}, Phe_{255}, Ser_{264} and Leu_{275} (see review in *23*). This information was essential in establishing the folding homology between the two subunits of the reaction center of photosystem II with that of the purple bacteria *(2, 11, 12)*. All mutations were in the hydrophobic pocket for herbicide binding already discussed above.

Many more mutations in the D1 protein leading to herbicide tolerance were identified via randomly induced mutagenesis or by site directed gene technology methods of the *psb*A gene in cyanobacteria and green algae (see *24, 25*). A summary is given Table I and in Figure 2. Double replacement and deletions of amino acids at several points in the Q_B binding niche of the D1 protein were checked in the cyanobacteria *Synechocystis* for herbicide resistance *(25)*. The results show the participation of further parts of the D1 protein in herbicide binding and tolerance. For example the two Ser_{221} and Ser_{222}, Tyr_{237}, Lys_{238}, Phe_{239}, and Ala_{250} are also contributing, i. e. these amino acids in the wild type allow high affinity herbicide binding where its substitution or deletion leads to herbicide tolerance *(25)*. The cross resistance of various herbicides in mutants in cell cultures *(27-29)* or of supersensitivity *(30-32)* gave important hints for modelling the orientation of the substituents of the chemical compounds towards the site change of the amino acids in the binding pocket *(15)*. Table II illustrates the principle of variations in positive and negative cross resistance for just four herbicides from the large amount of data in the literature on responses in site-selected (screened for tolerance) or site-directed (by genetic engineering) target site mutations in the D1 protein. For example there can be preferential tolerance to both diuron or atrazine or only to either one. In the Ser_{264} to Gly or Ala change there is never tolerance to a phenolic herbicide, like ioxynil and pyridate, but in the Val_{219} to Ile change there is. The data are used to point out that both classical and phenolic herbicides respond to such mutations and therefore are bound to the same binding niche *(13, 24, 25)* - though with different orientation *(15)* - the molecular basis of the old overlapping binding site model of 1979). The observed negative cross tolerance of pyridate was suggested to be possibly of importance for managing atrazine resistant weed *(26)*.

In addition to these principal and detailed knowledge of the mode of action and of tolerance there are three further interesting observations:

Table I. Amino acid Changes in Herbicide Tolerant Mutants of Higher Plants or Algae

Position in the D1 protein sequence	Organism
Single Replacement	
Phe_{211} to Ser	*Synechococcus*
Val_{219} to Ile	*Chlamydomonas*
Thr_{220} to Ala	*Chenopodium* cell culture
Ala_{251} to Val	*Chlamydomonas*
Phe_{255} to Tyr	*Chlamydomonas*
Gly_{256} to Asp	*Chlamydomonas*
Ser_{264} to Gly	*Amaranthus, Senecio* etc.
Ser_{264} to Ala	*Chlamydomonas*
Ser_{264} to Thr	*Nicotiana* cell culture
Ser_{264} to Asn	*Nicotiana* cell culture
Asn_{266} to Thr	*Synechococcus*
Leu_{275} to Phe	*Chlamydomonas*
Double Mutations	
Val_{219} to Ile + Ala_{251} to Thr	*Chenopodium* cell culture
Val_{219} to Ile + Glu_{229} to Gly + Ser_{270} to Phe	
Val_{219} to Ile + Ala_{251} to Thr + His_{272} to Arg	
Deletions	
$Ser_{221/222}$	*Synochocystis*
$YKF_{237-239}$	
$GQ_{240-241}$	

Table II. Cross Resistance of Herbicides in Photosystem II after Specific Amino Acid Substitutions in the D1 Protein

+ and ++ indicate small and large tolerance, - and -- no tolerance or even negative cross tolerance

Amino Acid Change	Atrazine	Diuron	Ioxynil	Pyridate
site-selected				
Ser_{264} to Gly	++	- -	- -	-
Ser_{264} to Ala	++	++	- -	-
site-directed				
Ser_{220} to Leu	+	++	+	

1. Only One Practical Mutation is Found in Weeds. In higher plants spontaneous herbicide resistance has been found in the field in various weed plants *(18-18, 20-22)*. In all it is only a Ser_{264} to Gly change that is responsible for the acquired tolerance. Only in herbicide tolerant cell cultures of higher plants, for example of *Nicotiana (27, 28)* or *Chenopodium (29)* Ser_{264} is changed also to other amino acids (Thr, Asn). There is even a triple amino acid change in cell cultures of *Chenopodium (29)* like Thr_{220} to Ala, Glu_{229} to Gly, Ser_{270} to Tyr and His_{272} to Arg. This dominance of the Ser_{264} to Gly change in tolerant plants is contrary to algae where site selection after random mutagenesis has led herbicide tolerance with many different changes of amino acids as described above (see Table I) not including site directed mutagenesis. The significance of this lack of different amino acid changes in acquired tolerance of weeds under field conditions is not clear.

2. Negatively correlated Resistance. A certain group of PS II herbicides called the phenol type family, like dinoseb or pyridate, also inhibit photosystem II, though with a somewhat different inhibitory pattern and a different structure-activity correlation (see *2*). The first D1 protein mutants with triazine or diuron tolerance with a Ser_{264} to Ala change in *Chlamydomonas* or Ser_{264} to Gly change in *Amaranthus* that became available showed surprisingly negative tolerance to these phenol type compounds, i. e. the plants became supersensitive to phenol herbicides *(30-32)*. This seemed to support the notion that these phenol type inhibitors occupy a different site on the PS II protein subunits. However, recently with a new set of highly efficient inhibitory nitrophenols, positive tolerance was observed in mutants that were also tolerant to the classical herbicides *(33)*, e.g., for example in the Val_{219} to Ile mutant. It appears then that both inhibitor families do occupy the same hydrophobic pocket in the D1 protein but orient themselves towards different amino acids. This was modelled in the protein folding *(34)*. Nevertheless the different properties of the phenol type PS II herbicides and their tolerance pattern may allow control of weeds that are tolerant to the classical PS II herbicides *(26)*.

3. D1 Protein Turnover and Light Regulation of Photosynthesis. It was mentioned above that the herbicide binding protein is a rapidly turning over protein, i. e., it is continuously degraded, resynthesized, and reassembled in the living plant *(7)*. The role of this behaviour is not entirely clear. There is a connection of this turnover to the response of a plant to high light intensities, i. e., in the repair of photoinhibition *(35)*. Recently a detailed study of the effect of herbicides on the turnover was presented *(36)*. It is speculated that the turnover may have to do with a role of the D1 protein in redox control of gene expression and protein activation via phosphorylation *(37)*. Although this is still unresolved, it is important to point out the observation that efficient commercial herbicides also block this D1 protein turnover. One could speculate from this that the herbicides affect control elements in light regulation of photosynthesis in a plant in addition to blocking photosynthetic electron flow. Inhibitors that block electron flow but not turnover do not effect this light control system and therefore do not become herbicides as well. It is possible therefore that it is actually this effect that is the mode of action of the herbicides *in vivo* rather than just arresting photosynthetic electron transport.

The mode of action of herbicides, the three dimensional structure of the protein that is the target of the herbicides, and the understanding of the basis of acquired as well as constructed herbicide tolerance by defined amino acid substitutions in this herbicide binding protein is obviously at an advanced level of understanding.

Literature Cited

1. Moreland, D. E. *Annu. Rev. Plant Physiol.* **1967**, 18, 365.
2. Trebst, A.; Draber, W. *Photosyn. Res.* **1986**, 10, 381.

3. Oettmeier, W. In *The Photosystems: Structure, Function and Molecular Biology*; Barber J. Ed.; Topics in Photosynthesis, 11; Elsevier Science Publishers B.V., 1992, Chapter 9, pp 349-408.
4. Hock, B.; Fedtke, C.; Schmidt, R. R. *Herbizide: Entwicklung, Anwendung, Wirkungen, Nebenwirkungen*; Georg Thieme Verlag: Stuttgart-New York, 1995.
5. Velthuys, B. R. *FEBS Lett.* **1981,** 126, 277.
6. Pfister, K.; Steinback, K. E.; Gardner, G.; Arntzen, C. J. *Proc. Natl. Acad. Sci. USA* **1981**, 78, 981.
7. Mattoo, A. K.; Pick, U.; Hoffmann-Falk, H.; Edelman, H. *Proc. Natl. Acad. Sci. USA* **1981**, 78, 1572.
8. Zurawski, G.; Bohnert, H. J.; Whitfield, P. R.; Bottomley, W. *Proc. Natl. Acad. Sci. USA* **1982**, 79, 7699.
9. Youvan, D. C.; Bylina, E. J.; Alberti, M. Begusch, H.; Hearst, J. E. *Cell* **1984**, 37, 949.
10. Deisenhofer, J.; Epp, O.; Miki, K., Huber, R.; Michel, H. *Nature* **1985**, 318, 618.
11. Trebst, A. *Z. Naturforsch.* **1986**, 41c, 240.
12. Trebst, A. *Z. Naturforsch.* **1987**, 42c, 742.
13. Draber, W.; Trebst, A.; Oettmeier, W., *Classical and 3 D QSAR in Agrochemistry and Toxicology*, Hansch, C.; Fujita, T., In *ACS Symposium Serie 606*, Hansch, C.; Fujita,T., Serie 606; American Chemical Society, Washington, DC 1995, Chapter 14, pp. 186.
14. Draber, W.; Fedtke, C. In *Advances in Pesticide Science, Part 2*; Geißbühler, H., Ed.; Pergamon, Oxford 1979, pp. 475.
15. Tietjen, K.; Kluth, J. F.; Andree, R.; Haug, M.; Lindig, M.; Müller, K. H.; Wroblowsky, H. J.; Trebst, *A. Pestic. Sci.* **1991**, 31, 65.
16. Radosevich, S. R.; De Viellier, O. T. *Weed Sci.* **1976**, 24, 229.
17. Pfister, K.; Arntzen, C. J. *Z. Naturforsch.* **1979**, 34c, 996.
18. Hirschberg, J.; McIntosh, L. *Science* **1983**, 222, 1346.
19. Holt, S. J.; Powles, S. B.; Holtum, J. A. M.*, Annu. Rev. Plant Physiol. Plant Mol. Biol.***1993**, 44, 203.
20. Van Oorschot, J. L. P. In *Herbicide Resistance in Weeds and Crops*; Caseley, J. C., Cussans, G. W., Atkin, R. K., Eds.; Butterworth-Heinemann Ltd., Oxford, GB, 1991, pp. 87-101.
21. Rubin, B. In *Herbicide Resitance in Weeds and Crops*; Caseley, J. C., Cussans, G. W., Atkin, R. K., Eds.; Butterworth-Heinemann Ltd., Osford, GB, 1991, Vol. II, pp. 387-414.
22. Gressel, J. in *Herbicide Physiology*; Duke, S. O., Ed.; CRC Press, Boca Raton, Fl, USA, 1985, Vol. II, pp. 159.
23. Rochaix, J.-D.; Erickson, J. *TIBS* **1988**, 13, 56.
24. Hirschberg, J.; Ohad, N.; Pecker, I.; Rahat, A. *Z. Naturforsch.* **1987**, 42c, 758.
25. Kless, H.; Oren-Shamir, M.; Malkin, S.; McIntosh, L.; Edelman, M. *Biochemistry* **1994**, 33, 10501.
26. Gressel, J.; Segel, L. A., *Z. Naturforsch.* **1989**, 45c, 470.
27. Sato, F.; Shigematsu, Y. Yamada, Y. *TIBS* **1988**, 13, 56.
28. Páy, A.; Smith, M. A.; Nagy, F.; Márton, L. *Nucleic Acids Res.* **1988**, 16, 8176.
29. Schwenger-Erger, C.; Thiemann, J.; Barz, W.; Johanningmeier, U.; Naber, D. FEBS Lett. **1993**, 329, 43.
30. Durner, J.; Thiel, A.; Böger, P., **1986**, *Z. Naturforsch.* 41 c, 881.
31. Wildner, G. F.; Heisterkamp, U.; Trebst, A. *Z. Naturforsch.* **1990**, 45c, 1142.

32. Oettmeier, W.; Masson, K.; Fedtke, C:; Konze, J.; Schmidt, R. R. *Pestic. Biochem. Physiol.* **1982**, 18, 357.
33. Trebst, A.; Hilp, U.; Draber, W. *Phytochemistry* **1993**, 33, 969.
34. Draber, W.; Hilp, U.; Schindler, M.; Trebst, A. In *ACS Symposium*; Hedin, P. A.; Menn, J. J.; Hollingworth, R. M., Eds.; Natural and Engineered Pest Management Agents, 551, American Chemical Society, Washington, DC 1994, pp. 449-472.
35. Schuster, G.; Timberg, R.; Ohad, I. *Eur. J. Biochem.* **1988**, 177, 403.
36. Jansen, M.A.K.; Depka, B.; Trebst, A.; Edelman, M. *J. biol. Chem.* **1993**, 268, 21246.
37. Allen, J. F. *Photosyn. Res.* **1993**, 36, 95

Chapter 6

Analysis of Sodium Channel Gene Sequences in Pyrethroid-Resistant Houseflies

Progress Toward a Molecular Diagnostic for Knockdown Resistance (*kdr*)

Martin S. Williamson, David Martinez-Torres, Caroline A. Hick, Nathalie Castells, and Alan L. Devonshire

Biological and Ecological Chemistry Department, Institute of Arable Crops Research-Rothamsted, Harpenden, Hertfordshire, AL5 2JQ, United Kingdom

Knockdown resistance (*kdr*) is a mechanism that confers nerve insensitivity to DDT and pyrethroid insecticides. This type of resistance is best characterized in the house fly where several *kdr* alleles, including the more potent *super-kdr* factor, have been identified. Resistance is thought to result from a modification of the voltage-sensitive sodium channel, the primary target site for these insecticides, and this is supported by genetic evidence that both *kdr* and *super-kdr* map closely to the house fly *para*-type sodium channel gene. To investigate the molecular basis of resistance, we have sequenced the full 6.3kb coding region of this gene from susceptible, *kdr* and *super-kdr* strains. Our results suggest that *kdr* is caused by a single amino acid substitution, leucine to phenylalanine, in the domain IIS6 segment of the channel; while an additional methionine to threonine change in the nearby IIS4-IIS5 linker is responsible for the enhanced resistance of *super-kdr*. Using this information, we have developed a PCR-based diagnostic technique for detecting the *kdr* mutation in individual house flies.

The intensive use of pyrethroid insecticides over the last 20 years has led to resistance in several important agricultural pests and this represents a significant threat to their continued effective use (*1*). The pyrethroids are potent neurotoxins and an important type of resistance is characterized by a marked reduction in the intrinsic sensitivity of the insect nervous system to these compounds. It confers resistance not only to all pyrethroids, but also to DDT 1,1'-(2,2,2-trichloroethylidene)bis[4-chlorobenzene] which shares a similar mode of action. This mechanism was first reported (2) over forty years ago in a strain of the house fly, *Musca domestica*, that withstood the normally rapid knockdown effects of DDT and was subsequently termed knockdown resistance (*kdr*). The *kdr* factor was isolated genetically and found to map to a single, recessive locus on chromosome III (*3*). Other *kdr* alleles were subsequently isolated from house fly populations in North America (*kdr*-Orlando) and Scandinavia (*kdr*-NPR) with similar resistance properties (*4, 5*). The *kdr* factor is characterized by a fairly uniform level of resistance (10 to 30-fold) to DDT and most pyrethroids (Table I), with no cross resistance to other insecticide classes. In 1978, Sawicki (*6*) reported

0097–6156/96/0645–0052$15.00/0

a similar factor conferring broad cross resistance to DDT and pyrethroids, but with a greatly enhanced resistance to the more active type II pyrethroids that are characterized by the presence of an α-cyano-3-phenoxybenzyl alcohol. This factor, termed *super-kdr*, mapped to the same region of chromosome III and so appeared to be allelic to *kdr*, but can confer up to 500-fold resistance to pyrethroids such as deltamethrin (Table I).

Table I. Resistance Factors of *kdr* and *super-kdr* House Flies To DDT,Type I and Type II Pyrethroids. Data from (*7, 8*)

Compound		House fly strain	
		kdr	*super-kdr*
DDT		17	53
Bioresmethrin	(I)	16	53
Cismethrin	(I)	13	59
Permethrin	(II)	13	68
Fenvalerate	(II)	20	170
Cypermethrin	(II)	28	250
Deltamethrin	(II)	31	560

The initial characterization and availability of well defined strains for both *kdr* and *super-kdr* in the house fly has meant that most studies aimed at understanding the physiological basis of resistance have focused on this insect. However, it should be noted that analogous *kdr*-type mechanisms showing similar cross resistance to DDT and pyrethroids have been reported in several other insect pest species (reviewed in *9*). *Kdr*-type resistance is ultimately defined as a mechanism that confers neuronal insensitivity to these compounds. This has been clearly established from comparative studies on nerve preparations from susceptible and *kdr (or super-kdr)* house flies that show the latter to be markedly (10^2 to 10^6 fold) less sensitive to the effects of pyrethroids (*10*). This suggests that the molecular basis of resistance resides in a modification of the nervous system that affects the normal mode of action of these insecticides. Although pyrethroids affect several neural processes, it is now generally accepted that their primary site of action is the voltage-sensitive sodium channel (see reviews *11, 12*). Electrophysiological studies using voltage- and patch-clamping techniques have shown that they alter the gating kinetics of the channel, particularly to slow inactivation, and thereby to prolong the Na^+ currents associated with membrane depolarizations. This results in uncontrolled bursts of action potentials leading to nerve exhaustion and death. Evidence that this type of resistance results from an alteration in the sodium channel that render it less sensitive to pyrethroids comes from cross resistance studies to certain sodium channel neurotoxins (*13*), and binding studies that indicate a reduced affinity for pyrethroids on the *super-kdr* sodium channel (*14*). This chapter reviews our recent progress in the cloning of sodium channel gene fragments from the house fly; the identification of single amino acid alterations in the *kdr (or super-kdr)* strains that correlate with resistance, and the development of a polymerase chain reaction (PCR)-based diagnostic technique for rapidly detecting the resistance mutations in individual flies.

Molecular Biology of the Voltage-Sensitive Sodium Channel

The voltage-sensitive sodium channel plays a key role in nerve signalling by generating the rising phase of action potentials in the neurons of vertebrates and invertebrates (*15*). The channel is sensitive to localized depolarization of the nerve membrane that cause it to open, allowing sodium ions to flow down a concentration gradient across the membrane and into the cell. The open (activated) state of the channel exists for less than a millisecond as it spontaneously closes (inactivates) before returning to the resting (also closed) state. This transient opening of the channel results in further depolarization of the membrane causing more channels to open and so generates a wave of depolarization down the axon that constitutes the nerve signal.

Sodium channels purified from mammalian brain comprise a large, glycosylated α subunit polypeptide (Mr 260 kD), together with two smaller associated β subunit polypeptides, β1 (36 kD) and β2 (33 kD) (*16*). The α subunit is the main structural component of the channel and is structurally conserved across a diverse range of vertebrate and invertebrate organisms; the ß subunits have so far only been identified in mammalian tissues. The first sodium channel α subunit to be cloned was that of electric eel, encoding a single polypeptide chain of over 1800 amino acids. Similar sequences were subsequently cloned from mammalian brain (3 different genes), mammalian skeletal and cardiac muscle, and from several invertebrates including *Drosophila melanogaster* (reviewed in *16, 17*). The cloned genes predict structurally-similar polypeptides with considerable sequence conservation, ranging from 45% amino acid homology between the vertebrate and invertebrate channel sequences to over 90% between the mammalian brain isoforms. The α subunit contains four homologous repeating domains (I-IV), with each domain containing six hydrophobic segments (S1-S6) that are thought to form membrane spanning α-helices (Figure 1). The predicted folding of these regions within the membrane has led to several related structural models of the channel with the four domains arranged as a square array about a central pore whose functional properties are formed by sequence elements within the transmembrane segments (*16, 17*).

The role of the α subunit in forming the functional channel protein has been confirmed by the expression of cloned α subunit sequences in heterologous expression systems such as the *Xenopus* oocyte (*18*). Both eel electroplax and mammalian brain α subunits form sodium-selective channels with normal activation properties in this system, although the brain channel inactivates slowly compared to the native form; a situation that is rectified by co-expression with the ß1 subunit (*19*). The availability of this system has enabled various aspects of the structural models to be tested directly by mutagenizing selected sequences and testing the activity of the modified channels. This has resulted in the characterization of sequence elements involved in voltage-dependent activation, ion selectivity and conductance, channel inactivation and receptor sites for the binding of certain neurotoxins (*16*). Since pyrethroids mainly interfere with channel inactivation, it is of interest that residues within the short intracellular peptide linking domains III and IV have been shown to be critically involved in this process. This peptide is thought to form a "ball" (the inactivation peptide) that binds to a hydrophobic site within the intracellular mouth of the activated (open) channcl, so blocking the ion flow and effecting inactivation. This highlights the inactivation peptide and sequences that form the intracellular mouth of the pore as likely candidates for interaction with DDT or pyrethroids.

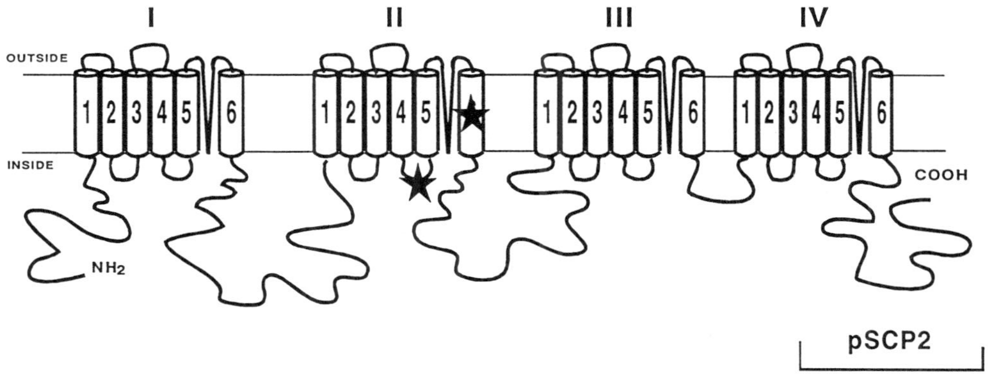

Figure 1. Diagram of the voltage-sensitive sodium channel showing its proposed membrane folding (*16,17*). The region covered by the house fly cDNA (pSCP2) that was used for RFLP linkage mapping is highlighted and positions of the two resistance-associated amino acid mutations Met_{918} to Thr and Leu_{1014} to Phe in the *kdr* or *super-kdr* channel sequences are marked (*). The full sequence of the house fly sodium channel will appear in the EMBL/GenBank databases under the accession number X96668.

Sodium Channel Sequences in Susceptible and *kdr* House Flies

The cloning of sodium channel sequences from vertebrates and *D. melanogaster* has provided the opportunity to study the corresponding sequences of susceptible, *kdr* and *super-kdr* house flies in order to look for changes that might confer resistance. In *Drosophila* two sodium channel genes have been cloned, *DSC1* located on chromosome II (*20*) and *para* on the X chromosome (*21*). The α subunit sequences encoded by these two genes are surprisingly divergent, being no more similar to each other (~ 50% amino acid homology) than they are to the vertebrate sequences. The *para* gene is known to encode a physiologically important sodium channel within the *Drosophila* nervous system and was cloned by selecting the mutant phenotypes that result from P element insertion at this locus (*21*). In contrast, *DSC1* was cloned by low stringency hybridization using a vertebrate sodium channel probe and since no neural mutants have been identified that map to this locus, the functional significance of this gene is at present unclear. For this reason, our studies of the *kdr* resistance mechanism have focused on the cloning and analysis of *para*-homologous gene sequences from the house fly.

In order to clone the house fly *para*-type gene we used a *para* fragment from *Drosophila* to screen a house fly adult head cDNA library at low stringency. A cDNA clone was recovered (pSCP2) that contained domain IV and C-terminal sequences of a House Fly sodium channel (Figure 1) with close homology to the published domain IV region of the *para* sequence (*22*). This cDNA was used to probe Southern blots of *Eco*R1-digested house fly genomic DNA and detected restriction fragment length polymorphisms (RFLPs) in the DNA of susceptible, *kdr* and *super-kdr* strains. These RFLPs serve as allele-specific DNA markers for the sodium channel gene and were therefore used to analyze the offspring of controlled genetic crosses involving these strains to establish linkage between the resistance phenotypes and the sodium channel locus. From the combined analysis of over 300 F_2 progeny we found that both *kdr* and *super-kdr* factors were closely linked to the sodium channel gene locus (*22*). These results consolidated physiological evidence that the sodium channel is the primary target of pyrethroid action and implicated the *para*-type sodium channel rather than *DSC1* as the site of resistance.

Similar studies in other insects have confirmed this linkage of *kdr*-type resistance to the *para* sodium channel. Taylor *et al.* (*23*) cloned a region extending across domains III and IV of the *para*-type sodium channel from the tobacco budworm, *Heliothis virescens*, and using a PCR-based RFLP technique showed evidence of linkage in a strain carrying multiple pyrethroid resistance factors. Dong and Scott (*24*) amplified a region from domain I of the *para*-type gene of German cockroach, *Blattella germanica*, and identified an RFLP in a *kdr*-type strain that also showed tight linkage to resistance. Finally, Knipple *et al.* (*25*) carried out an independent analysis using a *kdr* house fly strain, also using an RFLP within the domain I fragment of the *para*-type gene, and showed similar linkage to resistance. Taken together, these studies not only provide overwhelming support for the hypothesis that the *kdr* mechanism results from modification(s) of the *para*-type sodium channel, but also infer that this mechanism is likely to be highly conserved across a range of insect species.

To investigate further the molecular alterations that cause *kdr* and *super-kdr* resistance, we have cloned the full 6.3 kb coding region of the house fly *para*-type gene within five overlapping cDNA clones from the reference susceptible strain Cooper (*26*). The cDNAs predict a polypeptide of 2108 amino acids with 92% sequence identity to the *para* sodium channel of *Drosophila* and around 50% homology to vertebrate sodium channels. Using this (wild-type) sequence as a template, we designed sequence-specific oligonucleotide primers to selectively amplify the corresponding coding sequences of both *kdr* and *super-kdr* strains and carried out a full comparison of the wild-type and resistant channel sequences (*26*). The *kdr* strain showed only one amino acid change within the main body of the channel from the start of homology domain I through to the end of domain IV. This was the replacement Leu_{1014} to Phe (CTT to TTT) which was located within the transmembrane S6 segment of domain II (Figure 1). *Super-kdr* contained the same Leu_{1014} to Phe mutation together with an additional Met_{918} to Thr replacement (ATG to ATA) in the nearby S4-S5 intracellular loop of domain II (Figure 1). To assess the importance of these mutations in conferring the resistance phenotype, we then amplified the domain II region encompassing these changes from a second, unrelated *kdr* strain, six other *super-kdr* strains collected in different parts of the world, and five additional susceptible strains (*26*). The same Leu to Phe mutation was found in the second *kdr* strain, and both mutations (Leu to Phe and Met to Thr) in each of the six *super-kdr* strains. Neither mutation was present in the five pyrethroid-susceptible strains. The detection of only two amino acid changes within the main body of the sodium channel and their correlation across a range of *kdr* and *super-kdr* strains provide strong evidence that we have identified the molecular changes that underline these resistance phenotypes, and functional expression studies to confirm this are in progress. An important role for these mutations is further supported by their localization within the S4-S5 and S6 transmembrane regions of the channel protein. Recent studies of vertebrate channels suggest that these regions are located at the intracellular mouth of the channel pore where they form a receptor site for the inactivation peptide that blocks the channel during the inactivation process. This is consistent with the known physiological role of pyrethroids and DDT in delaying channel inactivation and suggests that the *kdr* mutations we have found define part of the binding site for these insecticides at the intracellular mouth of the channel. A more detailed discussion of this and the possible effects of these mutations on pyrethroid binding is given elsewhere (*26*).

The identification of these mutations has interesting implications for the evolution of the *kdr* and *super-kdr* resistance factors. The presence of the *kdr* (Leu to Phe) mutation in the *super-kdr* strains together with the additional (Met to Thr) mutation suggests that *super-kdr* arises sequentially from *kdr* rather than independently of it, since the likelihood of both mutations arising simultaneously from a wild type background is extremely low. *Super-kdr* would therefore only be predicted to arise from a population in which *kdr* is already established and would result from selection pressure with type II pyrethroids such as deltamethrin to which *super-kdr* offers a significant enhancement in the level of resistance. It will be interesting to determine whether the *super-kdr* (Met to Thr) mutation can also confer a level of resistance on its own, either by identifying field populations that contain only this mutation, or through site-directed mutagenesis and *in vitro* expression of the modified gene.

Development of a Molecular Diagnostic for Knockdown Resistance (*kdr*)

Our initial attempts to develop a DNA-based molecular diagnostic for resistance centered on exploiting the RFLPs associated with the sodium channel that were used as genetic markers in the linkage mapping (see above). Analysis of a wider range of *kdr* and *super-kdr* house fly strains did indeed reveal many similarities in their RFLPs patterns (*27*), however this technique was unreliable for diagnosing resistance in field populations since none of the bands occurred consistently in all strains. This is not surprising since the mutations associated with resistance are located in the central, domain II region of the sodium channel, whereas the probe used in these RFLP studies (pSCP2) detects changes in the C terminal and 3' non-coding regions of the gene (Figure 1). Hence, variability within this 3' region will not necessarily affect the resistance status of the gene.

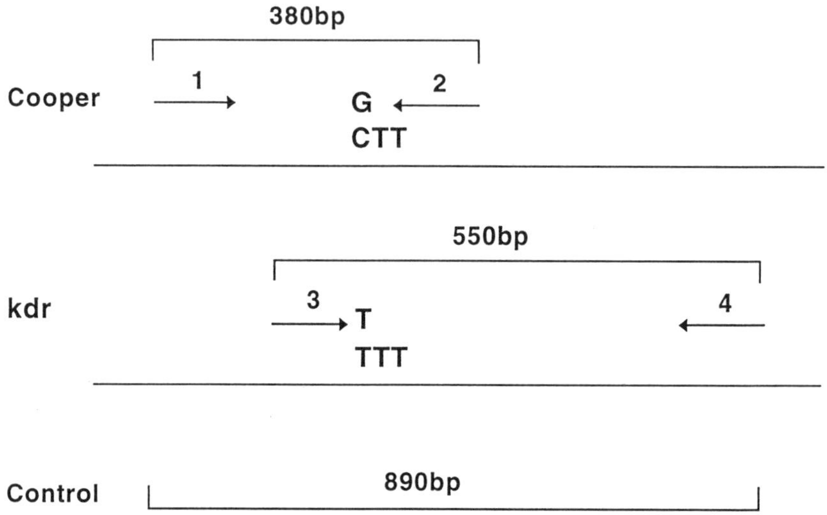

Figure 2. Diagram illustrating the PASA technique and the arrangement of primers used to amplify Cooper- and *kdr*-specific fragments; see text for details.

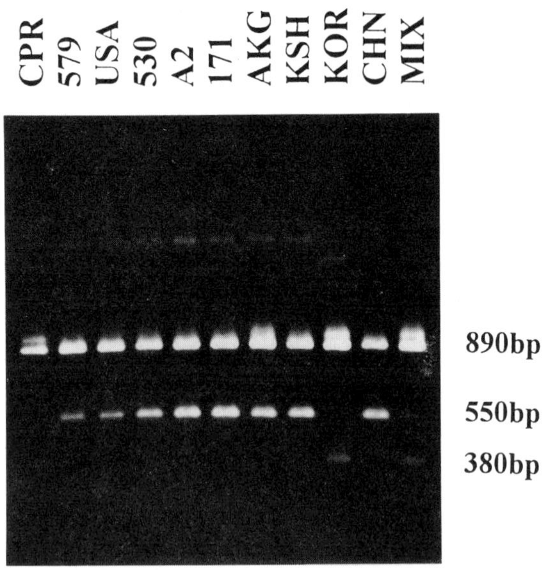

Figure 3. PASA banding patterns of 10 different House Fly strains. The PCR fragments were separated on a 1.5% agarose gel. Detailed descriptions of the different house fly strains are given elsewhere (*26,27*).

The identification of resistance-associated mutations within the domain II region of the gene not only allows the targeting of this region for a more reliable diagnostic of resistance, but also enables us to exploit the precise base changes that underlie resistance using rapid and sensitive polymerase chain reaction (PCR)-based methodologies. These techniques allow for direct detection of the mutant and wild-type alleles by selectively amplifying the regions in which the mutations occur (reviewed in *28*). One method, termed PCR Amplification of Specific Alleles (PASA) has found widespread application in clinical diagnostics (*29*) and has recently been applied to monitoring cyclodiene resistance in insects (*30*). The method relies on designing oligonucleotide primers that will preferentially amplify the mutant allele over the wild-type by matching one of the primers so that its 3' end corresponds to the base found in the mutant allele. Under stringent conditions, the PCR will then amplify the desired fragment from the mutant allele, but not from the wild-type allele because of the 3' mismatch. Although effective in detecting the presence of the mutant allele, this single primer pair will not discriminate between homozygous and heterozygous individuals. This is normally overcome by designing a third primer that matches the wild-type sequence at its 3' end and then carrying out two separate PCR reactions on each individual. Amplification of the fragment from both reactions then indicates that the individual is heterozygous.

We have adapted this technique for the detection of the house fly *kdr* mutation (Leu_{1014} to Phe; CTT to TTT), so that both homozygotes and heterozygotes are discriminated in a single PCR reaction (Figure 2). This involves using two distinct primer pairs; primers 1 and 2 select the wild-type allele (leucine) with primer 2 (antisense) ending in a G to match the CTT codon of the susceptible Cooper strain. Primers 3 and 4 select the *kdr* allele (phenylalanine; also present in *super-kdr*) with primer 3 (sense) ending in T to match the TTT. These primer pairs were designed so that they amplify fragments of different sizes (380 bp and 550 bp respectively) and can therefore be distinguished following their separation by agarose gel electrophoresis. Primers 1 and 4 are not selective and will therefore amplify a larger (890 bp) fragment from both wild-type and mutant alleles which serves as a useful internal control to ensure that PCR amplification has been successful. The assay is complicated by the presence of an intron close to the mutated base that makes it unsuitable for use on genomic DNA. To overcome this we have used cDNA reverse transcribed from RNA as the template for PCR (RT-PCR). Methods for RNA extraction from single flies and RT-PCR are described elsewhere (*26*); the stringency of the PCR reaction was optimized with 1.5mM $MgCl_2$ and an annealing temperature of 56°C.

The analysis of ten house fly strains by this method is shown in Figure 3. Cooper (CPR, lane 1) is the reference susceptible strain, 579 (derived from 538ge, ref *7*) and USA (lanes 2, 3) are strains with *kdr*-levels of resistance and the others (except for Kor, lane 9) are all strains that exhibit non-synergizable *super-kdr* levels of resistance to DDT and pyrethroids (Castells *et al.*, unpublished). The Korean strain (lane 9) is highly resistant to DDT but not to pyrethroids and the resistance appears to be metabolic (Castells, unpublished). The PASA banding patterns of these strains are fully consistent with these observed resistance phenotypes in that Cooper (lane 1) and Korean (lane 9) selectively amplify the 380 bp susceptible fragment, while the others give the 550 bp *kdr* fragment. As expected, all strains generate the 890 bp control fragment. The final track (lane 11, MIX) mimics a heterozygote in which Cooper and 579 DNA was mixed prior to amplification and from which both 550 bp and 380 bp fragments are generated. The low yield of the 380 bp Cooper fragment probably results from the lower melting temperature (Tm) associated with primer 2 and can be improved by reducing the stringency of the PCR reaction (not shown). The PASA designation that all the strains, except Cooper and Korean, contain the Leu_{1014} to Phe *kdr*-type mutation was subsequently confirmed by DNA sequencing of this region of the sodium channel from each strain. The sequence analysis also confirmed the presence of the additional Met to Thr mutation in each of the *super-kdr*-type strains (530, A2, 171, AKG, KSH and CHN). Although the assay does not distinguish between the *kdr* and *super-kdr* resistance factors, this could be addressed by designing similar primer pairs that are selective at the site of the Met to Thr mutation.

Future Work

The PASA assay described above offers a relatively simple and rapid method for detecting the *kdr*-associated sodium channel mutations in individual flies. A drawback is that it relies on cDNA rather than genomic DNA as the template, which reduces both the speed and sensitivity of the technique. We are sequencing the introns that lie in this region of the gene which will enable us to develop an improved version of the assay with primers that are suitable for use on genomic DNA. Such PCR-based assays offer several advantages over conventional bioassays in monitoring for insecticide resistance in pest populations (*31*). These include the direct detection and characterization of specific resistance mechanisms, more accurate estimates of resistance gene frequency in small sample populations and the potential to test different insect lifestages. Moreover, recent developments in the detection of PCR products using multiple fluorescent dyes (multiplexing) now make it possible to screen for several different mutations within a single PCR reaction. It is therefore feasible to develop assays that provide simultaneous screens for the *kdr* mutations as well as other major resistance genes for which the underlying molecular mutations are known.

A major goal of our work is to establish whether the mutations in the house fly are the same in other insect species that show *kdr*-type nerve insensitivity to pyrethroids. We have designed degenerate oligonucleotide primers against conserved sequences of the channel protein and are using these to selectively amplify and sequence the domain II region from a range of insect species. The data so far indicates that the *kdr* mutation (Leu_{1014} to Phe) appears to be conserved in at least two other insect orders, Lepidoptera and Homoptera. This further consolidates the role of this mutation in causing resistance and provides opportunities to develop similar PCR-based methodologies for detecting resistance in major agricultural pests.

Acknowledgments

IACR receives grant-aided support from the Biotechnology and Biological Sciences Research Council of the United Kingdom. D. M-T. was supported by a grant to ENIGMA (European Network for Insect Genetics in Medicine and Agriculture) under the EC Human Capital & Mobility Programme. N. C. was a visiting scientist from Dr. J. Bergé's group at INRA-Antibes, France and was supported by a Fellowship from the Ministère de la Recherche.

Literature Cited

1. Georghiou, G.P. In *Managing Resistance to Agrochemicals*; Green, M.B.; LeBaron, H.M. and Moberg, W.K. Eds; American Chemical Society, Washington, 1990, pp 18-41.
2. Busvine, J.R. *Nature* **1951**, *168*, 193-195.
3. Milani, R. *L. Riv. Parasitol.* **1954**, *15*, 513-542.
4. Hoyer, R.F.; Plapp, F.W. *J. Econ. Entomol.* **1966**, *59*, 495-501.

5. Farnham, A.W. *Pestic. Sci.* **1977**, *8*, 631-636.
6. Sawicki, R.M. *Nature* **1978**, *275*, 443-444.
7. Farnham, A.W; Murray, A.W.A.; Sawicki, R.M.; Denholm, I; White, J.C. *Pestic. Sci.* **1987**, *19*, 209-220.
8. Farnham, A.W.; Khambay, B.P.S. *Pestic. Sci.* **1995**, *44*, 269-275.
9. Soderlund, D.M.; Bloomquist, J.R. In *Pesticide Resistance in Arthropods*; Roush, R.T. and Tabashnik, B.E. Eds; Chapman and Hall, New York, 1990, pp 58-96.
10. Gibson, A.J.; Osborne, M.P.; Ross, H.F.; Sawicki, R.M. *Pestic. Sci.* **1990**, *30*, 379-396.
11. Soderlund, D.M.; Bloomquist, J.R. *Ann. Rev. Entomol.* **1989**, *34*, 77-96.
12. Narahashi, T. *Trends Pharmacol. Sci.* **1992**, *13*, 236-241.
13. Bloomquist, J.R.; Miller, T.A. *NeuroToxicology* **1986**, *7*, 217-224.
14. Pauron, D.; Barhanin, J.; Amichot, M.; Pralavorio, M.; Berge, J.B.; Lazdunski, M. *Biochemistry* **1989**, *28*, 1673-1677.
15. Hille, B. In *Ionic Channels of Excitable Membranes*. Sunderland, M.A. Sinauer Associates, 1984.
16. Catterall, W.A. *Annu. Rev. Biochem.* **1995**, *64*, 493-531.
17. Kallen, R.G.; Cohen, S.A.; Barchi, R.L. *Mol. Neurobiol.* **1993**, *7*, 383-428.
18. Noda, M.; Ikeda, T.; Kayano, T.; Suzuki, K.; Takeshima, H.; Kurasaki, M.; Takahashi, H.; Numa, S. *Nature* **1986**, *320*, 188-192.
19. Isom, L.L.; De Jongh, K.S.; Patton, D.E.; Reber, B.F.X. *Science* **1992**, *256*, 839-842.
20. Salkoff, L.; Butler, A.; Wei, A.; Scavarda, N.; Giffen, K.; Ifune, C.; Goodman, R.; Mandel, G. *Science* **1987**, *237*, 744-749.
21. Loughney, K.; Kreber, R.; Ganetzky, B. *Cell* **1989**, *58*, 1143-1154.
22. Williamson, M.S.; Denholm, I.; Bell, C.A.; Devonshire, A.L. *Mol. Gen. Genet.* **1993**, *240*, 17-22.
23. Taylor, M.F.J.; Heckel, D.G.; Brown, T.M.; Kreitman, M.E.; Black, B. *Insect Biochem. Mol. Biol.* **1993**, *23*, 763-775.
24. Dong, K.; Scott, J.G. *Insect Biochem. Mol. Biol.* **1994**, *24*, 647-654.
25. Knipple, D.C.; Doyle, K.E.; Marsella Herrick, P.A.; Soderlund, D.M. *Proc. Natl. Acad. Sci. USA* **1994**, *91*, 2483-2487.
26. Williamson, M.S.; Martinez-Torres, D.; Hick, C.A.; Devonshire, A.L. *Mol. Gen. Genet.* **1996** (in press).
27. Bell, C.A.; Williamson, M.S.; Denholm, I.; Devonshire, A.L. In *Molecular Action of Insecticides on Ion Channels*; Clark, J.M. Ed; ACS Symposium Series; American Chemical Society, Washington, DC, 1995, pp 86-96.
28. Prosser, J. *Tibtech* **1993**, *11*, 238-246.
29. Sommer, S.S.; Groszbach, A.R.; Bottema, C.D.K. *BioTechniques* **1992**, *12*, 82-87.
30. Steichen, J.C.; ffrench-Constant, R.H. *Pestic. Biochem. Physiol.* **1994**, *48*, 1-7.
31. ffrench-Constant, R.; Aronstein, K.; Anthony, N.; Coustau, C. *Pestic. Sci.* **1995**, *43*, 195-200.

Chapter 7

Molecular Genetics of Resistance in Fungi to Azole Fungicides

Maarten A. De Waard

Department of Phytopathology, Wageningen Agricultural University, P.O. Box 8025, Binnenhaven 9, 6700 EE Wageningen, Netherlands

Resistance in fungi to azole fungicides may result from a change in their target site, cytochrome P450-dependent sterol demethylase ($P450_{14DM}$), or from a decrease in accumulation of the fungicides in mycelium due to energy-dependent efflux. In order to study the first mentioned mechanism, the gene encoding $P450_{14DM}$ (*CYP51*) from *Penicillium italicum* has been cloned and characterized. It has an open reading frame of 1545 bp encoding a protein of 515 amino acids and is interrupted by three introns. The deduced protein sequence of this gene and $P450_{14DM}$'s from various yeast species share 46-47% identical amino acids. Laboratory isolates with a relatively high degree of resistance to azoles have the point mutation Tyr_{126} to Phe_{126}. A causal relation of this mutation with the resistance mechanism remains to be proven. Resistance to azoles based on energy-dependent efflux is probably mediated by enhanced P-glycoprotein activity. P-glycoproteins are plasma membrane ATP-ases, known to play an important role in secretion of exogenous toxic compounds. Several P-glycoprotein encoding genes from *Aspergillus nidulans* and *Botrytis cinerea* have been cloned and characterized. Differences in expression of these genes in sensitive and resistant isolates or differences in their base sequence should elucidate their relevance for fungicide resistance.

From 1969 onwards azoles and related compounds have been introduced as agricultural fungicides (*e.g.* fenarimol, imazalil and triadimefon) and antimycotics (*e.g.* miconazole, ketoconazole) (*1-3*). The mode of action of these fungicides is based on specific inhibition of a cytochrome P450-dependent sterol 14α-demethylase ($P450_{14DM}$), a key enzyme in sterol synthesis. Therefore, the fungicides are also referred to as sterol demethylase inhibitors (DMIs). Remarkable features of agricultural azoles are their high protective and curative properties, low use rates,

0097–6156/96/0645–0062$15.00/0

systemic and vapour phase activity, high chemical variability leading to diverse spectra of antifungal activity, and selective action between target and non-target organisms. At least 36 agricultural azoles have been registered for disease control in various crops and most of them have found wide spread applications. Hence, the introduction of azoles clearly represents a landmark in chemical disease control.

Development of resistance to fungicides with a specific mode of action is a major threat to effective chemical disease control. Azoles also have a site-specific mode of action. Therefore, it is understandable that these fungicides received timely attention with respect to the resistance risk. Fuchs and Drandarevski (1976) stated that development of resistance to azoles under practical conditions would be rather unlikely (*4*). History has demonstrated that this statement was premature since resistance to azoles has now been reported in various plant pathogens. However, resistance developed relatively slowly as compared with other classes of site-specific fungicides (*1-3*). The mechanisms of resistance involved are highly diverse and are seldom studied in field-resistant isolates but rather in azole-resistant laboratory-generated mutants. The reason for this is that pathogens in which resistance development in the field has been observed are often difficult to handle in biochemical studies.

The most common mechanisms of resistance reported are: an increased efflux of azoles from mycelium (*5,6*), a defect in sterol 14α-demethylation (*7*), circumvention of toxic sterol formation (*8*), overproduction of $P450_{14DM}$ (*9,10*) and decreased affinity of $P450_{14DM}$ (*11*). Overviews on these mechanisms have recently been published (*1-3*).

Molecular genetic analysis of resistance to azoles in filamentous fungi is very limited. Preliminary results have only been reported for the $P450_{14DM}$ encoding gene of *Erysiphe graminis* f.sp. *hordei* (*12*). This paper describes recent progress in the molecular analysis of resistance to azole fungicides. The first part describes the cloning and characterization of the $P450_{14DM}$ encoding gene (*CYP51*) from *Peni-cillium italicum*. Sequence differences of the gene from sensitive isolates and isolates with a high degree of resistance to imazalil suggest that azoles resistance can be due to point mutations. The second part of the paper reports on break-throughs to unrafle the molecular basis of resistance related to increased efflux of azoles from mycelium of *Aspergillus nidulans*. Evidence is provided that the driving force behind the increased fungicide secretion are membrane-bound P-glycoprotein pumps, encoded by *M(ulti)D(rug)R(esistance)* genes and that increased pump activity may relate to increased expression of these genes.

Biochemical Studies on Affinity of $P450_{14DM}$ to Azoles

The most frequently observed mechanism of resistance to fungicides is decreased affinity of the target site to the fungicide. However, reports on resistance to azoles caused by decreased affinity of $P450_{14DM}$ to these fungicides, without affecting functioning of the enzyme in sterol biosynthesis, are rare. Resistance to ketoconazole has been reported in a *Candida albicans* isolate obtained from patients with chronic mucocutaneous candidiasis who had relapsed after prolonged treatment (*11*). The reason for the reduced virulence of this isolate is not known. For plant pathogens the relevance of a target-site affinity mechanism is not known. The potency of imazalil to inhibit ergosterol biosynthesis in cell-free assays of laboratory-generated strains of *P. italicum* with different levels of resistance to imazalil is the same (*13*). These results suggest that azole resistance in the strains is not based on reduced affinity of $P450_{14DM}$. In fact, resistance based on changes in affinity of the target site has not been reported for any plant pathogen. The reason for this situation may be a technical one, since $P450_{14DM}$ is very unstable during preparation of cell-free extracts of filamentous fungi. So far, sterol demethylase activity has only been demonstrated in cell-free extracts of two filamentous fungi (*14,15*). Even for these fungi a proper comparison of the affinity of the target enzyme to azoles in isolates with a differential sensitivity remains difficult if not impossible, since no ways exist to determine specific activities of the enzyme and Ki values of the fungicides. Hence, biochemical approaches to test the relevance of resistance due to reduced affinity of $P450_{14DM}$ to azoles are difficult. Therefore, another strategy, based on cloning and characterization of the sterol 14α-demethylase encoding gene from sensitive isolates and azole-resistant mutants is much more attractive.

Molecular Genetics of *CYP51* Genes

Cytochrome P450 gene families are found in a wide range of pro- and eukaryotic organisms. The family *CYP51* encodes proteins catalyzing the 14α-demethylation of lanosterol ($P450_{14DM}$) in ergosterol biosynthesis of yeasts (*16*). The *CYP51* genes from the yeasts *Saccharomyces cerevisiae* (*9-10*), *Candida tropicalis* (*17-18*), and *C. albicans* (*19-20*) have been cloned and characterized. The isolation of *CYP51* genes from filamentous fungi has not yet been reported. The reason for this is not known but may relate to major differences in sequence homology of *CYP51* from yeasts and filamentous fungi. One difference is that the encoded enzyme in filamentous fungi uses eburicol as substrate instead of lanosterol (*2,3*). Therefore, the first step in the molecular analysis of azole resistance in plant pathogens is the cloning and molecular characterization of a *CYP51* gene. Recently, we have achieved this goal for the *CYP51* gene of *P. italicum* (Van Nistelrooy, J.G.M.; Van den Brink, J.M.; Van Kan, J.A.L.; Van Gorcum, R.F.M.; De Waard, M.A. *Molec. Gen. Genet.*; in press).

Cloning Strategy. Chromosomal DNA isolated from *P. italicum* W5 was digested with *Bam*HI, *Hin*dIII and *Eco*RI. Heterologous hybridisation with the 1.5 kb *Hin*dIII-*Sal*I fragment from pCC13 containing the $P450_{14DM}$ gene of *C. tropicalis* revealed under low stringency conditions hybridizing bands. In digests with *Bam*HI, a 3 kb band was detected. Next, 33,000 plaques from a non-amplified genomic library in a λ-EMBL3 vector were screened. Putative positive phages were selected, plated and purified. DNA from the selected phages was isolated, digested with *Bam*HI and *Sal*I, and analyzed in a Southern blot experiment using the 1.5 kb fragment from *C. tropicalis* as a probe. *BamH*I digests of purified clones identified two phages with a 3 kb hybridizing fragment co-migrating with the hybridizing fragment from the genomic DNA. A 1.3 kb *Bam*HI-*Sal*I subclone was sequenced and proved to contain the ORF of a HR2 region, which is characteristic for P450 genes. The protein sequence was about 70% identical to the HR2 regions of the *CYP51* genes of *C. albicans*, *C. tropica-*

lis and *S. cerevisiae*. Further subcloning of positive phages allowed the construction of a restriction map and a sequence strategy.

Sequence Analysis. The sequence of the subclones identified a P450 encoding gene of 1739 bp in which an open reading frame (ORF) and three putative introns were present. Since *CYP51* genes of yeasts lack intron sequences, it was decided to synthesize and clone a cDNA copy of the *CYP51* identified. The CDNA contained an ORF of 1545 bp and had a sequence identical to the predicted ORF of the genomic clone. The CDNA sequence data confirmed the presence of three introns at positions 217-276 (60 bp), 475-546 (72 bp) and 1624-1685 (62 bp), relative to the ATG initiation codon in the genomic clone. The deduced amino acid sequence of the ORF constitutes a protein of 515 amino acids with a molecular weight of 57.3 kDa. The protein contains two highly conserved cysteine-containing regions termed HR1 and HR2. The HR2 region corresponds to amino acid residues 452-473 near the C-terminal end, and is known to be the fifth ligand binding to the heme iron. Amino acid residues 121-142 correspond to the HR1 region, located in the N-terminal part. Its function is not known. Alignment of the deduced amino acid sequence of the *CYP51* gene from *P. italicum* with those of *S. cerevisiae* (*10*), *C. tropicalis* (*18*) and *C. albicans* (*20*) shows significant identity in the conserved HR1 (residues 121-142) and HR2 regions (residues 452-473). Identical sequences in other parts of the protein are shorter. Since the deduced amino acid sequence of the cloned P450 gene has more than 40% identity with *CYP51* genes from yeasts, it is concluded that the cloned gene is the *CYP51* from *P. italicum*. Obviously, the fact that $P450_{14DM}$ from *P. italicum* uses eburicol instead of lanosterol does not result in a new gene family. This conclusion may also be valid for other plant pathogens.

Functional Analysis. The cloned *CYP51* gene with an upstream promoter sequence of approximately 600 bp was subcloned in plasmid YEp24. *A. niger* N402 was co-transformed with this plasmid and plasmid pAN7-1 with the *hph* gene as a selectable marker (Hyg^R). Southern analysis revealed that transformant had >10 (D1) or less copies of the cloned P450 gene from *P. italicum* integrated in its genome. Northern analysis of the transformants showed that the gene is overexpres-sed. The sensitivity of the recipient isolate and the transformants to fenarimol was determined in radial growth tests. The EC_{95} value of fenarimol was 17 times higher for the transformant D1 than for the recipient isolate. The decrease in sensitivity correlated with the number of P450 copies in the transformed isolates. These results confirm the identity of the cloned gene as *CYP51*, since it is accepted that the reduced sensitivity of the transformants is due to functional overexpression of the transfected genes (Van Nistelrooy, J.G.M.; Van den Brink, J.M.; Van Kan, J.A.L.; Van Gorcum, R.F.M.; De Waard, M.A. *Molec. Gen. Genet.*; in press).

Analysis of *CYP51* in Azole-sensitive and -resistant Isolates.

Isolates of *P. italicum* with increasing levels of resistance to DMIs were obtained via stepwise mass selection of conidia of the wild-type isolate W5 (*21*). The mechanism of resistance in an isolate with a low degree of resistance (E300-3) was ascribed to decreased accumulation of the fungicides (*22*). Differential accumulation between isolates with a low (E300-3), medium (H17), and high degree (I33) of resistance was not observed, suggesting additional mechanisms of resistance may operate in isolates with a medium and high degree of resistance. Such mechanisms may be related to the target site and can either be caused by increased expression of *CYP51* or by mutations in this gene. Increased expression can be the result of gene amplification of *CYP51* or be due to changes in transcription. Both mechanisms would result in overproduction of $P450_{14DM}$, a "titration effect" of azoles, and hence, in a lower activity of azoles. These mechanisms have not been investigated yet. Mutations in *CYP51* may encode a $P450_{14DM}$ with a relatively low affinity to azoles. Analysis of *CYP51* from the resistant isolates mentioned above revealed that the sequence of the gene from isolate E300-3 was the same as the one from W5. In isolates H17 and I33, codon 126 of *CYP51* for tyrosine (TAC) was altered to a codon for phenylalanine (TTC) (unpublished results). The relevance of this observation for azole resistance remains to be established by functional analysis in *A. niger* as described for the gene from the wild-type isolate.

Biochemical Studies on Increased Efflux of Azoles

A well characterized mechanism for azole resistance in filamentous fungi is increased energy-dependent efflux of azoles from mycelium of laboratory-generated mutants. Increased efflux counteracts passive influx of azoles in mycelium and results in a relatively low and constant level of accumulation. This will reduce complex formation between azoles and their target site, $P450_{14DM}$, and hence may explain the relatively low levels of resistance observed. While efflux in the resistant mutants has a constitutive character, activity in wild-type isolates appears to be inducible, resulting in a transient accumulation pattern in time (*5,6*). Efflux activity in wild-type isolates was inducible by various azoles, but not by carbendazim, carboxin, and chloroneb, indicating a specific effect of azoles (*23*). In both wild-type isolates and resistant mutants, efflux activity could be inhibited by many metabolic inhibitors (*24*). These may be regarded as potential synergists of DMI toxicity (*25,26*). Increased energy-dependent efflux has now been described as a mechanism of resistance to various azoles (*27*) in *Aspergillus nidulans* (*5,6*), 1979), *Penicillium italicum* (*28*), *Candida albicans* (*29*), *Monilia fructicola* (*30*), and *Nectria haematococca* var. *cucurbitae* (*31*).

Multidrug Resistance. Increased efflux as a mechanism of resistance has also been observed in tumor cells resistant to antitumor drugs. This phenomenon is described in literature as multidrug resistance (MDR) since selection in tumor cells for resistance to one drug generally results in the simultaneous resistance to many chemically unrelated compounds (*32,33*). In most cases MDR is caused by overproduction of multidrug transport proteins, named P-glycoproteins (*34*). These multidrug transporters are plasma membrane ATPases which directly use the energy of ATP to secrete the drugs. The enhanced secretion capacity in MDR tumor cells results in a reduced cytoplasmatic accumulation of drugs and, in consequence, in a decrease of drug activity. In view of these properties, P-glycoproteins are regarded as "cytoplasmic vacuum cleaners".

In recent years, MDR has also been described in various other classes of organisms. Among fungi, the phenomenon has been reported most extensively in *Saccharomyces cerevisiae* under the name "pleiotropic drug resistance" (PDR; *35-36*). MDR in filamentous fungi has not been reported yet. However, phenotypic, genetic and biochemical evidence strongly indicates that this is the case for laboratory-generated mutants, resistant to DMIs (De Waard, M.A.; Van Nistelrooij, J.G.M.; Langeveld, C.R.; Van Kan, J.A.L.; Del Sorbo, G. In: *Modern Fungicides and Antifugal Compounds*; Lyr,H.; Russell, P.E.; Sisler, H.D., Eds; Intercept: Andover, U.K. 1996; in press). Phenotypic and genetic evidence has already been described by Van Tuyl (1977) who demonstrated cross resistance to DMIs and various antibiotics (*37*). Biochemical evidence is based on similarities in the biochemical mechanism of resistance in MDR tumor cells and azole resistant fungi (*5,6,32,33*).

P-glycoprotein. A model for the multidrug transporter in mammalian tumor cells has been well established. It consists of a plasma membrane bound efflux pump, identified as a P-glycoprotein. The most important feature of this P-glycoprotein is that drugs can be detected and expelled as they enter the plasma membrane. The pump activity accounts for the decreased accumulation in the cytosol. The second feature is that the transport occurs through a single barrel of the transporter. Common characteristics of the drugs involved are only their hydrophobic and amphipathic properties (*32,33*). The transporter itself is composed of one or more subunits of a 170 kD P-glycoprotein. The protein has two ATP-binding sites. The energy required for drug transport is derived directly from ATP hydrolysis. MDR cell lines possess relatively large amounts of P-glycoprotein in their cell membranes which in turn results in increased drug efflux. In many cell lines it has been demonstrated that the efflux activity increases with each selection step for a higher degree of multidrug resistance. Impairment of drug efflux by compounds may be a consequence of inhibition of ATP synthesis or a result of the potency to directly inhibit P-glycoprotein activity. Some compounds may mimic substrates and thus competitively inhibit multidrug transport. A similar model as described for mammalian tumor cells has widely been accepted for the mechanism of multidrug resistance in other organisms such as the budding yeast *S. cerevisiae* (*35,36*).

The Physiological Function of P-glycoproteins

P-glycoproteins can be classified as transporters of non-toxic or toxic substrates. The identity of the substrates of the first class is largely unknown but comprises a variety of endogenously produced compounds such as cortisol and phospholipids in mammals, and mating factors, peptides and enzymes in microorganisms. Substrates of the second class are, in general, toxins either endogenously produced (*e.g.* antibiotics in microorganisms) (*38*) or exogenously present (*e.g.* antibiotics, drugs, pesticides). It is believed that the natural function of the latter transporters in mammals is the secretion of natural cytotoxic compounds present in the diet (*e.g.* toxic plant compounds).

A physiological role of P-glycoproteins in plant pathogens may be, in analogy to the second class of P-glycoproteins mentioned above, the secretion of exogenous toxins, which naturally occur in nature. It is proposed that this may relate to antibiotics produced by other microorganisms as well as to toxins produced by host plants. In the latter case, P-glycoproteins can have a role in efflux of plant defense factors (phytoalexins, phytoncides) which accumulate in fungal cells upon colonization of the plant tissue. This hypothesis is supported by the observation that *N. haematococca* possesses an inducible and energy-dependent mechanism that secretes pisatin from its mycelium (*39*). This hypothesis also corroborates the simultaneous resistance to azoles and isoflavonoid phytoalexins in *Cladosporium cucumerinum* (*40*). Another role of P-glycoproteins in pathogenesis may be the secretion of pathogenicity factors (toxins, peptides and proteins) from plant pathogens. If these hypotheses are true, impaired P-glycoprotein activity can result in enhanced accumulation of plant defense factors in pathogens or reduced secretion of pathogenicity factors. This hypothesis also implies that inhibitors of P-glycoprotein activity can be regarded as new plant disease control agents by exploiting the natural plant defense response or by annulment of the action of pathogenicity factors. For a long time, the development of non-toxic disease control agents has been considered but could hardly be realized due to a lack of fundamental knowledge and rational leads. Inhibitors of P-glycoproteins may provide a concept to achieve this goal.

Molecular Genetics of *MDR* Genes

Genes encoding P-glycoproteins are often described as *MDR* or *PDR* genes. The human CDNA of the *MDR1* gene encodes a 1280-amino acid protein with 12 predicted transmembrane domains in two homologous halves, each containing six transmembrane regions and a large intracytoplasmic loop encoding an ATP site (*33*). Similar genes have now been isolated and characterized from different classes of organisms such as various mammalian species, plants, microorganisms and yeasts. In most of these organisms various genes encoding different P-glycoproteins have been detected. Conclusive evidence that at least some of these genes encode a multidrug transporter has been obtained from experiments in which these genes have been inactivated or transfected to parental sensitive cell lines. Inactivation resulted in increased and transfection in decreased sensitivity to drugs. The validity of this molecular approach has also been demonstrated for the *MDR* genes *PDR5* of *S. cerevisiae* and *CDR1* of *C. albicans* (*41,42*). Transfection of *CDR1* to a *PDR5*-disrupted mutant of *S. cerevisiae* resulted in multidrug resistance to several compounds including cycloheximide and the azole antimycotic miconazole. This is the first conclusive molecular evidence that resistance to azoles can be mediated by P-glycoprotein activity. The main mechanisms causing overexpression of P-glycoprotein in multidrug resistant organisms are amplification of the wild-type gene and alterations in gene regulation. Gene amplification is the most common mechanism in mammalian MDR tumor cells (*33*). The most common mechanism of PDR in yeast is based on mutations in the *PDR1* gene, encoding a transcription regulator which enhances transcription of P-glycoprotein encoding genes (41). Point mutations in structural genes encoding P-glycoproteins merely play a role in determining their substrate specificity (*43,44*).

Our current research tries to elucidate the presence of genes encoding P-glycoproteins in filamentous fungi and the role of the gene products in fungicide resistance and pathogenicity. The organisms selected are *A. nidulans* and *B. cinerea*. *A. nidulans* has been selected as a model organism because of the availability of genetically defined imazalil-resistant mutants and the wide experience with this organism in molecular genetics. *B. cinerea* has been selected because of its wide host range indicating that the pathogen is obviously able to cope with many different plant defense products, and because of its potency to secrete a wide variety of pathogenicity factors. The first aim of the research was to isolate genes encoding P-glycoproteins from both organisms. Significant progress in this respect has been made. Some of the cloned genes are almost fully sequenced and show a high degree of homology with *PDR5* and *SNQ2* of *S. cerevisiae*. In *S. cerevisiae* overexpression of both genes results in PDR. The aim of future studies is to construct transformants with decreased or increased expression of the P-glycoprotein genes isolated. The phenotype of these mutants, both with respect to fungicide resistance and pathogenesis, will be assessed to study the physiological function of the genes.

Concluding Remarks

Extensive efforts have been made to elucidate mechanisms of resistance to azole fungicides. The fragmentary evidence available so far indicates that different mechanisms operate and suggests that these mechanisms are not necessarily related to a decrease in affinity of the target site of azoles, $P450_{14DM}$ in sterol biosynthesis. The variety in mechanisms of resistance also corroborates with the preliminary observations that the genetic basis of resistance may vary for different organisms or even in the same organism (*31,45*). Differences in resistance mechanisms may particularly be relevant for laboratory-generated mutants of model fungi and field-resistant isolates of plant pathogens, since reductions in comparative fitness or pathogenicity are often restricted to the first category. Much additional knowledge is needed to fully understand the phenomenon.

Literature Cited

1. De Waard, M.A. In *Fungicide Resistance*. Heaney, J.; Slawson, D.; Hollomon, D.W.; Smith, M.; Russell, P.E.; Parry, D.W., Eds; BCPC Monograph 60; British Crop Protection Council: Farnham, U.K. 1994; pp 3-10.
2. Köller, W. In *Target Sites of Fungicides Action*; Köller, W., Ed.; CRC Press: Boca Raton, Florida, 1992; pp 119-206.
3. Vanden Bossche, H. In *Modern Selective Fungicides*; Lyr, H., Ed.; Gustav Fischer Verlag: Jena, Germany, 2nd ed., 1995; pp 431-484.
4. Fuchs, A; Drandarevski, C.A. *Neth. J. Pl. Pathol.* **1976**, *82*, 8[illegible]1.
5. De Waard, M.A.; Van Nistelrooy, J.G.M. *Pestic. Biochem. Physiol.* **1979**, *10*, 219-229.
6. De Waard, M.A.; Van Nistelrooy, J.G.M. *Pestic. Biochem. Physiol.* **1980**, *13*, 279-286.
7. Walsh, R.C.; Sisler, H.D. *Pestic. Biochem. Physiol.* **1982**, *18*, 122-131.
8. Taylor, F.R.; Rodriguez, R.J.; Parks, L.W. *J. Bacteriol.* **1983**, *155*, 64-68.

9. Kalb, V.F.; Loper, J.C.; Dey, C.R.; Woods, C.W.; Sutter, T. *Gene* **1986**, *45*, 237-245.
10. Kalb, V.F.; Woods, C.W.; Turi, T.G.; Dey, C.R.; Sutter, T.R.; Loper, J.C, *DNA* **1987**, *6*, 529-537.
11. Vanden Bossche, H.; Marichal,P.; Gorrens, J.; Bellens, D,; Moereels, H.; Janssen, P.A. *Biochem. Soc. Transact.* **1990**, *18*, 122-131.
12. Hollomon, D.W.; Butters, J.A., Hargreaves, J.A. In *Managing Resistance to Agrochemicals*; Green, M.B.; LeBaron; H.M.; Moberg, W.K., Eds; ACS Symposium Series 421; American Chemical Society: Washington, DC 1990; pp 199-214.
13. Guan, J.; De Waard, M.A. *Pestic. Biochem. Physiol.* **1993**, *46*, 1-6.
14. Guan, J; Stehmann, C.; Ellis, S.W.; Kerkenaar, A; De Waard, M.A. *Pestic. Biochem. Physiol.* **1992**, *42*, 262-270.
15. Stehmann, C.; Kapteyn, J.C.; De Waard, M.A. *Pestic. Sci.* **1994**, *40*, 1-8.
16. Nelson, D.R.; Kamataki, T.; Waxman, D.J.; Guengerich, F.P.; Estabrook, R.W.; Feyereisen, R.; Gonzalez, F.J.; Coon, M.J.; Gunsalus, I.C.; Gotoh, O., Okuda, K.; Nebert, D.W. *DNA and Cell Biol.* **1993**, *12*, 1-51.
17. Chen, C.; Turi, T.G.; Sanglard, D., Loper, J.C. *Biochem. Biophys. Res. Comm.* **1987**, *146*, 1311-1317.
18. Chen, C,; Kalb, V.F.; Turi; T.G.; Loper, J.C. *DNA* **1988**, *9*, 617-626.
19. Kirsch, D.R.; Lai, M.H.; O'Sullivan, J. *Gene* **1988**, *68*, 229-237.
20. Lai, M.H.; Kirsch, D.R. Nucleic Acids Res. **1989**, *17*, 804.
21. De Waard, M.A.; Van Nistelrooy, J.G.M. *Neth. J. Pl. Pathol.*,**1990**, *96*, 321-329.
22. Guan, J.; Kapteyn J.C.; Kerkenaar, A.; De Waard, M.A. *Neth. J. Pl. Path.* **1992**, *98*, 313-324.
23. De Waard, M.A.; Van Nistelrooy, J.G.M. *J. Gen. Microbiol.* **1981**, *126*, 483-489.
24. De Waard, M.A.; Van Nistelrooy, J.G.M. *Experim. Mycol.* **1987**, *11*, 1-10.
25. De Waard, M.A.; Van Nistelrooy, J.G.M. *Pestic. Sci.* **1982**, *13*, 279-286.
26. De Waard, M.A.; Van Nistelrooy, J.G.M. *Pestic. Sci.* **1984**, *15*, 56-62.
27. De Waard, M.A.; Van Nistelrooy, J.G.M. *Pestic. Sci.* **1988**, *22*, 371-382.
28. De Waard, M.A.; Van Nistelrooy, J.G.M. *Neth. J. Pl. Pathol.* **1984**, *90*, 143-153.
29. Ryley, J.F.; Wilson, R.G.; Barrett-Bee, K.J. *Sabouraudia* **1984**, *22*, 53-63.
30. Ney, C. *Untersuchungen zur Resistenz von Monilia fructicola (Wint.) Honey gegenüber Ergosterol-Biosynthese Hemmern.* Thesis University Basle, Basle, Switserland, 1988.
31. Kalamarakis, A.E.; De Waard, M.A.; Ziogas, B.N.; Georgopoulos, S.G. *Pestic. Biochem. Physiol.* **1991**, *40*, 212-220.
32. Gottesman, M.M.; Pastan, I. *J. Biol. Chem.* **1988**, *263*, 12163-12166.
33. Gottesman, M.M.; Pastan, I. *Ann. Rev. Biochem.* **1993**, *62*, 385-427.
34. Juliano, R.L.; Ling, V. *Biochim. Biophys. Act.* **1976**, *455*, 152-162.
35. Balzi, E.; Goffeau, A. *Biochim. Biophys. Act.* **1991**, *1073*, 241-252.
36. Balzi, E.; Goffeau, A. *Biochim. Biophys. Act.* **1994**, *1187*, 152-162.
37. Van Tuyl, J.M. *Meded. Landbouwhogeschool Wageningen* **1977**, *77-2*, 1-136.

38. Guilfoile, P.G.; Hutchinson, C.R. *Proc. Nat. Ac. Sci., USA* **1991**, *88*, 8553-8557.
39. Denny, T.P.; Matthews, P.S.; VanEtten, H.D. *Physiol. Mol. Pl. Pathol.* **1987**, *30*, 93-107.
40. Fuchs, A.; De Vries, F.W.; De Waard, M.A. *Neth. J. Pl. Pathol.* **1984**, *90*, 3-11.
41. Balzi, E.; Wang, M.; Leterme, S.; Van Dyck, L.; Goffeau, A. *J. Biol. Chem.* **1994**, *269*, 2206-2214.
42. Prasad, R.; De Wergifosse, P.; Goffeau, A.; Balzi, E. *Curr. Genet.* **1995**, *27*, 320-329.
43. Devine, S.E.; Ling, V.; Melera, P.W. *Proc. Nat. Ac. Sci. USA* **1992**, *89*, 4564-4568.
44. Devine, S.E.; Melera, P.W. *J. Biol. Chem.* **1994**, *269*, 6133-6139.
45. Kalamarakis, A.E.; Demopoulos, V.P.; Ziogas, B.N.; Georgopoulos S.G. *Pestic. Biochem. Physiol.* **1989**, *95 suppl.1*, 109-120.

Chapter 8

Expression of Amplified Esterase Genes in Insecticide-Resistant *Myzus persicae* (Sulzer)

Linda M. Field[1], Caroline A. Hick[1], Alan L. Devonshire[1], Naghmy Javed[2], Jennifer M. Spence[2], and Roger L. Blackman[2]

[1]Biological and Ecological Chemistry Department, Institute of Arable Crops Research-Rothamsted, Harpenden, Hertfordshire, AL5 2JQ, United Kingdom
[2]Department of Entomology, Natural History Museum, London SW7 5BD, United Kingdom

Insecticide resistance in *Myzus persicae* results from amplification of genes encoding insecticide-detoxifying esterases and from differential transcription of the amplified genes which may be mediated by changes in DNA methylation. Methylation is usually stable in resistant aphid clones and can be inherited during sexual reproduction. However, when a sudden loss of methylation occurs within a clone, it is accompanied by silencing of the amplified genes. When reselected with insecticides some recovery of expression can occur but this is not accompanied by methylation. Some aphid clones have amplified esterase genes arranged as tandem repeats at a single locus, whereas others have arrays dispersed around the genome. Thus, although resistance in *M. persicae* is dependent primarily on gene amplification, this may be modulated by other molecular and genetic factors.

There are now two well established examples of insecticide resistance resulting from the amplification of genes encoding insecticide detoxifying carboxylesterases (*1*). In the mosquito, *Culex pipiens quinquefasciatus*, resistant insects with more than 250 copies of the B1 esterase gene have been reported (*2*). In the aphid, *Myzus persicae*, variation in both copy number (*3*) and transcription (*4*) of amplified esterase E4 genes can affect the resistance status of an individual insect, the latter being associated with changes in the presence of 5-methylcytosine in and around the amplified genes. In *Culex* there is no evidence for changes in expression of the esterase genes (*5*). Here we bring together the current knowledge of DNA amplification, E4 gene expression and changes in DNA methylation during both asexual and sexual reproduction in *M. persicae* and attempt to assess the relative contributions these genetic phenomena make to insecticide resistance.

0097–6156/96/0645–0072$15.00/0

Quantification of Amplified Esterase Sequences in *Myzus persicae*

It is now more than seven years since the cloning of a cDNA, encoding the E4 esterase (carboxylester hydrolase, EC3.1.1.1), and the identification of amplified esterase genes were reported in insecticide-resistant *M. persicae* (*3*). At that time it was noted that although the esterase gene copy number increased with increasing levels of resistance, there was not the difference between susceptible and extremely resistant R_3 aphids which would be expected from their relative amounts of esterase enzyme. Figure 1 summarises the situation. Immunoassays of the E4 enzyme in R_3 aphids (A) can be accurately quantified in a microplate reader (*6*) and are in line with enzyme titration studies using radiolabelled ligands which showed an approximately 60-fold increase in esterase protein compared with susceptible aphids (*7*). This is reflected in dot blots of poly A^+ RNA probed with E4 cDNA (Figure 1B), which show *c.* 4-fold increases between S and R_1, R_1 and R_2, and R_2 and R_3, in line with their levels of esterase enzyme. However, probing of DNA dot blots (Figure 1C) shows much smaller differences between the aphid clones with only a *c.* 8-fold increase between susceptible and R_3. This has been attributed to non-specific binding of the probe to related sequences (*3*), but even in Southern blots of DNA digested with restriction enzymes, the binding to E4 fragments was difficult to quantify because of the band diversity and uncertainty regarding the extent of homology of the probe to amplified and susceptible esterase sequences.

Recent attempts to quantify the E4 gene copy number in an R_3 aphid clone, 794J, have used a cloned *Eco*RI/*Kpn*I fragment from the amplified E4 gene (known to be present in the DNA of susceptible aphids) to probe *Eco*RI/*Kpn*I digests of susceptible and 794J aphid DNA (Field, Devonshire and Tyler-Smith, *Biochem. J.* in press). This showed only 5-10 times more esterase sequences in 794J aphids, in which the amplification is known to be heterozygous at a single locus, suggesting the presence of a single array of 10-22 copies of the E4 gene (assuming two copies per diploid genome of susceptible aphids). In the same study, analyses of the repeat units (amplicons) containing the E4 genes, using pulsed-field gel electrophoresis, confirmed the presence of only *c.* 12 copies of the E4 gene, each on *c.* 24 kb amplicons, arranged as a tandem array of direct repeats. This is consistent with *in situ* hybridisation studies (*8*) and with crossing experiments (Blackman, Spence, Field, Javed and Devonshire, *Heredity*, in press) which confirm that amplified E4 genes are at a single heterozygous locus. Thus in the R_3 aphid clone, 794J, the increase in E4 copy number is not sufficient to account for the levels of E4 enzyme synthesized.

Methylation of amplified sequences in *M. persicae*

The presence of 5-methylcytosine in the amplified esterase genes and their flanking DNA has been detected in *M. persicae* using an *Msp*I/*Hpa*II diagnostic assay (*4,9*). This showed that esterase sequences in susceptible aphids are unmethylated, in line with most of the aphid genome. However, amplified expressed E4 genes in resistant aphids contain 5-methylcytosine, which is not

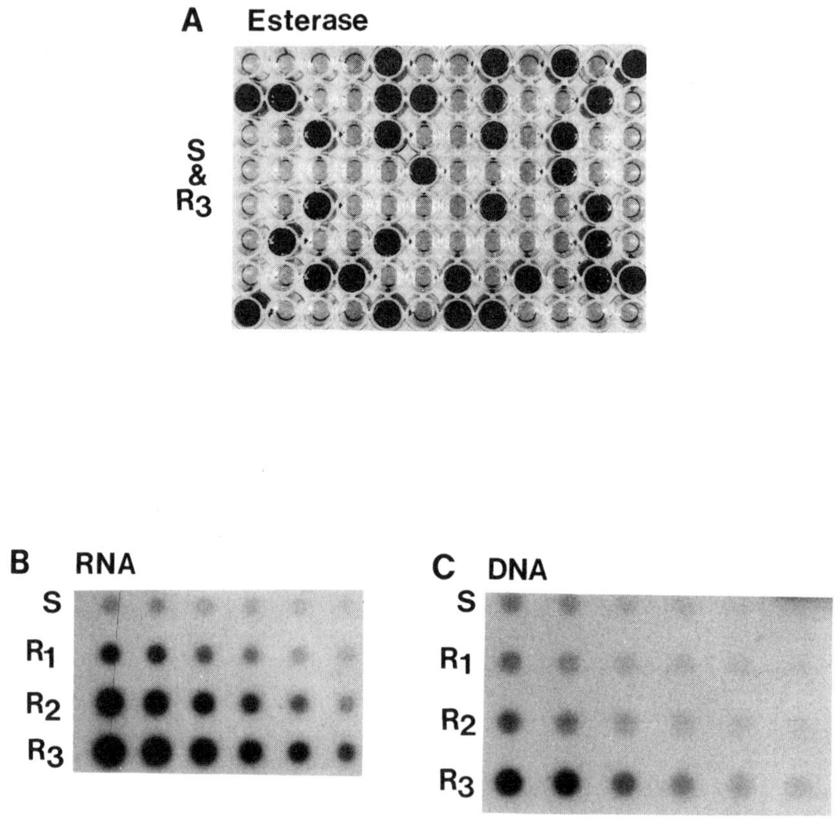

Figure 1. A: Immunoassay of the E4 enzyme present in susceptible (S, 64 light-coloured wells) and highly resistant (R_3, 32 dark wells) aphids and binding of an E4 cDNA probe to 2-fold serial dilutions of B: poly A^+ RNA and C: genomic DNA extracted from susceptible and increasingly resistant (R_1-R_3) aphids.

present in aphids that have lost resistance and are not expressing their amplified genes (revertants). These results were surprising since it was generally thought that insect DNA does not contain significant amounts of 5-methylcytosine (*10*) and in most studies of vertebrate gene expression the presence of DNA methylation prevents transcription (*11*).

We have recently monitored changes in DNA methylation during the loss of resistance in an aphid clone established in 1991 from a UK glasshouse population (Hick, Field and Devonshire, *Insect Biochem. & Mol. Biol.*, in press) and shown a concomitant loss of E4 gene transcription and 5-methylcytosine, as the aphids' progeny changed from R_3 to susceptible levels of E4 activity over three generations. When these revertant aphids were reselected for resistance with insecticide treatments, their E4 esterase content increased to $R_{1/2}$ levels without an accompanying change in DNA methylation. Thus there is a complex interaction between esterase copy number, DNA methylation and expression which is summarised in Table 1.

Table 1. Summary of molecular changes during loss and reselection of resistance in *M. persicae*

Resistance Status	Gene Amplification	DNA Methylation	Expression (E_4 levels)
Susceptible*	- (2x)	-	S (1x)
Resistant	+ (c12x)	+	R_3 (60x)
Revertant	+ (c12x)	-	S (1x)
Reselected	+ (c12x)	-	$R_{1/2}$ (4-16x)

*susceptible included for comparison, assumed diploid for esterase gene

These data suggest that resistance occurs primarily by E4 gene amplification, but with an increase in transcription, possibly as the result of DNA methylation. During reversion the methylation is lost and transcription of the amplified genes diminishes drastically. After reselection the amplified genes are again expressed but not re-methylated. It is interesting that we have not been able to reselect aphids to produce R_3 levels of E4 and it may be that the lack of methylation limits the E4 levels to reflect the gene copy number (i.e. 12 copies giving $R_{1/2}$ levels). This would support the view that it is the methylation of the amplified genes which is involved in their overexpression.

Inheritance of DNA methylation during sexual reproduction

The presence of 5-methylcytosine in the amplified genes of *M. persicae* indicates that, even if methylation is not common in the aphid genome, the aphid must have the enzyme systems necessary for both *de novo* methylation and its maintenance during asexual cell growth and development. It should be noted that the loss of methylation discussed earlier is a rare event and normally

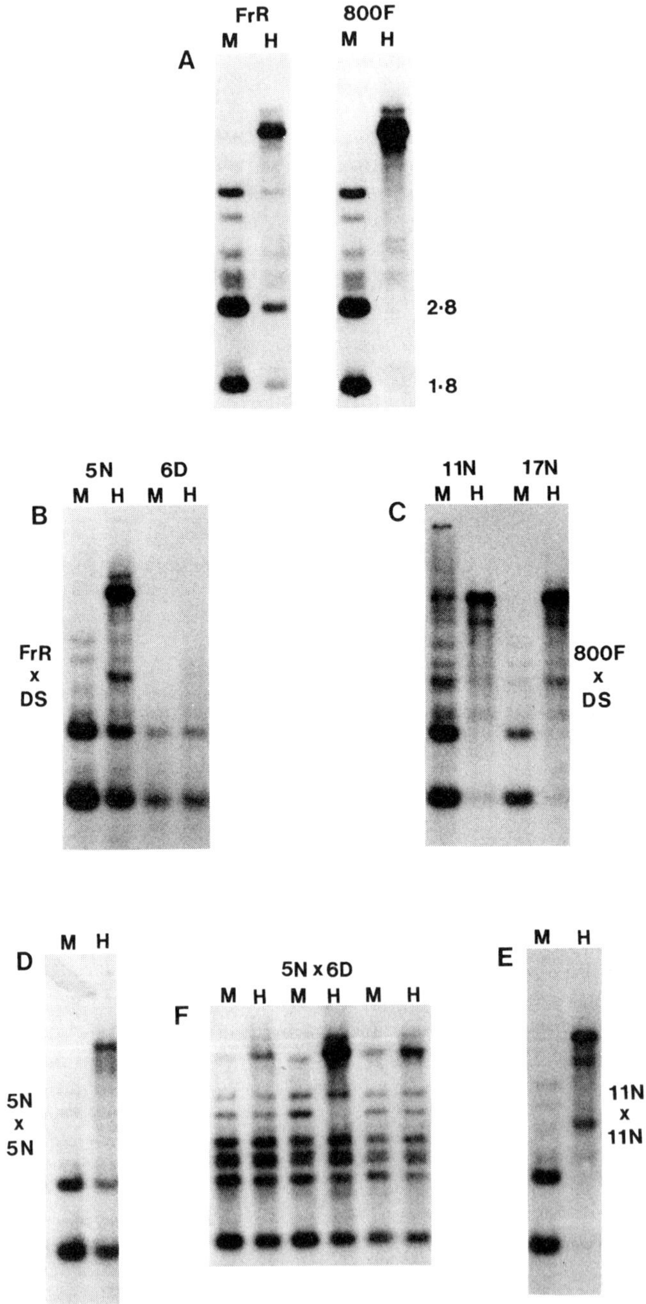

Figure 2. Binding of cloned esterase genomic sequences to *Msp*I (M) and *Hpa*II (H) digests of DNA extracted from resistant aphid clones FrR and 800F, and the offspring of 2 generations of sexual crosses.

resistance, E4 gene expression and the presence of 5-methylcytosine are stably inherited in asexual reproduction.

So are patterns of DNA methylation also stable during sexual reproduction, i.e. can they be maintained through meiosis? We have recently been studying the inheritance of amplified esterase genes through the sexual phase of *M. persicae* (Blackman, Spence, Field, Javed & Devonshire, *Heredity*, In press) and were able to monitor the methylation patterns in successive sexual generations. Figure 2 shows the *Msp*I and *Hpa*II esterase restriction fragments obtained over 3 generations. The presence of 2.8 and 1.8 kb bands in *Msp*I digests shows the presence of amplified genes (see ref 9); which encode FE4, an insecticide-detoxifying esterase very closely related to E4 (*11*). When these small bands are absent in *Hpa*II digests and replaced by larger fragments it shows that the amplified sequences are methylated (*9*).

The two parental clones FrR and 800F (Figure 2A) have amplified FE4 genes, but whereas the 800F sequences are fully methylated (i.e. they have no 2.8 and 1.8 kb bands in the *Hpa*II digest), the FrR sequences show "partial" methylation, with only some of the sites cut by *Hpa*II.

When 800F was crossed with susceptible (DS) aphids or selfed all amplified sequences inherited by the F1 clones were fully methylated (e.g. Figure 2C) and when one of these (11N) was selfed, the methylation was again inherited (Figure 2E).

For the FrR/susceptible cross a few aphids lost methylation completely (e.g. 6D Figure 2B) but most maintained the "partial" state (e.g. 5N of Figure 2B) which was again inherited during selfing (Figure 2D). When 5N was crossed with 6D the esterase genes in the offspring showed varying degrees of methylation (Figure 2F). Thus DNA methylation can be stable during sexual reproduction even if only some of the sites are methylated.

Future Prospects

This work has identified a so-called "partial" state of DNA methylation (*9*) where only some of the amplified esterase sequences in an aphid contain 5-methylcuytosine. Until recently it has only been possible to characterise methylation in homogenates of whole aphids since the *Msp*I/*Hpa*II technique requires enough DNA for two restriction digests. However, we have now developed a PCR-based assay (Field, Crick and Devonshire, in preparation) to diagnose the type of amplified gene present (E4 or FE4) and the presence of 5-methylcytosine in very small amounts of aphid DNA (<0.001 aphid). This now opens the possibility of studying DNA methylation in individual embryos and tissues during aphid development and reproduction.

Although amplified E4 genes are usually at a single heterozygous locus, *in situ* hybridization has shown that there are at least two other sites where amplified E4 genes can be found, and FE4 genes can also occur at several sites (*8*). This raises the possibility of an involvement of transposable elements in the development of insecticide resistance in *M. persicae* as has been suggested for mosquitoes (*13*), and this possibility will be explored further.

We now have aphid clones established from the offspring of the crosses having single sites of amplified genes and this creates the opportunity to establish the gene copy number, amplicon size and structure, DNA methylation and esterase expression at each of the loci. Thus a detailed picture of the genetic changes underlying resistance in this major agricultural pest can be achieved.

Acknowledgments

IACR receives grant-aided support from the Biotechnology and Biological Sciences Research (BBSRC) Council of the United Kingdom and this work was supported by a BBSRC Linked Research Groups grant.

Literature Cited

1. Devonshire, A.L.; Field, L.M. *Ann. Rev. Entomol.* **1991**, *36*, 1-23.
2. Mouchès, C.; Pasteur, N.; Bergé, J.B.; Hyrie, O.; Raymond, M.; De Saint Vincent, B.R.; De Silvestri, M.; Georghiou, G.P. *Science* **1986**, *233*, 778-780
3. Field, L.M.; Devonshire, A.L.; Forde, B.G. *Biochem. J.* **1988**, *251*, 309-312.
4. Field, L.M.; Devonshire, A.L.; ffrench-Constant, R.H.; Forde, B.G. *FEBS Letts* **1989**. *243*, 323-327.
5. Raymond, M; Poulin, E.; Boiroux, V.; Dupont,E.; Pasteur, N. *Heredity* **1993**, *70*, 301-307.
6. Devonshire, A.L.; Moores, G.D.; ffrench-Constant, R.H. *Bull. Ent. Res.* **1986**, *76*, 97-107.
7. Devonshire, A.L.; Sawicki, R.M. *Nature*, **1979**, *270*, 140-141.
8. Blackman, R.L.; Spence, J.M.; Field, L.M.; Devonshire, A.L. *Heredity*, **1995**, *75*, 297-302.
9. Field, L.M.; Devonshire, A.L. In *Molecular Mechanisms of Insecticide Resistance*; Mullin, C.A. and Scott, J.G. Eds; ACS Symposium Series; American Chemical Society, Washington DC, 1992, pp. 209-217.
10. Adams, R.L.P. *Biochem. J.* **1990**, *265*, 309-320.
11. Eden, S.; Cedar, H. *Current Opinions in Genetics and Development*, **1994**, *4*, 255-259.
12. Field, L.M.; Williamson, M.S.; Moores, G.D.; Devonshire, A.L. *Biochem. J.* **1993**, *294*, 569-574.
13. Mouchès, G.; Agarwal, M.; Campbell, K. In Insecticides: *Mechanisms of Action and Resistance*; Intercept Ltd Andover, Hants, UK, 1992; pp. 345-353.

Chapter 9

Molecular Genetics and Evolution of Copper Resistance in Bacterial Plant Pathogens

Donald A. Cooksey

Department of Plant Pathology, University of California, Riverside, CA 92521

Related copper resistance operons have been described in recent years from different genera of bacteria from agricultural environments where copper compounds are applied to plants for disease control and fed to livestock as dietary supplements. Although generally similar in overall structure, there is considerable divergence between the copper resistance determinants at the sequence level, in their functions, and at the level of metal-induced gene expression. Copper resistance operons are likely of ancient origin, related to indigenous bacterial multicopper oxidase systems. There is evidence for recent spread of copper resistance genes among closely-related bacteria, but not between different bacterial genera that contain highly specialized versions of distantly-related copper resistance operons.

Copper compounds have been used as antimicrobial agents in agriculture for many decades, but it was not until the 1980s that copper resistance was recognized in plant pathogens (*1*). Copper resistance was reported in *Xanthomonas campestris* first in Florida (*2*), and then in Mexico (*3*), Oklahoma (*4*), California (*5,6*), and at least two other countries (*7*). Plasmid-determined copper resistance was also described in a pathovar of *Pseudomonas syringae* that infected tomato in California and Mexico (*8*) and later in pathovars of *P. syringae* that infected impatiens (*9*), cherry (*10*), and ornamental trees (*11*). Copper resistance has also been characterized in epiphytic strains of *P. syringae* on citrus and almond (*12,13*) and saprophytic pseudomonads associated with copper-resistant pathogens on tomato plants (*5*). In addition to its use on plants, copper is fed to pigs as a dietary supplement, and enteric bacteria from such animals in different geographical locations have been shown to carry plasmid-determined copper resistance (*14,15*).

Early genetic studies suggested that the genes responsible for copper resistance in *Xanthomonas,* different strains of *P. syringae*, and the enteric bacteria from pigs were unrelated (*4,16,17*). The mechanisms of resistance were also considered to be

0097–6156/96/0645–0079$15.00/0

different, since some, such as *P. syringae* pv. tomato, were observed to accumulate copper (*18-20*), while others, such as *E. coli*, had a measurable decrease in copper accumulation when carrying copper resistance genes (*21-22*). More sensitive Southern hybridization experiments with cloned resistance genes (*5,9,13,15,23*), and DNA sequencing (*6,24*), subsequently revealed that several of the resistance determinants in different plant pathogenic and enteric bacteria from diverse agricultural environments are related. A better understanding of the origins and mechanisms of copper resistance may be useful in designing resistance management strategies for important copper-based agrochemicals.

Relatedness of Copper Resistance Determinants

Copper resistance has been described to date in only a few of the many host-specific pathovars of plant pathogenic bacterial species. Most of the resistance determinants in these species are related, but there are important structural and regulatory differences between them that argue against a recent dissemination of a common system between the different taxa.

Structure of the *cop* Operon. The basic structure, as described in *Pseudomonas syringae* pv. tomato (*25,26*), consists of four structural genes, *copABCD*, under the control of a copper-inducible promoter, and two regulatory genes, *copRS*, required for sensing copper and activating the copper-inducible promoter (Figure 1). This gene cluster occurs on a conserved plasmid in resistant strains of *P. syringae* pv. tomato and in the chromosome of certain other strains of that pathovar (*27*). Related copper resistance determinants have been described in *P. syringae* strains from impatiens, almond, and citrus (*9,13*), although DNA hybridization experiments suggest that some may be considerably diverged from the *cop* genes at the sequence level (*13*).

In the related plant pathogen *Xanthomonas campestris*, *cop*-related genes have been shown to confer resistance in pathovars from tomato, pepper, walnut, and crucifers (*5,6,23*). The resistance determinants from *X. campestris* have the same general *copABCD* structure as *P. syringae*, but with some differences in gene size and spacing, and considerable divergence at the sequence level (Figure 1). There also is no structural or functional evidence for *copRS* homologs in *X. campestris* (*6*), and as discussed later, there are functional differences at the level of expression for the *cop* genes from *Xanthomonas* vs. *Pseudomonas*.

A copper resistance system more closely related to the *cop* system of *P. syringae* has been described in enteric bacteria from pigs that were fed dietary supplements of copper sulfate (*15,24*). The *pcoABCDRSE* determinant was cloned and sequenced from *Escherichia coli* (*N. L. Brown, GenBank Accession No. X83541)* and is structurally similar to *copABCDRS*; however, considerable sequence divergence was observed. A general structural difference between the two systems is the presence of *pcoE* after *pcoABCDRS*, which is not found after *copABCDRS* (Mills, S. D. and Cooksey, D. A., unpublished data). An unrelated open reading frame in the opposite orientation is present downstream from *copABCDRS* (Figure 1), but there is no evidence that this open reading frame is important for copper resistance in *P. syringae*.

Relatedness of Protein Structure. *copA/pcoA* is the most conserved gene in these related operons, and it is the only one of the structural genes to be related to genes of known function from other organisms. The CopA product is related to a family of eucaryotic multi-copper oxidases, and more distantly to small blue copper proteins, such as azurin and plastocyanin (*28,29,30*). Many bacteria may contain *copA*-related genes, as shown by probing various species of *Pseudomonas* (*5*), and as revealed recently from sequencing the *E. coli* genome, where a *copA* homolog exists with no known function and without proximity to any other *cop* gene homologs (*31*). CopA contains conserved sites that are known to coordinate four copper atoms in the related eucaryotic multi-copper oxidases. Four amino acid residues coordinate a single type-1 copper site that is similar to the sites in small blue copper proteins (*29*). The type-1 copper is coordinated by two histidines, a cysteine, and usually a methionine present near the carboxyl end of these proteins (His-X_{48}-Cys-X_4-His-X_4-Met). In addition, a trinuclear copper center consists of one type-2 copper bound by two histidine ligands and two type-3 copper atoms with three histidine ligands each (*32*). The histidines that contribute to this trinuclear copper center are symetrically supplied from two conserved domains near the beginning and end of these proteins. CopA from *P. syringae* and *X. campestris,* as well as PcoA from *E. coli,* contain these conserved sites (*6,25,28*), which presumably bind copper atoms. However, it is not yet known whether these proteins are functional oxidases.

The similarity of CopA sequences with other known proteins can be used in studying the phylogeny of the copper resistance determinants. CopA and PcoA from the copper resistance operons are more closely related to each other than to the CopA homolog that is present in the *E. coli* chromosome (Figure 2). Sequence analysis also confirms, as noted by comparison of overall operon structure (Figure 1), that the plasmid-borne *pco* copper resistance system of *E. coli* and the *cop* system of *P. syringae* are more closely related to each other than to the *Xanthomonas* copper resistance system. However, *Pseudomonas* and *Xanthomonas* are considered to be closely related aerobic bacteria, with *E. coli* belonging to a separate family of facultatively anaerobic bacteria. This preliminary phylogenetic analysis is consistent with exchange of copper resistance determinants in the past between bacterial taxa rather than recent independent evolution of resistance in each group under copper selection in agriculture (*20*). However, as discussed later in relation to divergent functional and regulatory mechanisms, the sequence data do not support recent exchange of copper resistance determinants between these particular genera.

Mechanisms of Copper Resistance

Protein Characterization. In addition to analysis of protein sequence, products of the *cop* genes of *P. syringae* have been characterized functionally. CopA (72 kDa) and CopC (12 kDa) are periplasmic copper-binding proteins (18). CopC was purified and shown to bind a single copper atom per polypeptide, while CopA was estimated to bind 11 copper atoms per polypeptide. In addition to the amino acid ligands for binding the four copper atoms conserved in multicopper oxidases, CopA contains internal repeated sequences, rich in methionine and histidine, which may bind the additional copper atoms. Both CopA and CopC are produced abundantly in copper-

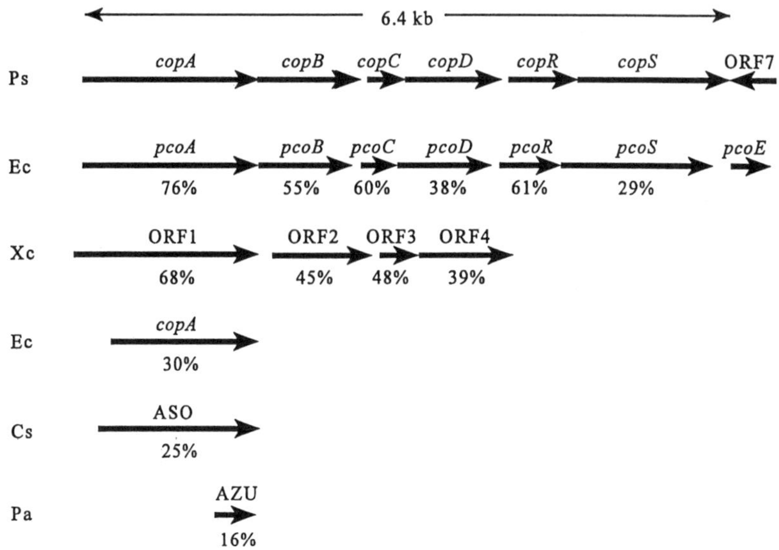

Figure 1. Comparative structure of copper resistance operons and related oxidase genes from *Pseudomonas syringae* (Ps), *Escherichia coli* (Ec), *Xanthomonas campestris* (Xc), *Cucumis sativus* (Cs), and *P. aeruginosa* (Pa). Percent identities of amino acid sequences with the corresponding *P. syringae* gene products are indicated.

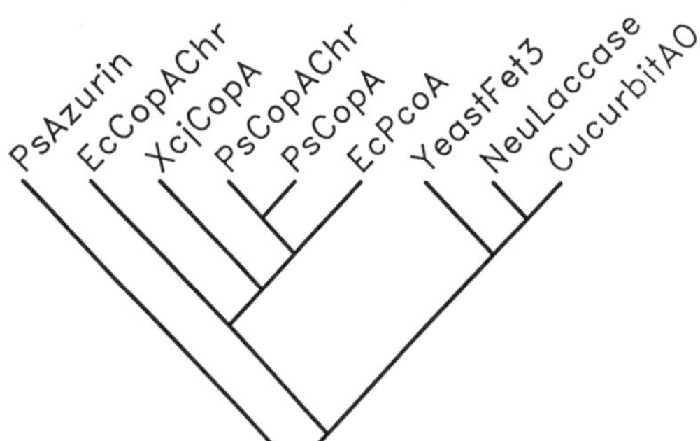

Figure 2. Phylogeny of CopA and related protein sequences derived from parsimony analysis (PHYLIP, Version 3.5c) of aligned amino acid sequences of azurin from *Pseudomonas aeruginosa* (*33*), CopA from the chromosome of *Escherichia coli* (*31*), CopA from the copper resistance operon of *Xanthomonas campestris* pv. juglandis (*6*), CopA from the chromosome of *P. syringae* (*34*), CopA from the plasmid-borne *cop* operon of *P. syringae* (*25*), PcoA from the plasmid-borne *pco* operon of *E. coli* (*GenBank Accession No. X83541*), Fet3 from yeast (*35*), laccase from *Neurospora crassa* (*36*), and ascorbate oxidase from *Cucumis sativus* (*37*). Azurin was designated as the outgroup sequence for the analysis.

induced cells, accounting for 3% and 1%, respectively, of total protein on a dry weight basis (*38*).

CopB is a 39-kDa protein located in the outer membrane of *P. syringae* (*18*), which may bind or channel copper through the outer membrane. Outer membrane fractions containing CopB are blue with bound copper, but methods used to isolate the protein from the membrane released copper into solution. Therefore, there is no direct evidence that CopB binds copper, but there are several tandem repeats in CopB similar to the internal repeats of CopA, with the consensus Asp-His-X-X-Met-X-X-Met (*25*). A similar sequence, His-X-X-Met-X-X-Met, also appears in PcoE of *E. coli* (*N. L. Brown, GenBank Accession No. X83541*), involved in plasmid-borne copper resistance, and in the chromosomally-encoded CutE of *E. coli* (*39*), which may bind copper and play a role in copper transport. The sequence Asp-His-X-X-Met-X-X-Met is also repeated near the amino terminus of a P-type copper-transporting ATPase of the Gram-positive bacterium *Enterococcus hirae* (*40*), although CopB of *P. syringae* has no other apparent similarities with the ATPase. In addition, a copper-transporting protein CTR1, from yeast, contains 11 repetitions of a Met-X-X-Met motif that was compared with the 12 repetitions of Met-X-X-Met in CopA of *P. syringae* (*41*). CopB has six copies of Met-X-X-Met within its repeated region (*25*). It therefore seems likely that the repeated domains of CopB interact with copper. Dancis et al. (*41*) noted that these repeated sites are distinct from the ligands of copper-containing electron transport proteins where copper must be tightly bound in a fashion to allow reversible oxidation and reduction. The repeated sites may be more consistent with a reversible binding of copper, which would be expected for a role in copper transport and release to other cellular components.

CopD has not been purified, but is likely to be an inner membrane protein with multiple membrane-spanning regions (*25,42*). Interestingly, expression of CopD and CopC, without CopA or CopB, resulted in 26-fold-greater sensitivity to copper than wild-type cells. This hypersensitivity was associated with increased cellular copper accumulation, suggesting that CopD, together with CopC, function in copper uptake (*42*).

Proposed Functions of Proteins in Copper Resistance. Analysis of the location and copper-binding properties of Cop proteins has led to the suggestion of sequestration of copper in the periplasmic space and outer membrane as a mechanism of resistance to copper in *P. syringae* pv. tomato (*18*). This model is supported by cellular copper accumulation studies and the observation that colonies of resistant strains turn blue when grown on copper-containing media (*19,20,42*). However, the amount of copper predicted to bind to Cop proteins measured in cells accounted for only a fraction of the total copper accumulated by *P. syringae*, and the concentration of the proteins did not continue to increase at higher levels of copper, while total accumulated copper did increase (*18*). Other cellular components, such as lipopolysaccharide, may be involved in the further accumulation of copper. The Cop proteins are required for growth at high levels of copper, and they may therefore function in transient binding and delivery of copper ions to other binding components of the cell wall.

In contrast to the system described from *P. syringae* pv. tomato, neither *E. coli*, *X. campestris*, or certain epiphytic strains of *P. syringae*, all with *cop*-related resistance determinants, turn blue or appear to accumulate copper as a resistance mechanism (*13,23,28*). A decreased uptake of copper has been measured in *pco*-containing *E. coli* cells, suggesting an efflux mechanism for copper resistance (*17,28*). However, none of the Pco or Cop proteins from Gram-negative bacteria resembles known copper-transporting ATPases, which show considerable conservation from bacteria to humans (*40*). A copper-transporting ATPase, also called Cop, has been described from the Gram-positive bacterium *Enterococcus hirae* (*43*), but no such protein has been shown to be involved in copper resistance in *E. coli*. Clearly, the mechanisms of copper resistance in the *pco* and *cop* systems have yet to be fully defined, but it is apparent that functional differences between them exist. Further work is needed to define the possible role of structural differences in the observed functional divergence between these systems. These include the presence of *pcoE* after the plasmid-borne *pcoABCDRS* of *E. coli*, while no such gene is present after *copABCDRS* of *P. syringae*. In addition, the repeated units in CopB of *P. syringae* are absent in PcoB of *E. coli* and CopB of *X. campestris* (*6,28*).

Regulation of Copper Resistance

Copper and Zinc Sensing. All of the copper resistance operons that have been studied at the regulatory level are induced by copper ions. The *cop* operon from *P. syringae* pv. tomato is induced only by copper (*1,44*), but recent work has shown that a related operon cloned from an epiphytic strain of *P. syringae* is also induced by zinc (*13*). Both determinants only provide resistance to copper.

Unlike most other metal resistance systems (*45*), induction of the *cop* operons, at least in *P. syringae* and *E. coli*, involves a two-component system for sensing copper (CopS) and activating transcription of *cop* promoters (CopR). By analogy to related two-component systems (*46*), CopS is thought to sense copper in the region that probably loops out into the periplasm, but this topology, and the copper-interactive sites, have not been defined experimentally. Comparison of sequences between CopS and PcoS, which share only 29% amino acid identity overall, but which presumably sense copper similarly, reveals several conserved amino acid residues in the probable periplasmic domain that could potentially interact with copper. These sites are being investigated through mutagenesis. If CopS is produced by the *cop* homolog from the epiphytic strain of *P. syringae* that also responds to zinc (*13*), then sequence analysis and mutagenesis of that clone may provide insight into specificity for copper vs. zinc sensing.

The hypothetical model for copper induction, derived from related two-component regulatory systems, is that the membrane-spanning CopS protein senses copper ions in the periplasm and phosphorylates CopR in the cytoplasm, which converts CopR to an active state for induction of the *cop* operon. The sequence of CopS is similar to other bacterial sensors such as PhoR and EnvZ of *E. coli*, including conservation at the histidine kinase autophosphorylation site and the relative positions of two hydrophobic regions that may span the cytoplasmic membrane (*26*).

Transcriptional Activation. The cytoplasmic response regulatory protein, CopR, has been purified and shown to bind to a specific DNA sequence, referred to as the *cop* box (*34*), that spans the -35 region with respect to the start of transcription of the *cop* operon of *P. syringae*. Similar sites are present in the promoter/operator region of a chromosomal homolog of the *cop* operon from *P. syringae* and in front of both *pcoA* and *pcoE* of the *E. coli* copper resistance determinant (Figure 3). No *cop* box is present after *copS* in the *P. syringae* copper resistance determinant, confirming the uniqueness of the *pcoE* gene in the *E. coli* version (Figure 1). As discussed for different mechanisms of resistance conferred by related copper resistance operons, there appears to be considerable divergence in the function of these copper-responsive regulatory elements, in spite of the apparent structural similarities of two-component systems and CopR binding sites present in copper-inducible promoters. There is no expression in *E. coli* or in *X. campestris* from the *cop* promoter of *P. syringae* (*26,44*), and the *X. campestris* resistance genes do not function in *E. coli* or *P. syringae* (*23*).

Some copper-sensitive strains of *P. syringae* and of other pseudomonads that did not contain plasmid-borne resistance genes were shown to support copper-inducible activation of an introduced *cop* promoter that was fused to a reporter gene (*26*). This suggested that functional *copRS* homologs were present on the chromosome, possibly regulating other chromosomally-encoded genes involved in copper metabolism. In one strain, spontaneous mutations, apparently in the chromosomal *copRS* genes, resulted in increased transcription of *cop*-related chromosomal genes and an elevated level of copper resistance (*27*). Thus, the ability to efficiently regulate expression of copper resistance genes may have evolved from indigenous regulatory pathways already involved in sensing copper, just as the functional copper resistance proteins may have evolved from common copper oxidase systems.

copA	CAAGCTTACAGAAATGTAATCGCGCCGC
copAH	CCAGCTTACGGAAATGTAATTACCTCGT
pcoA	GAAGATGACGTAAATGTAATACAGCTAT
pcoE	GAAGGTGACAAAATTGTCATCATTCAGT

Figure 3. Conservation of the binding site of the CopR regulatory protein in the promoter/operator region of plasmid-borne *copA* from *P. syringae* (*34*), chromosomal *copAH* from *P. syringae* (*34*), and *pcoA* and *pcoE* from *E. coli* (*GenBank Accession No. X83541*). Shading indicates nucleotides conserved between the plasmid-borne *copA* gene promoter and other the promoters. Horizontal lines were placed above palindromic sequences, which may indicate binding of CopR as a dimer.

Conclusions

Considering the level of divergence in sequence, function, and expression, between the different *cop*-related copper resistance systems, it is not likely that a single system evolved and was disseminated among bacterial taxa recently, in response to mobilization of copper from human activities. Like other metal resistance determinants, it seems more likely that copper resistance had an ancient origin. The highly specialized expression of the copper resistance determinants that have been described in different genera may even create a barrier to their dissemination between taxa. There is evidence suggesting that closely-related strains, pathovars or species of plant pathogens and of enteric bacteria from livestock have exchanged copper resistance plasmids recently (*9,15,23,47*), but it appears that different genera of bacteria from the agricultural environment probably obtained copper resistance independently from related species. Except in one type of strain of *P. syringae* pv. tomato (*27*), there is no evidence that copper resistance, determined by *cop*-related operons, can evolve under selection from indigenous chromosomal genes. Chromosomal homologs of *copA*, probably with an oxidase function, may be widespread in bacteria, but at least in the *E. coli* chromosome (*31*), the other essential genes of the *cop* operon are not known to be present in indigenous copper-sensitive populations.

With the continued widespread use of antimicrobial copper compounds in agriculture, copper resistance is likely to spread further among populations of related plant pathogenic bacteria. Long-distance transport of plant propagative materials, which can be contaminated with plant pathogens, may contribute to further spread of copper resistance geographically. Since most pathovars of plant pathogenic bacteria are not yet reported to be resistant to copper, increased efforts to monitor for resistance in the field would be important in any attempts to manage resistant populations. In some plant pathogens with copper resistance determined by *cop* genes, the level of copper resistance is only 2-4 times the normal sensitive levels, and disease control may still be achieved with certain modifications of copper formulations that enhance copper toxicity (*1,2,48,49*). Such enhanced formulations may only have a temporary benefit, however, since higher levels of copper resistance can be selected easily from some of the resistant bacteria, at least in the laboratory (*27*). Further monitoring of resistance levels in the field and investigation of adaptive mechanisms may help to predict the success of such resistance management strategies.

Acknowledgment

This work was supported by National Science Foundation Grant DEB-9306559.

Literature Cited

1. Cooksey, D. A. *Annu. Rev. Phytopathol.* **1990**, *28*, 201-19.
2. Marco, G. M.; Stall, R. E. *Plant Dis.* **1983**, *67*, 779-81.
3. Adaskaveg, J. E.; Hine, R. B. *Plant Dis.* **1985**, *69*, 993-96.
4. Bender, C. L.; Malvick, D. K.; Conway, K. E.; George, S.; Pratt, P. *Appl. Environ. Microbiol.* **1990**, *56*, 170-75.
5. Cooksey, D. A.; Azad, H. R.; Cha, J.-S.; Lim, C.-K. *Appl. Environ. Microbiol.* **1990**, *56*, 431-35.
6. Lee, Y.-A; Hendson, M.; Panopoulos, N. J.; Schroth, M. N. *J. Bacteriol.* **1994**, *176*, 173-88.
7. Jones, J. B.; Stall, R. E.; Scott, J. W.; Somodi, G. C.; Bouzar, H.; Hodge, N. C. *Plant Dis.* **1995**. *79*, 395-98.

8. Bender, C. L.; Cooksey, D. A. *J. Bacteriol.* **1986**, *165*, 534-41.
9. Cooksey, D. A. *Appl. Environ. Microbiol.* **1990**, *56*, 13-16.
10. Sundin, G.W; Jones, A. L.; Fulbright, D. W. *Phytopathology* **1989**, *76*, 861-65.
11. Sundin, G. W.; Bender, C. L. *Appl. Environ. Microbiol.* **1993**, *59*, 1018-24.
12. Andersen, G. L.; Menkissoglu, O.; Lindow, S. E. *Phytopathology* **1991**, *81*, 648-56.
13. Rogers, J. S.; Clark, E.; Cirvilleri, G.; Lindow, S. E. *Phytopathology* **1994**, *84*, 891-97.
14. Tetaz, T. J.; Luke, R. K. J. *J. Bacteriol.* **1983**, *154*, 1263-68.
15. Williams, J. R.; Morgan, A. G.; Rouch, D. A.; Brown, N. L.; Lee, B. T. O. *Appl. Environ. Microbiol.* **1993**, *59*, 2531-37.
16. Bender, C. L.; Cooksey, D. A. *J. Bacteriol.* **1987**, *169*, 470-74.
17. Rouch, D.; Lee, B. T. O.; Camakaris, J. In *Metal Ion Homeostasis: Molecular Biology and Chemistry*; Hamer, D. H.; Winge, D. R., Eds.; Alan R. Liss, Inc., New York, 1989, pp 439-46.
18. Cha, J.-S.; Cooksey, D. A. *Proc. Natl. Acad. Sci. USA* **1991**, *88*, 8915-19.
19. Cooksey, D. A.; Azad, H. R. *Appl. Environ. Microbiol.* **1992**, *58*, 274-78.
20. Cooksey, D. A. *Mol. Microbiol.* **1993**, *7*, 1-5.
21. Rouch, D. A.; Camakaris, J.; Lee, B. T. O.; Luke, R. K. J. *J. Gen. Microbiol.* **1985**, *131*, 939-43.
22. Rouch, D.; Camakaris, J.; Lee, B. T. O. In *Metal Ion Homeostasis: Molecular Biology and Chemistry*; Hamer, D. H.; Winge, D. R., Eds.; Alan R. Liss, Inc., New York, 1989, pp 469-77.
23. Voloudakis, A. E.; Bender, C. L.; Cooksey, D. A. *Appl. Environ. Microbiol.* **1993**, *59*, 1627-34.
24. Silver, S.; Lee, B. T. O.; Brown, N. L.; Cooksey, D. A. In *The Chemistry of the Copper and Zinc Triads*; Welch, A. J., Chapman, S. U., Eds.; Royal Society of Chemistry, London, 1993, pp 38-53.
25. Mellano, M. A.; Cooksey, D. A. *J. Bacteriol.* **1988**, *170*, 1879-83.
26. Mills, S. D.; Jasalavich, C. A.; Cooksey, D. A. *J. Bacteriol.* **1993**, *175*, 1656-64.
27. Lim, C.-K.; Cooksey, D. A. *J. Bacteriol.* **1993**, *175*, 4492-98.
28. Bryson, J. W.; O'Halloran, T. V.; Rouch, D. A.; Brown, N. L.; Camakaris, J.; Lee, B. T. O. In *Bioinorganic Chemistry of Copper*; Karlin, K. D; Tyeklar, Z., Eds.; Chapman and Hall, New York, 1993, pp 101-109.
29. Ouzonis, C.; Sander, C. *FEBS Lett.* **1991**, *279*, 73-78.
30. Cooksey, D. A. *FEMS Microbiol. Rev.* **1994**, *14*, 381-86.
31. Fujita, N.; Mori, H.; Yura, T.; Ishihama, A. *Nucl. Acids Res.* **1994**, *22*, 1637-39.
32. Messerschmidt, A. *Biochem. So. Transact.* **1992**, *20*, 364-68.
33. Hoitink, C. W.; Woudt, L. P.; Turenhout, J. C.; Van de Kamp, M.; Canters, G. W. *Gene* **1990**, *90*, 15-20.
34. Mills, S. D.; Lim, C.-K.; Cooksey, D. A. *Mol. Gen. Genet.* **1994**, *244*, 341-51.
35. Askwith, C.; Eide, D.; Van Ho, A.; Bernard, P. S.; Li, L.; Davis-Kaplan, S.; Sipe, D. M.; Kaplan, J. *Cell* **1994**, *76*, 403-10.
36. German, U. A.; Müller, G.; Hunziker, P. E.; Lerch, K. *J. Biol. Chem.* **1988**, *263*, 885-96.

37. Ohkawa, J.; Okada, N.; Shinmyo, A.; Takano, M. *Proc. Natl. Acad. Sci. USA* **1989**, *86*, 1239-43.
38. Cha, J.-S. Ph.D. Thesis, University of California, Riverside, CA, 1990, 74 pp.
39. Rogers, S. D.; Bhave, M. R.; Mercer, J. F. B.; Camakaris, J.; Lee, B. T. O. *J. Bacteriol.* **1991**, *173*, 6742-48.
40. Solioz, M.; Odermatt, A.; Krapf, R. *FEBS Lett.* **1994**, *346*, 44-7.
41. Dancis, A.; Yuan, D. S.; Haile, D.; Askwith, C.; Eide, D.; Moehle, C.; Kaplan, J.; Klausner, R. D. *Cell* **1994**, *76*, 393-402.
42. Cha, J.-S.; Cooksey, D. A. *Appl. Environ. Microbiol.* **1993**, *59*, 1671-74.
43. Odermatt, A.; Suter, H.; Krapf, R.; Solioz, M. *J. Biol. Chem.* **1993**, *268*, 12775-79.
44. Mellano, M. A.; Cooksey, D. A. *J. Bacteriol.* **1988**, *170*, 4399-401.
45. Silver, S.; Walderhaug, M. *Microbiol. Rev.* **1992**, *56*, 195-228.
46. Parkinson, J. S.; Kofoid, E. C. *Annu. Rev. Genet.* **1992**, *26*, 71-112.
47. Sundin, G. W.; Demezas, D. H.; Bender, C. L. *Appl. Environ. Microbiol.* **1994**, *60*, 4421-31.
48. Conlin, K. C.; McCarter, S. M. *Plant Dis.* **1983**, *67*, 639-44.
49. Lee, Y.-A.; Schroth, M. N.; Hendson, M.; Lindow, S. E.; Wang, X.-L.; Olson, B.; Buchner, R. P.; Teviotdale, B. *Phytopathology* **1993**, *83*, 1460-65.

Evolution and Population Genetics

Chapter 10

Evolution of Insecticide Resistance in the Mosquito *Culex pipiens*: The Migration Hypothesis of Amplified Esterase Genes

M. Raymond and N. Pasteur

Laboratoire Génétique et Environnement, Institut des Sciences de l'Evolution, Unité de Recherche Associée, Centre National de la Recherche Scientifique 327, Université de Montpellier II (C.C. 065), F–34095 Montpellier, Cedex 05, France

Resistance to organophosphorus insecticides has been studied at the gene and the population levels in *Culex pipiens* in various geographic areas. Only three loci have developed major resistance alleles in this species, including *Est-2* (or esterase B), at which resistance occurs through gene amplification. Gene amplification involving a same particular haplotype has been found at the esterase B locus of mosquitoes from various continents. This situation, which has been explained by a unique amplification event followed by migration and selection by OP insecticides, has been sometimes questioned. A clarification of the hypotheses proposed is presented, and how it is possible to prove or disprove them. Recent data on the extent of polymorphism at the esterase B locus in susceptible populations provide a strong support of the migration hypothesis.

The wide use of organic insecticides to control medically important pest species has been a powerful agent of selection in natural populations of many insect species which have developed various degrees of resistance (*1,2*). In a few species, such as the mosquito *Culex pipiens*, it is possible to identify each gene conferring resistance to organophosphorous insecticides in single individuals.

This mosquito, common in temperate and tropical countries, is subjected to insecticide control in many places. World-wide surveys of resistance to organophosphorus insecticides have disclosed that only three loci have developed major resistance alleles (*3-6*). The first two loci, *Est-2* (or esterase B) and *Est-3* (or esterase A), code for detoxifying carboxylester hydrolases (EC 3.1.1.1), and resistance alleles correspond to an esterase overproduction (*4,7,8*). Six distinct electromorphs have been described so far at the *Est-2* locus (named B1, B2, B4, B5, B6 and B7) and four at the *Est-3* locus (A1, A2, A4 and A5) (*3,4,6,9-11*). In the case of esterase B, overproduction corresponds to the amplification of a DNA segment containing the structural gene (*4,10,12*). The third locus, *Ace*, codes the acetylcholinesterase

0097–6156/96/0645–0090$15.00/0

(insecticide target), and insensitive alleles have been reported in various places (*e. g. 13-15*) but it is not known how many Ace^R alleles have occurred independently.

Analysis by Restriction Endonuclase Digestion

How resistance to organophosphorus insecticides evolved in natural populations of *Culex pipiens* could be studied at the molecular level for the esterase B locus, for which molecular tools have been developed for field studies. It is possible to build a restriction map of the DNA area within and around the esterase B structural gene, in susceptible mosquitoes with a single copy of the gene, as well as in mosquitoes with an amplified gene. When such maps are compared, large differences are observed. For example, the map found in S-Lab, a susceptible reference strain from California, and the map from Tem-R, a strain also from California possessing the B1 amplification, have only 21% of their restriction sites in common (*16*). Similar results are found in comparing maps from different susceptible strains, or in comparing strains with distinct overproduced electromorph (*4,16,17*). However, when strains with the B2 electromorph are compared, restriction maps are strictly identical (*16*), independently of their geographical origins (Table I). A similar situation is found for B1 electromorph, which possesses the same restriction map in mosquitoes from various parts within the Americas and in China (*18*). How can such similarity be explained ?

How to Explain the Similarity of the Restriction Maps?

A large part of the polymorphism detected by restriction enzymes around the esterase B structural gene is probably neutral. The identity of the restriction maps of B1 or B2 haplotypes in many geographic areas indicates therefore that these alleles are identical by descent. There are two possibilities: either they were first amplified in a particular place, and have then spread (Figure 1A), or they have first spread and then been independently amplified in various places (Figure 1B).

The first scenario has been proposed by Raymond et al. (*16*) and Qiao and Raymond (*18*), based on the argument that the amount of divergence between distinct restriction maps (such as between S-Lab and Tem-R) could indicate a large amount of polymorphism in natural populations, so that the probability of independently amplifying a same allele is very low. In addition, the selective advantage provided by the amplification itself promotes its spread in places treated with organophosphate insecticides. The multiple and independent amplification of B2 has been favoured by Hemingway et al. (*19*) and Ketterman et al. (*20*), based on variation in the kinetics of esterases studied on partially purified enzymes.

Only an analysis of the polymorphism of susceptible populations could discriminate between these two possibilities. Under the first scenario, the polymorphism at the esterase B locus in non-treated populations should be extensive, and the probability of sampling a non-amplified allele already amplified elsewhere should be very low. On the other hand, the second scenario predicts that either the non-amplified B1 or B2 allele (which are amplified most commonly world-wide) is present at a detectable frequency in susceptible populations.

Table I. Geographic distribution of amplified B esterase identified by restriction fragment length polymorphism analysis from published information. The year of collection of the material studied is indicated.

Country	Year	Allele	RE[a]	References
North America:				
USA California	1974	B1	12	(*23*)
USA California	1974	B1	13	(*16*)
USA California	1986	B1	1	(*10*)
USA Illinois	1986	B1	1	(*10*)
USA Texas	1986	B2	13	(*16*)
Latin America:				
French Guiana	1991	B1	1	(*18*)
French Guiana	1991	B2	1	(*18*)
Venezuela	1991	B1	1	(*18*)
Puerto Rico	1992	B1	13	(*18*)
Cuba	1986	B8[b]	1	(*22*)
Europe:				
France	1984	B4	6	(*4*)
Corsica	1988	B4	1	(*24*)
Cyprus	1987	B5	6	(*4*)
France	1986	B2	3	(*25*)
France	1991	B4	1	(*26*)
Italy	1992	B4	1	(*21*)
Africa:				
Egypt	1987	B2	13	(*16*)
Congo	1988	B2	13	(*16*)
Ivory coast	1986	B2	13	(*16*)
Asia:				
Sri Lanka	1986	B2	1	(*22*)
China	1992	B1	13	(*18*)
China	1992	B1	2	(*11*)
China	1992	B2	2	(*11*)
China	1992	B6	2	(*11*)
China	1992	B7	2	(*11*)
Pakistan	1985	B2	13	(*16*)

[a]RE: number of restriction enzymes used
[b]Unnamed by Vaughan et al. 1995.

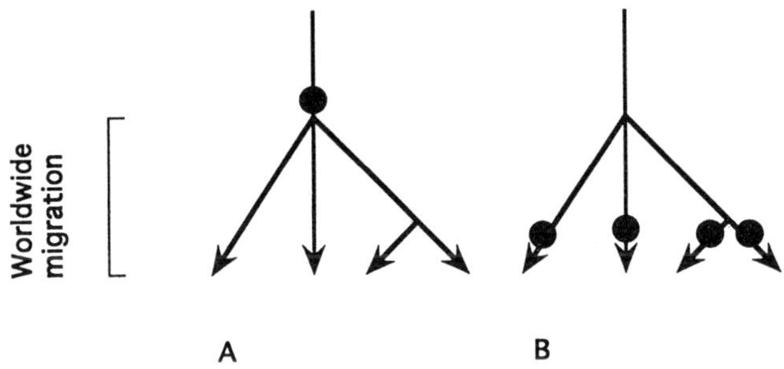

Figure 1. The two possible scenarios for the identity of amplified alleles in various geographical areas. A) amplification occurs once and extensive migration is promoted through organophosphorus resistance. B) migration of the non-amplified allele occurs first, and then is amplified independently in several organophosphorous treated areas. The circle represents an amplification event.

Study of Polymorphism at the Esterase B Locus in Susceptible Populations.

Culex pipiens susceptible populations still exist in northern France and northern Europe. Three such susceptible populations were sampled and analyzed for esterase electromorph and DNA polymorphism (*17*). At the protein level, 16 alleles were found for esterase B in one French population (N = 74), and 14 in an English one (N = 50). One electromorph had the same mobility as B2, but it was never associated with A2 (a characteristic of the amplified B2 throughout the world), and it was concluded that this similar migration is probably coincidental. At the DNA level, 24 alleles at the esterase B locus were identified in a sample of 72 mosquitoes from one population, with the use of only one restriction enzyme (Figure 2). Restriction maps of two non-amplified alleles randomly sampled from a single breeding site in Belgium were built with 6 restriction enzymes (Figure 2). 60% of the sites are different among the two maps. In addition, these two maps were not more related than a pair drawn at random from a pool containing other European alleles (Figure 2).

The huge polymorphism found in susceptible populations considerably strengthens the hypothesis that amplification of an allele occurs before it is spread. It is still possible that non amplified B2 or B1 alleles exist at very low frequencies in every susceptible populations, but how such a situation would be created and maintained requires specific explanation before further considerations.

Conclusion

The unique amplification event prior extensive migration of esterase B haplotype (Figure 1A) seems the most likely hypothesis to explain the existing data. This hypothesis is based on 1) the existence of a large neutral polymorphism around the esterase B structural gene in susceptible mosquitoes, and 2) the presence of the same amplified haplotype in populations from distant geographical areas. The second point

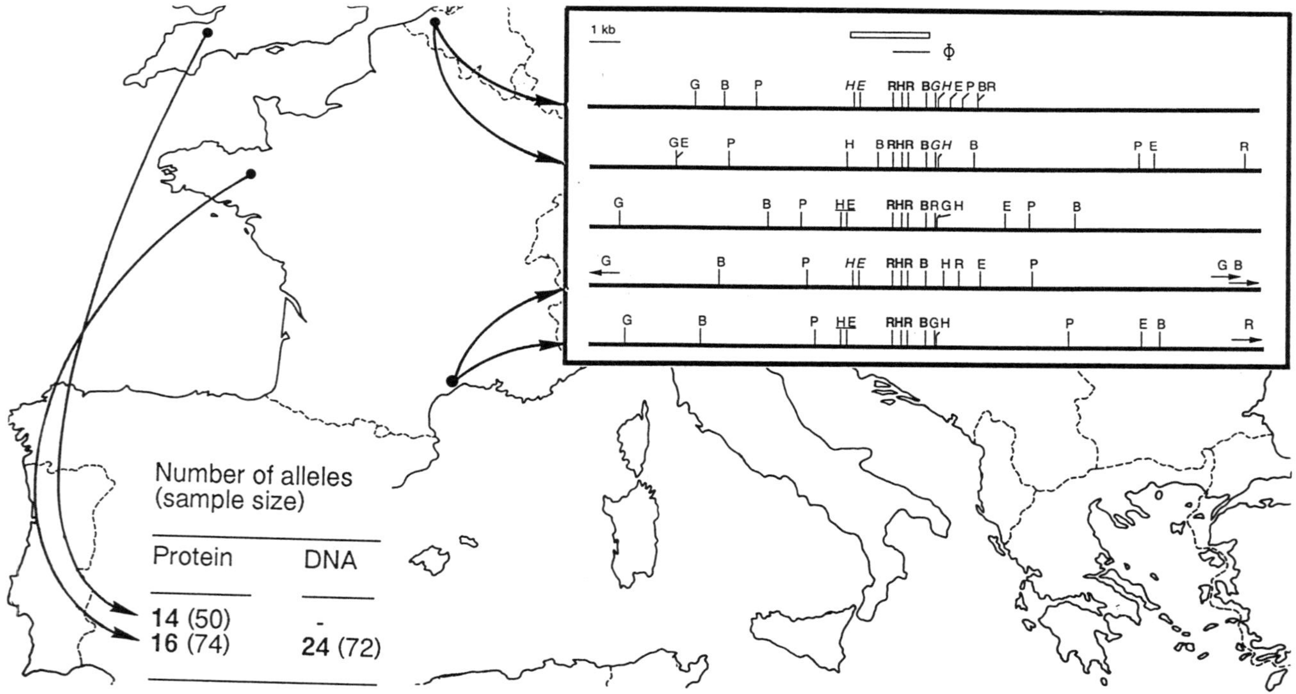

Figure 2. Analysis of susceptible populations in Brittany (France), Bruges (Belgium) and Bristol (England) at the esterase B locus. The French and the English population have been analysed for the number of alleles at the protein and DNA level (bottom left inset). The Belgium population has been studied for two esterase B alleles taken at random, for which a restriction map has been built and compared (top right inset) with restriction maps of B5 and B4 (3rd and 4th from top), and MSE (bottom). Restriction enzymes are abbreviated as: B (*Bam*H1), G (*Bgl*2), E (*Eco*R1), R (*Eco*RV), H (*HindIII*) and P (*Pst*1). SOURCE: Adapted from ref.17.

has been substantially documented (Table I), and the first point is now supported by extensive studies from susceptible populations from northern Europe (*17*), as well as from small samples studies from Portugal (*16*), Italy (*21*) and Venezuela (*18*).

When distinct restriction maps are found for the same overproduced electromorph, as described by Poirié *et al.* (*4*) in the Mediterranean region or by Vaughan *et al.* (*22*) in Cuba, this indicates that independent amplification of distinct alleles coding co-migrating electromorphs has occurred. This situation does not contradict the migration hypothesis.

Acknowledgments
We are very grateful to C. Bernard, M. Marquine and G. Pistre for technical help. Research was supported in part by the "Programme Environnement du CNRS" (G.D.R. 1105) and INRA. This is paper no. ISEM 95.090 of the Institut des Sciences de l'Evolution.

Literature Cited

1. Georghiou, G. P.; Mellon, R. B. In *Pesticide resistance to pesticides,* G. P. Georghiou; T. Saito, Eds; Plenum Press: New York, N.Y., 1983, p. 1-46.
2. Georghiou, G. P.; Lagunes-Tejeda, A. *The occurrence of resistance to pesticides in arthropods;* Food and Agriculture Organization: Rome, 1991.
3. Pasteur, N.; Iseki, A.; Georghiou, G. P. *Biochem. Genet.* **1981**, *19*, 909-919.
4. Poirié, M.; Raymond, M.; Pasteur, M. *Biochem. Genet.* **1992**, *30*, 13-26.
5. Wirth, M.; Marquine, M.; Georghiou, G. P.; Pasteur, N. *J. Econ. Entomol.* **1990**, *27*, 202-206.
6. Georghiou, G. P. In *Insecticides: mechanism of action and resistance,* D. Otto; B. Weber, Ed; Intercept: Andover, 1992, p. 407-408.
7. Fournier, D.; Bride, J.-M.; Mouchès, C.; Raymond, M.; Magnin, M.; Bergé, J.-B.; Pasteur, N.; Georghiou, G. P. *Pestic. Biochem. Physiol.* **1987**, *27*, 211-217.
8. Mouchès, C.; Magnin, M.; Bergé, J.-B.; De Silvestri, M.; Beyssat, V.; Pasteur, N.; Georghiou, G. P. *Proc. Natl. Acad. Sc., USA* **1987**, *84*, 2113-2116.
9. Pasteur, N.; Sinègre, G.; Gabinaud, A. *Biochem. Genet.* **1981**, *19*, 499 508.
10. Raymond, M.; Beyssat-Arnaouty, V.; Sivasubramanian, N.; Mouchès, C.; Georghiou, G. P.; Pasteur, N. *Biochem. Genet.* **1989**, *27*, 417-423.
11. Xu, J.; Qu, F.; Liu, W. *J. Med. Coll. PLA* **1994**, *9*, 20-23.
12. Mouchès, C.; Pasteur, N.; Bergé, J. B.; Hyrien, O.; Raymond, M.; Robert de Saint Vincent, B.; De Silvestri, M.; Georghiou, G. P. *Science* **1986**, *233*, 778-780.
13. Bisset, J. A.; Rodriguez, M. M.; Hemingway, J.; Diaz, C.; Small, G. J.; Ortiz, E. *Med. Vet. Entomol.* **1991**, *5*, 223-228.
14. Raymond, M.; Fournier, D.; Bride, J.-M.; Cuany, A.; Bergé, J.; Magnin, M.; Pasteur, N. *J. Econ. Entomol.* **1986**, *79*, 1452-1458.
15. Bourguet, D.; Capela, R.; Raymond, M. *J. Econ. Entomol.* **1995**, Submitted.
16. Raymond, M.; Callaghan, A.; Fort, P.; Pasteur, N. *Nature* **1991**, *350*, 151-153.
17. Raymond, M.; Qiao, C. L.; Callaghan, A. *Genet. Res.* **1995**, *66*, in press.
18. Qiao, C.-L.; Raymond, M. *Heredity* **1995**, *74*, 339-345.

19. Hemingway, J.; Ketterman, A. J.; Karunaratne, S. H. P. P.; Jayawardena, K. G. I.; Vaughan, A., In *First International Conference on Insect Pests in Urban Environment,* Willey, K. B.; Robinson W. H., Eds.; BPCC Wheatons Ltd: Exeter, 1993, p. 319-328.
20. Ketterman, A.; Karunaratne, S. H. P. P.; Jayawardena, K. G. I.; Hemingway, J. *Pestic. Biochem. Physiol.* **1993**, *47*, 142-148.
21. Severini, C.; Marinucci, M.; Raymond, M. *J. Med. Entomol.* **1994**, *31*, 496-499.
22. Vaughan, A.; Rodriguez, M.; Hemingway, J. *Biochem. J.* **1995**, *305*, 651-658.
23. Mouchès, C.; Pauplin, Y.; Agawarl, M.; Lemieux, L.; Herzog, M.; Abadon, M.; Beyssat-Arnaouty, V.; Hyrien, O.; Robert de Saint Vincent, B.; Georghiou, G. P.; Pasteur, N. *Proc. Natl. Acad. Sc., USA* **1990**, *87*, 2574-2578.
24. Raymond, M.; Marquine, M. *J. Evol. Biol.* **1994**, *7*, 315-337.
25. Rivet, Y.; Marquine, M.; Raymond, M. *Biol. J. Linn. Soc.* **1993**, *49*, 249-255.
26. Chevillon, C.; Pasteur, N.; Marquine, M.; Heyse, D.; Raymond, M. *Evolution* **1995**, *49*, 997-1007.

Chapter 11

Evolution and Selection of Antibiotic and Pesticide Resistance: A Molecular Genetic Perspective

George W. Sundin

Department of Microbiology and Immunology, University of Illinois—Chicago, 835 South Wolcott Avenue, Chicago, IL 60612

A workshop was held at the American Chemical Society Special Conference VI aimed at comparing various features of antibiotic and pesticide resistance. Contributors and participants discussed topics related to the evolution of bacterial antibiotic resistance and debated the potential relevance of these phenomena to currently known information concerning the evolution of pesticide resistance in fungi, insects, and plants. This chapter will focus on important characteristics of bacterial antibiotic resistance and include examples of pesticide resistance systems which may be similar in nature.

Initial comparisons of bacterial antibiotic resistance and pesticide resistance in eukaryotes indicate that there are two obvious differences. One is the genetic basis of resistance itself and the other is the involvement of gene transfer in the acquisition of resistance genes. Resistance to important classes of antibiotics is usually conferred by genes which encode enzymes that hydrolyze or alter the antibiotic molecule (*1*). By contrast, in addition to enhanced detoxification, pesticide resistance is often conferred by mutations which alter one or more amino acids in a protein target thereby affecting the pesticide-target interaction. The evolution and dissemination of antibiotic resistance (Ab^r) is driven by gene transfer; i.e. bacterial populations tend to encode Ab^r genes on genetic elements which can be efficiently transferred in a rapid and seemingly indiscriminate manner (for excellent reviews, see *2,3*). The evolution of pesticide resistance, however, involves the selection and amplification of resistant individuals, and gene transfer has apparently not played a key role in this process to date.

Despite these major differences, an examination of resistance evolution from a bacterial perspective could shed light on possibilities which might be confronted in the future of pesticide resistance management. Throughout this chapter, I have attempted to cite reviews and other articles which would be of broad interest to readers from many disciplines.

0097–6156/96/0645–0097$15.00/0

Introduction to Antibiotic Resistance

When antibiotics were initially introduced in clinical medicine, the propensity for resistance development in target bacterial pathogens was assessed from the standpoint of spontaneous mutation. The spontaneous mutation rate of resistance to streptomycin, one of the first antibiotics, is quite low (ca. $\leq 10^{-9}$ per generation); therefore, researchers did not anticipate that the development of streptomycin resistance would become a serious clinical problem. However, researchers also did not anticipate the presence of streptomycin-resistance genes in bacterial populations and the chance that these genes eventually would find their way to important clinical pathogens. This problem of prediction of resistance potential to a drug still exists today. A single pathogenic bacterium may not have the genetic capacity to evolve resistance to a particular antibiotic. However, the presence of a resistance gene, even in an obscure unrelated bacterium, allows for the eventuality of the acquisition of the gene by the pathogen.

The widespread, sudden deployment of a variety of Ab^r genes in bacterial populations leads to the question, where did they come from? Sequence analyses of Ab^r genes suggest that they originated from a variety of sources and in some cases from the producer organisms themselves. Organisms such as *Streptomyces griseus*, which produces streptomycin, encode a streptomycin-resistance gene as a self-protection mechanism. However, it has also been suggested that other genes including protein kinases, sugar kinases, and acetyltransferases may have evolved into aminoglycoside Ab^r genes (*4,5*). These Ab^r genes may have evolved in soil organisms in response to natural antibiotic selection pressure from Streptomycetes and other producer organisms.

The current state of antibiotic resistance illustrates important points about the genetic cooperativity of bacterial species. Although bacterial communities are characterized by extreme competition, individual species are also unwittingly cooperative and exchange DNA sequences due to the activity of conjugative plasmids. The close nucleotide sequence identity of specific Ab^r genes isolated from unrelated species is evidence for recent gene transfer events (*6*). The wide dissemination of resistance genes suggests that a gene pool is readily accessible to a multitude of bacteria under selection pressure from antibiotic usage. The acquisition of a resistance gene in a recipient organism does not guarantee its expression as promoter elements may be incompatible. However, bacteria typically contain an array of mobile genetic elements (MGEs) called insertion sequence (IS) elements, some of which encode outwardly directed promoter sequences (*7*). Selection pressure for the expression of the Ab^r gene dictates the insertion upstream of an IS element mobile promoter, thereby effecting gene expression.

Mobile Genetic Elements

The horizontal transfer of Ab^r genes is thought to occur quite commonly in the microbial world (*8,9*). Homologous Ab^r gene sequences have been detected among bacterial inhabitants of animals, humans, plants, and soil (*10*) suggesting that there are no real barriers preventing gene exchange between organisms from distinct environmental niches. The critical traffickers of Ab^r genes are MGEs (plasmids, transposable elements, and integrons).

Plasmids. Plasmids are obligate, intracellular DNA molecules which are replicated by host cells and transferred vertically through the lineage of the cell. Plasmids are normally nonessential components of a bacterial genome and thus can undergo radical alterations without affecting cell viability. Rare events such as gene acquisition, amplification, and rearrangements may be selected when they confer a fitness benefit. Compared to eukaryotes, this selection phenomenon may be magnified in bacteria due to comparatively immense population sizes. This may be a reason for the very efficient genetic systems for resistance which have evolved in bacteria. Plasmids are visitors which arrive but do not leave; once established, most plasmids are highly stable even without selection pressure. In fact, some plasmids even encode poison/antidote systems which are capable of the killing of plasmid-free segregants (*11*). However, plasmids do bring gifts, in the form of resistance and other genes which enhance the ability of the host to respond to environmental stresses (*12*). Whether these entities are purely selfish or beneficial symbionts, their widespread nature is testament to their promiscuity and broad adaptability to different bacterial hosts.

Conjugative plasmids encode and regulate all of the genetic determinants which are necessary for the transfer process. Plasmid transfer is a replicative process which results in both donor and recipient cells containing a plasmid copy (*13*). Plasmid transfer by conjugation has been demonstrated in many environmental habitats (for examples, see *14,15*); also, plasmid transfer between distinctly unrelated bacteria has been observed *in vitro* (*6*). Thus, the transfer of plasmids between bacteria provides an effective bridge for the exchange of DNA sequences and for the trafficking of Ab^r genes.

Transposable Elements. Three main forms of transposable elements are known in bacteria; the simplest is comprised of a gene encoding a transposase enzyme, which functions in the transposition of the element, flanked by short inverted repeat (IR) sequences (Figure 1A). This MGE, termed an IS element, typically ranges in size from 0.8 to 3.0 kb (*7*). Elements more commonly referred to as transposons are divided into two classes (*16*): class I transposons contain DNA sequences flanked by two copies of an IS element (Figure 1B) whereas class II transposons are flanked by short IR sequences (Figure 1C). Thus, in a class I transposon, transposition of the element is accomplished by the IS elements while the class II transposon encodes genes which function in the transposition process. Both of these classes of transposons may contain additional genes which encode traits such as antibiotic resistance, heavy metal resistance, and novel catabolic functions (*17*). Transposons and IS elements may reside within the host chromosome or on plasmids. The interplasmid mobilization of transposons increases the possibility of their association with superior genotypes. Transposons may also be disseminated horizontally on conjugative plasmids or on "suicide" plasmids, i.e. a conjugative plasmid which could not be established in the recipient host. In such a situation, the transposon simply excises from the sinking ship to reinsert itself within the new host's genome.

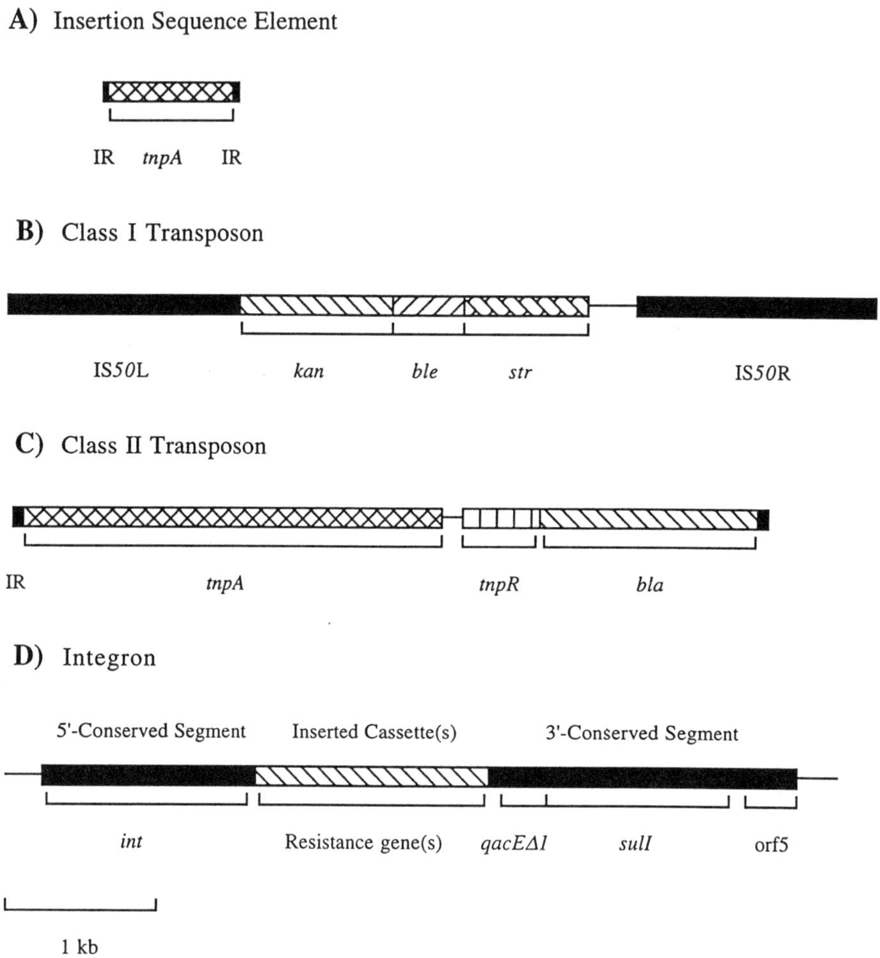

Figure 1. Examples of mobile genetic elements present in bacteria: A) insertion sequence element -- IS*6100* (*21*); B) class I transposon -- Tn*5* (*22*); C) class II transposon - Tn*3* (*23*); D) integron basic structure (*19*). Abbreviations: *bla*, β-lactamase; *ble*, bleomycin-resistance gene; *int*, integrase; *kan*, kanamycin-resistance gene; *orf5*, open reading frame of unknown function; *qacE*Δ*1*, defective quarternary ammonium compound exporter; *str*, streptomycin-resistance gene; *sulI*, sulfonamide-resistance gene; *tnpA*, tranposase; *tnpR*, transposon resolvase gene.

There are even transposons which are capable of catalyzing horizontal transfer on their own; these "conjugative transposons" do not require the actions of a plasmid for their dissemination (*18*).

Integrons. Integrons are recently discovered genetic elements which are thought to be responsible for much of the multiple antibiotic resistance observed today. These elements, presumed to be mobile themselves, consist of two conserved segments which flank a central region that contains a varying number of gene cassettes (Figure 1D) (*19,20*). The 5' conserved segment encodes an integrase enzyme which inserts gene cassettes via site-specific recombination at a specific insertion site within the integron directly downstream from a strong promoter sequence. Each gene cassette is always inserted in the correct orientation relative to the promoter due to the presence of a palindromic 59-bp element near its 3' end. The 3' conserved segment appears to have become associated with the integrase sequence early in the evolutionary history of the integron; it contains a defective gene homologous to the antiseptic-resistance gene *qacE1* and the sulfonamide-resistance gene *sulI* (*20*). The number of different gene cassettes found within naturally-occurring integrons is increasing and includes genes encoding resistance to at least seven important antibiotics (*19*). There also is no apparent limit to the number of gene cassettes inserted within a single integron and there are two examples of integrons which contain four gene cassettes within the central region (*19*). Sequence analyses of multiple-resistant integrons indicates that the inserted gene cassettes are markedly different in codon usage and therefore, in origin. Thus, the integron is an example of natural genetic engineering by bacteria providing a mechanism for the rapid acquisition of useful genes.

Mobile Genetic Elements and the Evolution and Dissemination of Multiple Antibiotic Resistance in Bacteria. As stated above, the critical traffickers of Ab^r genes are the MGEs plasmids, transposons, and integrons. Although both integrons and transposons can be found within bacterial chromosomes, the location of these elements on plasmids facilitates their transfer among a broader range of bacterial hosts. The evolution of multiple antibiotic resistance in bacteria typically involves the accumulation of resistance genes on plasmids. This can be accomplished through the acquisition and accumulation of transposons encoding different Ab^r genes; there are many examples of multiresistance plasmids which have evolved in this manner (*24,25*). Alternatively, multiple antibiotic resistance can evolve through the acquisition and incorporation of Ab^r genes within an integron. The integrons would appear to have a selective advantage over multiple transposons in that the Ab^r genes are contained in a smaller unit of DNA sequence (i.e. multiple transposons would each encode other genes functioning in transposition in addition to the Ab^r genes). However, the type of multiresistance strategy which is more prevalent today is currently unknown.

Transposable Elements and Pesticide Resistance in Eukaryotes

As shown by the previous examples, the evolution of antibiotic resistance in bacteria illustrates the extremely complex, interwoven nature of bacterial communities. Could similar processes be involved in the evolution of pesticide resistance in eukaryotes?

Transposons themselves may be good candidates for involvement in pesticide resistance processes. Transposons occur naturally in fungi, insects, and plants (*17*). An important biological activity of transposons in eukaryotes is the influence of transposon insertion on gene activity. For example, *Ty* elements in yeast seem to selectively insert in the 5' regulatory region of genes, thereby affecting expression (*26*). In plants, the activity of transposons may enable populations to respond with more flexibility to changes in the environment (*27*). There are numerous examples of plant transposons influencing gene regulation through insertion into 5' transcriptional control regions of genes or through insertion into an exon within a gene (*28*). Transposons in *Drosophila* such as *P*, *hobo*, and *gypsy* have also been shown to turn genes on or off following insertion (*26*). Thus, one possible mechanism for the evolution of insecticide resistance may be through gene disruption (a hypothetical example would be the disruption of the gene encoding the insect midgut receptor for *Bacillus thuringiensis* toxin). In the mosquito *Culex pipiens*, transposable elements were suspected to be involved in the amplification of a region containing the esterase B1 gene. The amplification was associated with a transposon-induced recombination event and resulted in resistance to organophosphate insecticides (*29*). Others have also speculated on the role of transposons in inducing insecticide resistance (*30*).

P elements are an example of a transposon which have been recently introduced into *Drosophila* and have progressively invaded populations to a point where P^- individuals are rarely found today (*31*). *P* elements are responsible for hybrid dysigenesis which results in an extremely high frequency of transposition in the progeny of a cross between a P^+ male and a P^- female (*32*). Although most of the progeny may carry deleterious mutations, the mutator activity of *P* could provide adaptive benefits for some variants fixing these genotypes within populations. It has been stated that the invasion of *P* elements in *Drosophila* populations occurred following the widescale human usage of organophosphate insecticides. Perhaps the mutator activities of *P* are involved in the selection of genotypes which are more competent to adapt to environments under insecticide stress.

Thus, the potential roles for transposons in inducing insecticide resistance include gene disruption, increasing gene expression, gene amplification, or general mutator activities. However, in all of these cases, the capacity for evolving resistance rests solely within the genome of the individual. What are the possibilities for horizontal gene transfer which, as described above, is a major contributor in the evolution of antibiotic resistance in bacteria?

While horizontal gene transfer may occur at low frequencies in eukaryotic populations, the main observations of transfer have occurred on an evolutionary time scale (*33,34*). The lateral transfer of MGEs is believed to occur at higher frequencies than that of other sequences (*33*). The widespread distribution of *mariner* transposable elements in species of *Drosophila* and other unrelated insects indicates that horizontal gene transfer events have occurred among insects (*35,36*). Organisms thought to mediate such transfers include viruses, retroviruses, or even semiparisitic mites (*37,38*). The *P* element is thought to have been introduced to *Drosophila melanogaster* from *Drosophila willistoni* approximately 50 years ago (*39*). Thus, it is clear that a gene transfer event, coupled with normal rates of gene flow, can facilitate the invasion of a species with a novel genetic sequence. However, to date, the horizontal transfer of an actual pesticide resistance gene has not been observed or inferred to have occurred in nature. Also, since transposons such as *P* or *mariner* have not been shown to encode additional genes (unlike the Ab^r genes present within some bacterial transposons), the chances for the horizontal transfer of a pesticide resistance gene may be limited.

Concluding Perspective of Antibiotic and Pesticide Resistance

The deployment of antibiotics in medicine and agriculture was initially seriously debilitating to bacterial populations and placed them in a life or death struggle. Obviously, bacteria have survived this onslaught, and their war chest, containing an arsenal of Ab^r genes and MGEs, has provided them with a great defense and flexibility. The unwitting cooperative interaction of these extremely adaptible organisms has further compromised man's attempts to destroy them. Resistance in other pests such as insects and weeds is on the rise worldwide. The potential role of transposons in the evolution of insecticide resistance is gaining increased attention. Although MGEs and gene transfer are not currently considered as important in the pesticide resistance field, observations of trends in antibiotic resistance could foreshadow a plethora of future problems in pesticide resistance management.

Acknowledgments

I thank Henri Darmency, Keith Hughes, Alan Jones, and Francois Tardif for contributing to the workshop, and Tom Brown, Don Cooksey, and Keith Hughes for helpful discussions.

Literature Cited

1. Hayes, J.D., and Wolf, C.R. Molecular mechanisms of drug resistance. *Biochem. J.* **1990**, *272*, 281-295.
2. Davies, J. Inactivation of antibiotics and the dissemination of resistance genes. *Science* **1994**, 264, 375-382.
3. Fekete, T. Bacterial genetics, antibiotic usage, and public policy: the crucial interplay in emerging resistance. *Perspect. Biol. Med.* **1995**, *38*, 363-382.
4. Udou, T., Mizuguchi, Y., and Wallace, R.J.Jr. Does aminoglycoside-acetyltransferase in rapidly growing mycobacteria have a metabolic function in addition to aminoglycoside inactivation? *FEMS Microbiol. Lett.* **1989**, *57*, 227-230.
5. Rather, P.N., Orosz, E., Shaw, K.J., Hare, R., and Miller, G. Characterization and transcriptional regulation of the 2'-*N*-acetyltransferase gene from *Providencia stuartii*. *J. Bacteriol.* **1993**, *175*, 6492-6498.
6. Mazodier, P., and Davies, J. Gene transfer between distantly related bacteria. *Annu. Rev. Genet.* **1991**, *25*, 147-171.

7. Galas, D.J., and Chandler, M. Bacterial insertion sequences. In *Mobile DNA*; Berg, D.E., and M.M. Howe, Ed.; American Society for Microbiology: Washington, D.C., 1989, pp. 109-162.
8. *Gene Transfer in the Environment*; Levy, S.B., and Miller, R.V., Ed.; McGraw-Hill, New York, 1989.
9. Amábile-Cuevas, C.F., and Chiculrel, M.E. Bacterial plasmids and gene flux. *Cell* **1992**, *70*, 189-199.
10. Sundin, G.W., and Bender, C.L. Dissemination of the *strA-strB* streptomycin-resistance genes among commensal and pathogenic bacteria from humans, animals, and plants. *Mol. Ecol.* **1995**, *4*, in press.
11. Thirsted, T., and Gerdes, K. Mechanism of postsegregational killing by the *hok/sok* system of plasmid R1. *J. Mol. Biol.* **1991**, *223*, 41-54.
12. Tschäpe, H. The spread of plasmids as a function of bacterial adaptability. *FEMS Microbiol. Ecol.* **1994**, *15*, 23-32.
13. Frost, L.S. Bacterial conjugation: everybody's doin' it. *Can. J. Microbiol.* **1992**, *38*, 1091-1096.
14. van Elsas, J.D. Antibiotic resistance gene transfer in the environment: an overview. In *Genetic Interactions among Microorganisms in the Natural Environment*; Wellington, E.M.H., and van Elsas, J.D., Ed.; Pergamon Press: Oxford, 1992, pp. 17-39.
15. Trevors, J.T., Barkay, T., and Bourquin, A.W. Gene transfer among bacteria in soil and aquatic environments: a review. *Can. J. Microbiol.* **1987**, *33*, 191-198.
16. Kleckner, N. Transposable elements in prokaryotes. *Annu. Rev. Genet.* **1981**, *15*, 341-404.
17. *Mobile DNA*; Berg, D.E., and Howe, M.M., Ed.; American Society for Microbiology: Washington, D.C., 1989.
18. Scott, J.R., and Churchward, G.G. Conjugative transposition. *Annu. Rev. Microbiol.* **1995**, *49*, 367-397.
19. Hall, R.M., and Stokes, H.W. Integrons: novel DNA elements which capture genes by site-specific recombination. *Genetica* **1993**, *90*, 115-132.
20. Hall, R.M., and Collis, C.M. Mobile gene cassettes and integrons: capture and spread of genes by site-specific recombination. *Mol. Microbiol.* **1995**, *15*, 593-600.
21. Sundin, G.W., and Bender, C.L. Expression of the *strA-strB* streptomycin resistance genes in *Pseudomonas syringae* and *Xanthomonas campestris* and characterization of IS*6100* in *X. campestris*. *Appl. Environ. Microbiol.* **1995**, *61*, 2891-2897.
22. Berg, D.E. The transposon Tn*5*. In *Mobile DNA*; Berg, D.E., and M.M. Howe, Ed.; American Society for Microbiology: Washington, D.C., 1989, pp. 185-210.
23. Heffron, F., McCarthy, B.J., Ohtsubo, H., Ohtsubo, E. DNA sequence analysis of the transposon Tn*3*: three genes and three sites involved in transposition of Tn*3*. *Cell* **1979**, *18*, 1153-1163.

24. Labigne-Roussel, A., Witchitz, J., and Courvalin, P. Modular evolution of disseminated Inc7-M plasmids encoding gentamicin resistance. *Plasmid* **1982**, *8*, 215-231.

25. Saunders, J.R., Hart, C.A., and Saunders, V.A. Plasmid-mediated resistance to β-lactam antibiotics in Gram-negative bacteria: the role of in-vivo recyclization reactions in plasmid evolution. *J. Antimicrob. Chemother.* **1986**, *18*(Suppl. C), 57-66.

26. Syvanen, M. The evolutionary implications of mobile genetic elements. *Annu. Rev. Genet.* **1984**, *18*, 271-293.

27. Gierl, A., Saedler, H., and Peterson, P.A. Maize transposable elements. *Annu. Rev. Genet.* **1989**, *23*, 71-85.

28. Flavell, A.J., Pearce, S.R., and Kumar, A. Plant transposable elements and the genome. *Curr. Opin. Genet. Develop.* **1994**, *4*, 838-844.

29. Salvado, J.C., Bensaadi-Merchermek, N., and Mouchès, C. Transposable elements in mosquitoes and other insect species. *Comp. Biochem. Physiol.* **1994**, *109B*, 531-544.

30. Wilson, T.G. Transposable elements as initiators of insecticide resistance. *J. Econ. Entomol.* **1993**, 86:, 645-651.

31. Kidwell, M.G. Evolution of hybrid dysigenesis determinants in *Drosophila melanogaster*. *Proc. Natl. Acad. Sci. USA* **1983**, *80*:1655-1659.

32. Rio, D.C. Molecular mechanisms regulating *Drosophila P* element transposition. *Annu. Rev. Genet.* **1990**, *24*, 543-578.

33. Kidwell, M.G. Lateral transfer in natural populations of eukaryotes. *Annu. Rev. Genet.* **1993**, *27*, 235-256.

34. Syvanen, M. Horizontal gene transfer: evidence and possible consequences. *Annu. Rev. Genet.* **1994**, *28*, 237-261.

35. Robertson, H.M. The *mariner* transposable element is widespread in insects. *Nature(London)* **1993**, *362*, 241-245.

36. Robertson, H.M., and Lampe, D.J. Recent horizontal transfer of a *mariner* transposable element among and between Diptera and Neuroptera. *Mol. Biol. Evol.* **1995**, *12*, 850-862.

37. Houck, M.A., Clark, J.B., Peterson, K.R., and Kidwell, M.G. Possible horizontal transfer of *Drosophila* genes by the mite *Proctolaelaps regalis*. *Science* **1991**, *253*:1126-1129.

38. Lerch, R.A., and Friesen, P.D. The baculovirus integrated retrotransposon TED encodes *gag* and *pol* proteins that assemble into virus-like particles with reverse transcriptase. *J. Virol.* **1992**, *66*:1590-1600.

39. Clark, J.B., Altheide, T.K., Schlosser, M.J., and Kidwell, M.G. Molecular evolution of *P* transposable elements in the genus *Drosophila*. I. The *saltans* and *willistoni* species groups. *Mol. Biol. Evol.* **1995**, *12*, 902-913.

Chapter 12

Single Versus Multiple Origins of Insecticide Resistance: Inferences from the Cyclodiene Resistance Gene *Rdl*

Richard H. ffrench-Constant, Nicola M. Anthony, Dmitri Andreev, and Kate Aronstein

Department of Entomology, University of Wisconsin, Madison, WI 53706

The number of independent origins of insecticide resistance alleles is currently the subject of intense debate. Support for the importance of a single point of origin and spread of resistance through insect populations comes from studies of amplified esterases and insensitive acetylcholinesterase in *Culex* mosquitoes. Here we argue that it is difficult to determine precisely the number of origins of resistance alleles due to the complexity of the two mechanisms studied in *Culex*. The repeated replacement of the same amino acid in the *Resistance to dieldrin* (*Rdl*) gene, conferring resistance to cyclodiene insecticides, offers a model system within which to examine the diversity and origins of resistance alleles. By comparing *Rdl* alleles in two *Drosophila* species, two beetle species and the *Bemisia tabaci* whitefly complex we present repeated evidence for multiple independent origins of resistance. Evidence for independent origins comes not only from the finding of different replacements of this same amino acid but also flanking sequence data supporting multiple origins of the same amino acid replacement. Further, we emphasize that the life history of the insect under consideration can play a major role in determining the likely origin and spread of different resistance alleles.

Determining the number of independent origins of insecticide resistance associated mutations is of central importance to both to our understanding of the evolution of pesticide resistance and our attempts to delay or contain its spread. Estimates of the initial frequencies of resistance associated mutations vary enormously from as high as 10^{-3} to as low as 10^{-13} (1). Therefore we are still uncertain if resistance commonly arises independently in isolated populations of insects or if resistance appears once, or a limited number of times, and subsequently spreads globally by migration. Due to the difficulties of both predicting when resistance associated mutations will first

0097–6156/96/0645–0106$15.00/0

appear and at what frequency, initial mutation rates associated with resistance in most insect pests therefore remain obscure.

However, we *can* determine the number of resistance alleles present in populations *after* the advent and spread of resistance and attempt to construct a phylogeny of their origin(s). This chapter examines the current evidence for a hypothesis promoting the importance of a single point of origin, and subsequent spread, of amplified esterase loci in mosquitoes and contrasts this situation with studies on a single cyclodiene resistance locus *Resistance to dieldrin* or *Rdl* in a range of insects with very different life histories and dispersal capabilities. We propose that *Rdl* resistance alleles have arisen more than once in a range of insects and that their subsequent patterns of distribution are dictated largely by the dispersal ability and life cycle of the pest insect under examination. Caution should therefore be taken in extrapolating from a highly mobile pest such as the mosquito to other insect pests with very different life histories.

Evidence for Single Origin of Resistance in *Culex pipiens*

The debate on the origin of amplified esterase genes in mosquitoes centers upon the relative importance of mutation versus migration. That is to say, the relative frequency of independent gene amplification events versus the capacity for migration to drive the spread of different alleles within and between countries. The original hypothesis for a single origin of resistance was derived from a comparison of the restriction pattern of flanking DNA from the amplified B2 esterase locus in *Culex pipiens* (2). As the restriction enzyme patterns of flanking DNA from amplified resistant alleles from widely differing locations were identical, the results were argued to be consistent with a single initial amplification event and subsequent spread via migration. Secondary evidence to support single points of origin of amplified esterase alleles also lies in the strong linkage disequilibrium observed between the A2-B2 and A1- $Est\text{-}2^{0.64}$ (Esterase-2, mobility 0.64) alleles (2, 3). Thus, the fact that these pairs of alleles are always found together was argued to indicate the spread of a single allelic combination. These arguments are based therefore upon either the observed similarity of alleles at a single locus or the similarity of allelic combinations found across a large geographical area, and both are taken to indicate that mutational events are a rate limiting step in the origin of resistance genes.

Further evidence for mutation being a rare event was derived from two studies of the appearance of resistance to organophosphorus insecticides in *C. pipiens* in Europe (3). The first study was of resistance to chlorpyrifos in Southern France. In this case, the rapid selection of an amplified A1 esterase locus was taken to suggest that resistance was either originally present at low frequency or that it appeared (by mutation) or arrived (via migration) very early in the history of insecticide use in the region. In contrast, the slower appearance of the more effective resistance mechanism, insecticide insensitive acetylcholinesterase or Ace^R, was used to argue that the rate of appearance of resistance associated mutations within the *Ace* gene itself was a constraint in selection. The second study concerns resistance to temephos in Corsica. In this case it was argued that the failure of the widespread B1 resistance allele to appear in Corsica, following 15 years of high selection pressure, suggests than this allele failed to arrive via migration (3).

Problems with *Culex pipiens* as a Model System

There are several complicating factors in the analysis of the origins of insecticide resistance alleles in *Culex* mosquitoes, both with respect to amplified esterase and to insensitive acetylcholinesterase alleles. Firstly, amplification of more than one esterase locus (i.e. B1, B2) can confer resistance. Secondly, recent evidence from the green peach aphid *Myzus persicae* suggests that duplicated esterase genes themselves can, in some aphid clones, reside at a number of locations in the genome (see Field et al. this volume). Each separate cluster of gene copies would therefore be expected to behave as an independent locus. In this respect, a similar 6-cutter restriction pattern of flanking DNA could still be obtained from such identical amplicons at *different* locations in the genome (without a precise definition of the size of the amplicon). Further, in *Myzus* the number of amplified gene copies correlates poorly with levels of enzyme production observed. This suggests that alteration of transcriptional control by mutation(s) either within E4 itself or within a completely different upstream gene may play a key role in resistance. Thus in the strictest interpretation amplification of, or mutations within, the structural gene for E4 may not be the sole mechanism responsible for resistance, as other gene(s) upregulating E4 transcription may also have a major effect. In regard to mosquitos, no rigid correlation of gene copy number with amplified esterase content has been performed in *Culex* and we are currently unable to assess the relative roles of gene duplication and altered transcriptional control.

Finally, with respect to the origins of resistance associated mutations in the *Ace* gene, resistance in *Drosophila* appears to arise from the assembly of a number of 'weak' mutations within the same allele (4). It is therefore presently unclear as to whether it is the accumulation (via recombination) of pre-existing weak mutations within the same resistant allele or the rate of mutation itself that is rate limiting in the evolution of Ace^R mediated resistance.

Rdl as a Case Study

Several factors make *Rdl* a useful model system in which to address the question of the origin of insecticide resistance alleles.

(1) Despite the decline in the use of cyclodienes, resistance to cyclodienes still accounted for over 60% of reported cases of insecticide resistance at the time of the last survey (5). Many of these cases of cyclodiene resistance show the highly characteristic semi-dominant resistance phenotype and cross-resistance spectrum associated with *Rdl* resistance (6). *Rdl* mediated resistance has therefore been a widespread mechanism of resistance in the past and will probably remain important with the continued use of cyclodiene type compounds such as endosulfan.

(2) Resistance appears to be conferred by replacement of a single amino acid (Ala_{302} in *Drosophila*) within the *Rdl* encoded γ-aminobutyric acid (GABA) receptor (7) by either a serine or a glycine residue (8). Replacements of this same amino acid are found in cyclodiene resistant strains of house flies, red flour beetles,

cockroaches (9), yellow fever mosquitoes (10), whiteflies (11) and the coffee berry borer (12). Further, this mutation has been shown to be genetically linked to dieldrin resistance in the coffee berry borer (13, 14), the red flour beetle (15) and the yellow fever mosquito (our unpublished results). The *Rdl* cDNA has been shown to rescue the susceptible phenotype of dieldrin resistance following germline transformation of resistant *Drosophila* (16), proving that *Rdl* is indeed the resistance gene.

Table I. Different replacements of Ala_{302} found in cyclodiene resistant strains. Nucleotide sequences for both alternative codon usages of Ala to Ser (A>S) and the Ala to Gly (A>G) replacement are shown. Mutated base(s) are underlined and the number of resistant strains or clones* examined by sequencing or PCR diagnostic is given (n).

Species	Strains (n)	Resistance associated replacement		
		A>S codon 1	A>S codon 2	A>G
DIPTERA				
Drosophila melanogaster	(48)	GCG > TCG	-	-
D. simulans	(10)	GCG > TCG	-	GCG > GGG
Musca domestica	(1)	GCT > TCT	-	-
WHITEFLIES				
Bemisia tabaci (non-B biotype)	(7)	GCC > TCC	-	-
B. tabaci (Sudanese strain)	(1)	-	GCC > AGC	-
B. argentifolii (B biotype)	(7)	GCC > TCC	-	-
APHIDS				
Myzus persicae	(52)*	GCT > TCG	-	GCT > GGT
BEETLES				
Tribolium castaneum	(1)	GCT > TCT	-	-
Hypothenemus hampei	(3)	GCG/T > TCT	-	-
COCKROACHES				
Blatella germanica	(1)	GCC > TCC	-	-
LEPIDOPTERA				
Helicoverpa armigera	(2)	CAG (Q in both R and S strains)		

To date we have only examined a single case where cyclodiene resistance was not correlated with replacement of Ala_{302}. Thus both endosulfan resistant and susceptible *Helicoverpa armigera* had a glutamine residue at the equivalent position to Ala_{302} (17). This suggests that the *Rdl* receptor of lepidoptera may be more variable at this residue than that of diptera or coleoptera, and that resistance to endosulfan in this particular strain of *H. armigera* may be associated with increased insecticide metabolism rather than target site insensitivity.

Interestingly, Ala302 appears to play a unique role in both directly altering the binding site for cyclodienes within the proposed chloride ion channel pore of the GABA receptor, and in allosterically reducing the amount of time spent by the receptor in the insecticide-preferred desensitized state (18). Since this is the only mutation observed to date, it appears that only replacements of this amino acid are able to preserve sufficient channel function whilst conferring adequate levels of resistance. Within the context of replacements of a *single* amino acid within a *single* gene, the question of the number of origins of resistance is therefore vastly simplified. That is, does the single point mutation occur only once and spread through populations by migration? Or, does the same mutation occur repeatedly at the same position in the protein but in multiple locations throughout the geographic range of a given pest species? In order to address this question, we have examined the nucleotide sequence diversity of *Rdl* alleles in a range of several insects of different dispersal capabilities and life histories. By detailed analysis of the number of *Rdl* alleles and their distribution in different insect species we hope to determine whether cyclodiene resistance typically arises once and then spreads globally or if multiple origins of resistance are also equally common.

The Number of *Rdl* Alleles in Different Insect Species

We have cloned and sequenced resistant and susceptible *Rdl* alleles from a range of insect species with very different life histories. Firstly, two species of *Drosophila*, *D. melanogaster* and *D. simulans*. Secondly, two very different beetles, the coffee berry borer *Hypothenemus hampei* and the red flour beetle *Tribolium castaneum*. Thirdly, whiteflies of the *Bemisia tabaci* group.

***Drosophila* spp.** The *Rdl* gene was cloned from a mutant of *D. melanogaster* resistant to cyclodiene insecticides and picrotoxinin (19). Following the original description of the Ala_{302} to Ser replacement in resistant *D. melanogaster* (7) we surveyed resistant strains of both *D. melanogaster* and *D. simulans* worldwide using two polymerase chain reaction (PCR) based diagnostics REN/PCR (PCR amplification followed by restriction enzyme digest) (8) and PASA (PCR amplification of specific alleles) (20). Interestingly only the Ala_{302} to Ser replacement was found in resistant strains of *D. melanogaster* worldwide but the same alanine was replaced by either a serine or a glycine in resistant *D. simulans* (8). A preliminary survey of flanking restriction enzyme variation in *D. melanogaster* using REN/PCR quickly uncovered strong linkage disequilibrium between an *Eco*RI site within the *Rdl* gene and the resistance associated mutation (8). Sequence analysis revealed that this *Eco*RI site was only 700 bp away from the resistance associated mutation (in exon 7) in a flanking intron. As in the case of the amplified esterase loci in mosquitos, these results again suggest that a single allele has been spread globally. Subsequent detailed repetitive sequencing by C.

Aquadro and R. Roush has confirmed that the pattern of variability in *Rdl* alleles of *D. melanogaster* is consistent with a single recent origin and spread. In contrast, the finding of both Ala_{302} to Ser and Ala_{302} to Gly replacements in resistant *D. simulans* alleles is clearly evidence itself for more than one independent origin of resistance in this species. Further, sequence analysis of both allele classes (Ala_{302} to Ser and Ala_{302} to Gly) in *D. simulans* does not support a single point of origin for both alleles (C. Aquadro, personal communication).

Coffee berry borer. Although the biology of *D. melanogaster* is well studied, the high rates of dispersal, relationship to man and diplo-diploid status of this insect make it unrepresentative of many pest insects in terms of life history. For example, the scolytid beetle the coffee berry borer (*Hypothenemus hampei*) has a dramatically different lifecycle. Mated females of *H. hampei* enter individual coffee berries and lay large broods. Their progeny show a highly biased sex ratio (10 females : 1 male) and there is obligate sib mating between the daughters and their dwarf, flightless brothers. The male progeny therefore never leave the coffee berry in which they were born.

This beetle can infest a high percentage of coffee berries and is the major insect pest of coffee worldwide. Control is best achieved by endosulfan whose fumigant action can penetrate the coffee berry. The recent finding of endosulfan resistance in the South Pacific island of New Caledonia is thus a major threat to the international coffee industry. Although the insect itself has limited active dispersal capabilities, resistant beetles could be widely distributed in infested berries.

We were interested in investigating the molecular basis of endosulfan resistance as resistance appears to be confined to the South Pacific island of New Caledonia (21), where the beetle itself was introduced only after the second world war. Thus, resistance may have arisen on, or been introduced to, this island very recently. Examination of several resistant strains using PASA confirmed that resistance was again associated with replacement of Ala_{302} by Ser (12). Interestingly, simple studies of the inheritance of resistance showed that paternally derived copies of both R and S alleles were not phenotypically expressed in male progeny. Resistance therefore appeared to be 'sex-linked' (22). As haplo diploidy is often found in insects showing distorted sex ratios and inbreeding, like *H. hampei*, we postulated that this was consistent with males being haploid and females being diploid. However, cytological examination revealed that both males and females are diploid. Furthermore, the paternally derived chromosome set is condensed in the male soma. This condensed chromosome set is then discarded after the first meiotic division (14). Therefore, there is only one meiotic division and it is essentially mitotic. We termed this mode of inheritance of resistance "functional haplo-diploidy" as the effects on inheritance are similar to haplo-diploidy but the mechanism is very different (13).

The unique life-cycle of the coffee berry borer may provide some insight into the recent appearance and rapid spread of resistance in New Caledonia. Firstly, when resistance is expressed in males they are effectively hemizygous or *R*/- (where - is a copy of *Rdl* that is not expressed) and will thus behave phenotypically as resistant homozygotes *R*/*R* and survive higher doses of insecticide (e.g. hemizygous

Drosophila are equivalent to *R/R* rather than *R/S* insects (6)). Secondly, if inbreeding is complete, following the original mutational event each resistance allele will be maintained in an individual line derived from one female. Thirdly, within these lines the continuous maternal backcrossing will promote the rate at which resistance becomes homozygous. Thus, in possibly an even more extreme case than *Culex* or *D. melanogaster*, we can speculate that a single resistance allele may only have arisen once in *H. hampei* and that it is being maintained in a *single* inbreeding line founded by a *single* mated female.

Red Flour Beetle. The coffee berry borer clearly illustrates how different life histories, chromosome cycles and dispersal abilities are likely to influence the spread of resistance genes. We were therefore also interested in looking at a diplo-diploid beetle the red flour beetle *Tribolium castaneum* with a more conventional life history. Again this is a species that has the potential for being widely and passively distributed in stored grain but is also one in which lindane resistance is historically very widespread (23). Following the original description of *lindane Resistance* or *linR* (24), we cloned the *Rdl* homolog from *T. castaneum* and again showed that resistance was associated with the equivalent Ala_{302} to Ser replacement (15).

We have subsequently repeated a similar exercise as that conducted for the *Drosophila* alleles by sequencing out from the resistance associated mutation in exon 7 to examine flanking nucleotide variation in one of the adjacent introns. In contrast to *D. melanogaster* we have been able to identify variation between resistance alleles very close to the resistance associated mutation itself (within the adjacent codon). This will allow us to largely rule out the possibility that the different resistance alleles have arisen from a single ancestor by simple recombination. Preliminary analysis of sequence data from a range of both resistant and susceptible *Rdl* alleles from across the world has already provided us with strong evidence for multiple origins of resistance in this species.

The Sweet Potato Whitefly. The sweetpotato whitefly *Bemisia tabaci* is an important crop pest and vector of viral pathogens (25-27) and although highly mobile itself, also shows tremendous potential for passive distribution worldwide on infested plant material. Following the apparent recent emergence and spread of the novel 'B biotype', proposed to be a separate species *Bemisia argentifolii* (28), *Bemisia* has risen to primary pest status in many agroecosystems and caused unprecedented damage to cotton and vegetable crops in the United States and Caribbean basin (26, 29, 30). Due to the extreme success of the B biotype (29, 31), its rapid recent spread (32) and its reported resistance to insecticides (31, 33, 34) we were interested in examining the number of *Rdl* alleles in *Bemisia* in the light of three questions. 1) Does the B biotype show an apparently monophyletic origin of *Rdl* mediated resistance, as might be expected from a recent single point of origin of the biotype? 2) Is such resistance unique to the B biotype and thus could it have played a part in the dramatic ability of this new strain to replace existing endemic whitefly populations? 3) Is resistance found in the less dispersive endemic populations and if so, does it show evidence for multiple points of origin as might be expected within isolated populations?

Preliminary sequence analysis of the *Rdl* gene from several whitefly strains again revealed replacements of the same alanine residue with a serine. We then employed two molecular diagnostics, PASA and SSCP, to survey resistance associated mutations and their associated flanking polymorphisms in a range of strains of B and non-B biotype status (11). Results from the PASA diagnostic showed that the same Ala to Ser replacement was present in both B and non-B strains. SSCP analysis was used to confirm the resistance status of the same strains and also to examine nucleotide variation in the region encompassing the resistance associated mutation. Interestingly, of all the B strains examined, none showed any nucleotide variation other than the presence or absence of the resistance associated point mutation. In contrast, analysis of non-B strains revealed considerable variation in SSCP banding patterns, which were confirmed by sequence analysis to reflect one or more flanking nucleotide substitutions. Interestingly, one highly variable strain from the Sudan showed the alternative codon usage for the resistance associated serine (Table I) which we have not found in any other insect, suggesting an independent origin of resistance in this region. This hypothesis is supported by a phylogenetic analysis using parsimony (35) of *B. tabaci* *Rdl* nucleotide sequences (Figure 1) which suggests that the Sudanese resistance allele is widely separated from alleles from other localities.

We can draw several conclusions from this preliminary analysis. Firstly, within the *B. tabaci* complex we can find clear evidence for more than one origin of resistance. For example, in the case of the clearly very different Sudanese resistance allele. Secondly, the absence of nucleotide variation in the B biotype is consistent with the recent widespread introduction and rapid spread of this novel biotype worldwide. In contrast, non-B (indigenous) whitefly populations are considerably more heterogeneous than populations conforming to the B biotype. Finally, the same resistance associated mutation is present in both B and non-B strains. Therefore cyclodiene resistance is not uniquely associated with the B biotype and is therefore not likely to have accelerated its spread. Thus, populations of any biotype exposed to pesticides in a crop system seem equally likely to display resistance.

Conclusions and Future Directions

We have examined the allelic diversity of the *Rdl* gene in a range of species in order to clarify the relative roles of mutation and migration in the spread of resistance and also to highlight the importance of differing insect life histories. The *Rdl* system has the advantage of examining replacements of a single amino acid within a single locus of major effect, rather than alterations of a range of independent or interacting loci.

Several broad conclusions supporting multiple origins of resistance can be drawn at this stage in our work. Firstly, although only a single amino acid is replaced (the equivalent of Ala_{302} in *Drosophila*), at least two different types of replacements have been found in a number of different insects. 1) The alternative Ala to Gly replacement has been found in both *D. simulans* and *Myzus persicae* (our unpublished data) and 2) an alternative codon usage for the serine replacement

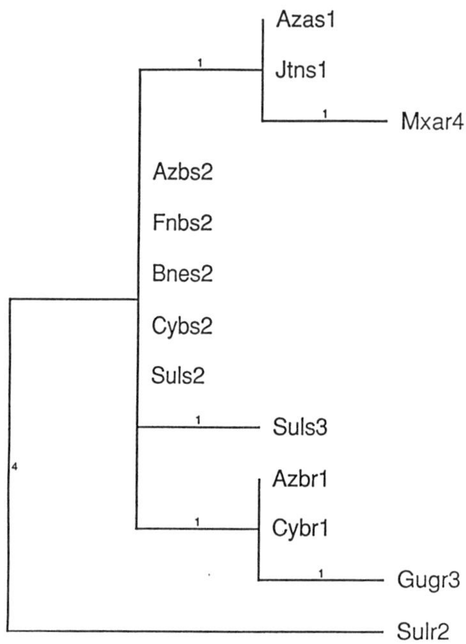

Figure 1. Inferred phylogenetic relationship between *Bemisia tabaci Rdl* alleles obtained using parsimony methods from the computer program PAUP 3.1 (35). Note that the Sudanese resistance allele (Sulr2), containing the alternative codon usage for serine, is the most divergent (see text). Support for particular phylogenies was assessed on the basis of 100 bootstrap runs. The numbers on the branches of the consensus tree indicate the number of single nucleotide differences between alleles. For individual alleles, the first two letters indicate the origin of the strain: Az, Arizona, USA; Jt, Puerto Rico; Mx, Mexico; Fn, Florida; Bn, Benin, Africa; Cy, Cypress; Su, Sudan and Gu, Guatemala. The third letter denotes the biotype status (32) . The last two characters indicate if the allele is resistant (r) or susceptible (s) and the number refers to their original description (11).

has also been observed in a Sudanese strain of *Bemisia*. Secondly, analysis of flanking nucleotide variation only seems to support a single point of origin in *D. melanogaster*. In both *D. simulans* and *T. castaneum* sequence analysis provides strong support for the independent origin of alleles coding for the same amino acid replacement. Thirdly, insect life history is obviously critical when considering the observed patterns of distribution of resistance alleles. For example, both *Culex*, *D. melanogaster* and the 'B biotype' of *B. tabaci* can be widely passively dispersed by man (as indeed can *Tribolium*) and accordingly all seem to show evidence for single origins of resistance alleles. In contrast, obligate inbreeding in the coffee berry borer may be confining resistance alleles to inbreeding lines derived from single females. Finally, it is important to consider that the complex population structure of some insects may also influence the distribution of resistance alleles. Thus the apparent diversity of sedentary indigenous whitefly biotypes may be responsible for maintaining the diversity of *Rdl* alleles observed in this species group.

Our future work in this area will therefore focus upon more detailed comparisons of flanking nucleotide variation and of the derived allelic phylogenies. In this fashion we hope to compound the evidence for the multiple origin of resistance alleles and gain an insight into how often resistance has arisen, how far resistance alleles can spread and therefore how often they are likely to arise in the future. These observations will not only provide valuable inputs into our models for understanding and managing the emergence of resistance, but will also help us manage resistance through the exclusion of new resistance alleles via quarantine procedures.

Literature cited

1. Roush, R. T.; McKenzie, J. A. *Annu. Rev. Entomol.* **1987**, *32*, 361-380.
2. Raymond, M.; Callaghan, A.; Fort, P.; Pasteur, N. *Nature* **1991**, *350*, 151-153.
3. Raymond, M.; Marquine, M.; Pasteur, N. In *Resistance 91: achievements and developments in combating pesticide resistance;* I. Denholm, A. L. Devonshire and D. W. Hollomon, Ed.; Elsevier Applied Science: 1992, pp 19-27.
4. Mutero, A.; Pralavorio, M.; Bride, J.-M.; Fournier, D. *Proc. Natl. Acad. Sci. USA* **1994**, *91*, 5922-5926.
5. Georghiou, G. P. In *Pesticide Resistance : Strategies and Tactics for Management;* National Academy of Sciences, Ed.; National Academy Press: Washington DC, 1986, pp 14-43.
6. ffrench-Constant, R. H.; Roush, R. T. *Genet. Res., Camb.* **1991**, *57*, 17-21.
7. ffrench-Constant, R. H.; Rocheleau, T. A.; Steichen, J. C.; Chalmers, A. E. *Nature* **1993**, *363*, 449-451.
8. ffrench-Constant, R. H.; Steichen, J.; Rocheleau, T. A.; Aronstein, K.; Roush, R. T. *Proc. Natl. Acad. Sci. USA* **1993**, *90*, 1957-1961.
9. Thompson, M.; Steichen, J. C.; ffrench-Constant, R. H. *Ins. Mol. Bio.* **1993**, *2*, 149-154.
10. Thompson, M.; Shotkoski, F.; ffrench-Constant, R. *FEBS Lett.* **1993**, *325*, 187-190.

11. Anthony, N. M.; Brown, J. K.; Markham, P. G.; ffrench-Constant, R. H. *Pestic. Biochem. Physiol.* **1995**, *51*, 220-228.
12. ffrench-Constant, R. H.; Steichen, J. C.; Brun, L. O. *Bull. Ent. Res.* **1994**, *84*, 11-16.
13. Brun, L. O.; Borsa, P.; Gaudichon, V.; Stuart, J. J.; Aronstein, K.; Coustau, C.; ffrench-Constant, R. H. *Nature* **1995**, *374*, 506.
14. Brun, L. O.; Stuart, J.; Gaudichon, V.; Arontein, K.; ffrench-Constant, R. H. *Proc. Natl. Acad. Sci. USA* **s 1995**, in press.
15. Andreev, D.; Rocheleau, T.; Phillips, T. W.; Beeman, R.; ffrench-Constant, R. H. *Pestic. Sci.* **1994**, *41*, 345-349.
16. Stilwell, G. E.; Rocheleau, T.; ffrench-Constant, R. H. *J. Molec. Biol.* **1995**, *253*, 223-227.
17.
18. Zhang, H.-G.; ffrench-Constant, R. H.; Jackson, M. B. *J. Physiol.* **1994**, *479*, 65-75.
19. ffrench-Constant, R. H.; Mortlock, D. P.; Shaffer, C. D.; MacIntyre, R. J.; Roush, R. T. *Proc. Natl. Acad. Sci. USA* **1991**, *88*, 7209-7213.
20. Steichen, J. C.; ffrench-Constant, R. H. *Pestic. Biochem. Physiol.* **1994**, *48*, 1-7.
21. Brun, L. O.; Marcillaud, C.; Gaudichon, V.; Suckling, D. M. *J. Econ. Entomol.* **1989**, *82*, 1311-1316.
22. Brun, L. O.; Suckling, D. M.; Roush, R. T.; Preisler, H.; Gaudichon, V.; Robertson, J. L. *J. Econ. Entomol.* **1995**, *88*, 470-474.
23. Georghiou, G. P.; Mellon, R. B. In *Pest Resistance to Pesticides;* G. P. Georghiou and T. Saito, Ed.; Plenum Press: New York, 1983, pp 1-46.
24. Beeman, R. W.; Stuart, J. J. *J. Econ. Entomol.* **1990**, *83*, 1745-1751.
25. Brown, J. K. *FAO Plant Prot. Bull.* **1990**, *30*, 5-23.
26. Brown, J. K.; Bird, J. *Plant Disease* **1992**, *76*, 220-225.
27. Byrne, D. N.; Bellows, T. S.; Parrella, M. P. In *Whiteflies: their bionomics, pest status and management;* D. Gerling, Ed.; Intercept: Andover, UK, 1990, pp 227-261.
28. Perring, T. M.; Cooper, A. D.; Rodriguez, R. J.; Farrar, C. A.; Bellows, T. S. *Science* **1993**, *259*, 74-77.
29. Perring, T. M.; Cooper, A.; Kazmer, D. J.; Shields, C.; Shields, J. *Calif. Agric.* **1991**, *45*, 10.
30. Perring, T. M.; Farrar, C. A.; Bellows, T. S.; Cooper, A. D.; Rodriguez, R. J. *Calif. Agric.* **1993**, *47*, 7-8.
31. AgriculturalReseachService In *Second Annual Review;* T. J. Henneberry, N. C. Toscano, R. M. Faust and J. R. Coppedge, Ed.; United States Department of Agriculture: Orlando, Florida, **1994**, pp 224.
32. Brown, J. K.; Coats, S.; Bedford, I. D.; Markham, P. J. *Biochem. Gen.* **1994**, in press.
33. Byrne, F. J.; Devonshire, A. L. *Pestic. Biochem. Physiol.* **1991**, *40*, 198-204.
34. Costa, H. S.; Brown, J. K.; Sivasupramaniam, S.; Bird, J. *Insect. Sci. Applic.* **1993**, *14*, 127-138.
35. Swofford, D. L.; Begle, D. P. In Ed.; Laboratory of Molecular Systematics Smithsonian Institution: **1993**, pp 257.

Chapter 13

Multiple Mechanisms and Multiple Herbicide Resistance in *Lolium rigidum*

Christopher Preston, Francois J. Tardif, and Stephen B. Powles

Co-operative Research Centre for Weed Management Systems and Department of Crop Protection, Waite Campus, University of Adelaide, PMB 1, Glen Osmond, South Australia 5064, Australia

Herbicide resistance in the grass weed *Lolium rigidum* is widespread across the cropping regions of southern Australia creating a serious practical problem for the control of this species. One biotype shows resistance to herbicides from nine chemical classes with five modes of action. This biotype contains two herbicide-resistant target sites, acetolactate synthase (ALS) and acetyl-coenzyme A carboxylase (ACCase); however, resistant ALS is restricted to about 5 % of the population. In addition, this population has enhanced capacity for detoxification of the herbicides chlorotoluron, simazine, chlorsulfuron, tralkoxydim and diclofop-methyl. This enhanced herbicide metabolism can be decreased by pre-treatment of plants with inhibitors of cytochrome P450 monooxygenases. There are differences in the synergistic effectiveness of these inhibitors between herbicides. For example, piperonyl butoxide inhibits metabolism of chlorotoluron and simazine, but not the other herbicides. These patterns indicate that while a single isozyme may be responsible for enhanced metabolism of chlorotoluron and simazine, different isozymes are responsible for enhanced metabolism of chlorsulfuron, diclofop-methyl, and tralkoxydim. Therefore, multiple resistance in this biotype is due to the accumulation of multiple mechanisms of resistance, probably encoded by several genes.

Lolium rigidum (annual ryegrass or rigid ryegrass) is widespread throughout the cropping regions of southern Australia. This species is a valuable pasture plant and was deliberately planted across many millions of hectares in Australia. However, *L. rigidum* is also a serious weed of cropping causing substantial yield losses (*1-2*). With the increasing intensity of cropping across southern Australia, there is increased reliance on herbicides to control *L. rigidum*. *L. rigidum* is a highly-variable, outcrossing species adapted to Mediterranean-type climates and has been a major target of herbicide use in Australia over the last two decades (*3*). The advent of the cereal-selective aryloxyphenoxypropanoate and sulfonylurea herbicides in the late 1970s and early 1980s allowed control of *L. rigidum* within the crop. These two herbicide groups in particular have greatly increased the herbicide selection pressure on *L. rigidum*, and it was not long before failures became evident. The first case of herbicide-resistant *L.*

0097–6156/96/0645–0117$15.00/0

rigidum appeared in 1980 (*4*), and now more than two thousand properties have herbicide-resistant populations of *L. rigidum* (*3*). This rapid development and proliferation of herbicide-resistant populations of *L. rigidum* has had a serious negative impact on the cost and ease of weed control.

One striking feature of herbicide resistance in *L. rigidum* is the propensity of this species to develop widespread cross and multiple resistance to herbicides. Frequently, cross resistance occurs to herbicide groups to which the population has never been exposed. Currently populations of *L. rigidum* demonstrate resistance across sixteen different herbicide chemical classes with eleven different modes of action (*5*). There is considerable variation between resistant populations in the extent of multiple resistance. Populations may show resistance to one or a few herbicides within a chemical class, or may have varying levels of resistance across many chemical classes.

To date we have identified target site-based resistance mechanisms to acetyl-coenzyme A carboxylase (ACCase)- and acetolactate synthase (ALS)-inhibiting herbicides in *L. rigidum* (*6-8*). In addition, enhanced metabolism of ACCase-, ALS-, and Photosystem II (PS II)-inhibiting herbicides has been documented in various populations of *L. rigidum* (*9-14,* Preston, Tardif, Christopher and Powles, unpublished). Resistant populations may contain one or several of these mechanisms, and a case of a population resistant to numerous chemical classes will be used to illustrate the accumulation of resistance mechanisms.

Herbicide Multiple Resistance in a Single Biotype of *L. rigidum*

A multiple-resistant *L. rigidum* biotype, designated VLR 69, collected from a property producing perennial ryegrass seed has wide cross resistance to at least nine herbicide chemistries with five modes of action as described in Table I. This biotype had an extensive history of herbicide use including 19 applications of diuron, 6 of chlorsulfuron, 5 of atrazine, 2 of diclofop-methyl, and 3 of a mixture of paraquat and diquat over 22 years (*15*).

Table I. Herbicide Classes and Target Sites to which *Lolium rigidum* Biotype VLR 69 Displays Resistance

Herbicide Class	*Target Site*
Aryloxyphenoxypropanoate	ACCase
Chloroacetamide	mitosis
Cyclohexanedione	ACCase
Imidazolinone	ALS
Substituted urea	PS II
Sulfonylurea	ALS
Thiocarbamate	ACC elongase
Triazine	PS II
Triazinone	PS II

From (*15*) and Preston, unpublished data.

We have extensively studied the mechanisms of resistance in this biotype over the last few years, concentrating on herbicides that inhibit ACCase, ALS and PS II. When grown in pots, the resistant biotype is resistant to several, but not always all, members of each of these classes. The resistant biotype displays variable resistance to the herbicides simazine, chlorotoluron, chlorsulfuron, tralkoxydim and diclofop-methyl compared to a standard susceptible biotype (Table II). Resistance to these five herbicides, representing the triazine, substituted urea, sulfonylurea, cyclohexanedione and aryloxyphenoxypropanoate classes respectively, will be considered in further detail.

Table II. Multiple Resistance to Herbicides of Three Modes of Action in Biotype VLR 69 of *Lolium rigidum*

Herbicide	*Susceptible*	*Resistant*	*R/S*[a]
	LD_{50} (kg ha^{-1})[b]		
Simazine	0.91	5.95	6.5
Chlorotoluron	0.69	8.31	12.1
Chlorsulfuron	0.007	0.47	67.1
Tralkoxydim	0.020	0.154	7.7
Diclofop-methyl	0.061	> 16	>266

[a]R/S is the ratio of the LD_{50} for the resistant biotype compared to the susceptible biotype. [b]LD_{50}s were determined by a logit procedure fitted to the log-transformed data using a maximum likelihood program (*16*). Confidence intervals (95 %) of LD_{50}s did not overlap between biotypes.

Target Site-Based Resistance

PS II. The vast majority of cases of triazine resistance are endowed by a resistant PS II (*17*). In contrast, resistance to the PS II-inhibiting herbicides in *L. rigidum* biotype VLR 69 is not target site based. Methyl viologen-dependent O_2 consumption of thylakoids isolated from the resistant and susceptible biotypes was equally sensitive to all PS II-inhibiting herbicides tested (*9,10,18*). From this data an I_{50}, the concentration of herbicide required to inhibit O_2 consumption by 50 %, was calculated. The I_{50} for inhibition of PS II activity in isolated thylakoids of the susceptible biotype of *L. rigidum* ranged from 0.03 to 0.4 μM, with various PS II-inhibiting herbicides. Similar values were obtained with thylakoids from the resistant biotype for the same herbicides (Table III). Clearly a resistant PS II target site does not endow resistance to these herbicides in this biotype of *L. rigidum.*

Table III. Inhibition of Photosystem II Activity in Isolated Thylakoids from Susceptible and Resistant Biotypes of *L. rigidum*

Herbicide	*Susceptible*	*Resistant*	*R/S*[a]
	I_{50} (μM)[b]		
Diuron	0.03	0.02	0.67
Metribuzin	0.07	0.08	1.14
Simazine	0.41	0.64	1.56

[a]R/S is the ratio of the I_{50} of the R biotype over that of the S biotype. [b]I_{50}s were estimated from previously published data (*9,10,18*).

ALS. The resistant biotype is resistant to a range of sulfonylurea and imidazolinone herbicides, but is conspicuously susceptible to sulfometuron-methyl and imazapyr (*15*). These two herbicides are lethal to wheat as this species is unable to metabolize these herbicides sufficiently rapidly (*19,20*). If susceptible *L. rigidum* seed is germinated on agar containing sulfometuron-methyl, all seeds fail to produce viable seedlings. This inhibition of germination by sulfometuron-methyl is also observed in the resistant population, except for about 5 % of the individuals that germinate normally on 27 μM sulfometuron-methyl (*14*). This procedure allows the selection of a sulfometuron-methyl-resistant sub-population, designated selected VLR 69. ALS extracted from the susceptible and resistant biotypes of *L. rigidum* and partially purified by $(NH_4)_2SO_4$ precipitation and gel filtration, was inhibited by low concentrations of chlorsulfuron, whereas that from the sulfometuron-methyl-resistant sub-population was not (*14*).

Median inhibitory concentrations calculated from this data demonstrate that ALS activity of both the susceptible and resistant biotypes was sensitive to chlorsulfuron (Table IV). In contrast, the ALS extracted from the sulfometuron-methyl-resistant sub-population was highly resistant to chlorsulfuron with an I_{50} 7 to 9-fold greater than that of the susceptible and bulk resistant populations respectively.

Table IV. Median Inhibitory Concentrations (I_{50}) of Chlorsulfuron Against ALS Activity from Susceptible and Resistant Biotypes of *L. rigidum*

Biotype	*Chlorsulfuron I_{50} [a], (μM)*
Susceptible (VLR 1)	30
Resistant (VLR 69)	23
Selected VLR 69[b]	200

[a]I_{50}s were estimated from previously published data (*14*). [b]Selected by germination on 27 μM sulfometuron-methyl.

Although a small sub-population contains a resistant ALS target site, the bulk of the resistant population has a sensitive ALS. In contrast, resistance to ALS-inhibiting herbicides in many other weed species is due to a modified target site (*21*). *L. rigidum* populations may display resistance to ALS-inhibiting herbicides due to a modified target site or through non-target site mechanisms (*13*). Our observations have been that many populations of ALS resistant *L. rigidum* have a susceptible target site (*22*), however studies in Western Australia have shown a high correlation between resistance to triasulfuron, a cereal-selective sulfonylurea, and sulfometuron-methyl (*23*) indicating many populations there contain a resistant target site.

ACCase. The resistant biotype is resistant to all aryloxyphenoxypropanoate herbicides and the cyclohexanedione herbicide tralkoxydim, but is susceptible to sethoxydim (*15*). The target site for these herbicides, ACCase, was extracted from the shoot meristematic tissue of both resistant and susceptible biotypes, partially purified by $(NH_4)_2SO_4$ precipitation and gel filtration, and assayed for acetyl-coenzyme A-dependent CO_2 fixation. The enzyme was challenged with a range of ACCase-inhibiting herbicides. The ACCase from the resistant biotype proved to be resistant to all the aryloxyphenoxypropanoate herbicides, but not to the cyclohexanedione herbicides sethoxydim and tralkoxydim (Table V).

Table V. Median Inhibitory Concentrations of Herbicides Against ACCase Activity from Susceptible and Resistant Biotypes of *L. rigidum*

Herbicide	*Susceptible*	*Resistant*	*R/S*
	(I_{50} μM)		
Diclofop acid	0.3 ± 0.1[a]	9.1 ± 1.2	30
Fluazifop acid	13 ± 2	52 ± 2	4
Haloxyfop acid	1.2 ± 0.1	23 ± 5	19
Sethoxydim	3.3 ± 0.2	3.9 ± 0.5	1.2
Tralkoxydim	0.5 ± 0.1	0.5 ± 0.1	1.0

[a]Data are means ± SEM of three experiments.

A herbicide-resistant ACCase accounts for resistance to all aryloxyphenoxypropanoate herbicides in this biotype; however, this mechanism cannot explain resistance to tralkoxydim. On the other hand, susceptibility to sethoxydim is the result of a sethoxydim-sensitive target site. A resistant target site is the most common mechanism of resistance to ACCase-inhibiting herbicides although other mechanisms have also been observed (*24*). In *L. rigidum* we have observed resistance due to both target site and non-target site mechanisms (*6-8,11*). We have also observed considerable variation in the responses of the ACCase from different resistant biotypes to different herbicide classes, ranging from high-level resistance to both groups to resistance to the aryloxyphenoxypropanoate herbicides only (*5*). This indicates that there are several possible mutations within ACCase that can provide resistance to herbicides without impairing the normal functioning of this enzyme.

Resistance due to Enhanced Herbicide Detoxicative Metabolism

Although the resistant biotype contains two resistant target sites, possession of a herbicide-resistant target site cannot explain resistance to many herbicides, such as the PS II-inhibiting herbicides and tralkoxydim. Enhanced detoxicative metabolism of herbicides appears to be an important mechanism of resistance in this biotype (*9-14,* Preston, Tardif, Christopher and Powles, unpublished) and is a mechanism of resistance for five different herbicides from five different chemical groups (Table VI). Metabolism of herbicides was determined with soil-grown plants, except for simazine and chlorotoluron where plants grown in liquid culture were used. Plants were exposed to ^{14}C-labeled herbicides either through the leaf (diclofop-methyl) (*11*), root (simazine and chlorotoluron) (*9-10*), or cut shoot (chlorsulfuron and tralkoxydim) (*13,25*). Following exposure to herbicides, shoot tissue was harvested, ground in liquid N_2 and extracted with 80 % methanol. Metabolites and parent herbicide were separated in the extracts by HPLC.

Table VI. Detoxification of Herbicides by Susceptible and Resistant Biotypes of *L. rigidum*

Herbicide	*Susceptible*	*Resistant*
	$T_{1/2}$ (h)[a]	
Simazine	18	6
Chlorotoluron	48	11
Chlorsulfuron	6	3
Tralkoxydim	5.5	2.5
Diclofop-methyl	42	30

[a]$T_{1/2}$ (time taken to metabolise 50% of applied simazine, chlorotoluron, chlorsulfuron, and tralkoxydim and to convert 50% of applied diclofop-methyl to products other than diclofop acid) were estimated from data in (*9,10,14*), and Preston, unpublished data.

The resistant biotype has enhanced metabolism of PS II-, ALS- and ACCase-inhibiting herbicides. The rate of metabolism of herbicides in the resistant biotype varies from 1.4-fold as fast as the susceptible for diclofop-methyl to 4-fold as fast for chlorotoluron. What cannot be determined from this data is the tyoe or number of herbicide-degrading enzymes involved in enhanced metabolism of herbicides.

Cytochrome P450 Inhibitors as Synergists of Herbicide Action in the Resistant Biotype

Research in our laboratory has shown that specific inhibitors of cytochrome P450 enzymes can reduce the rate of enhanced metabolism of herbicides in herbicide-resistant *L. rigidum* biotypes (*9,10,25*). This suggests that cytochrome P450 monooxygenase

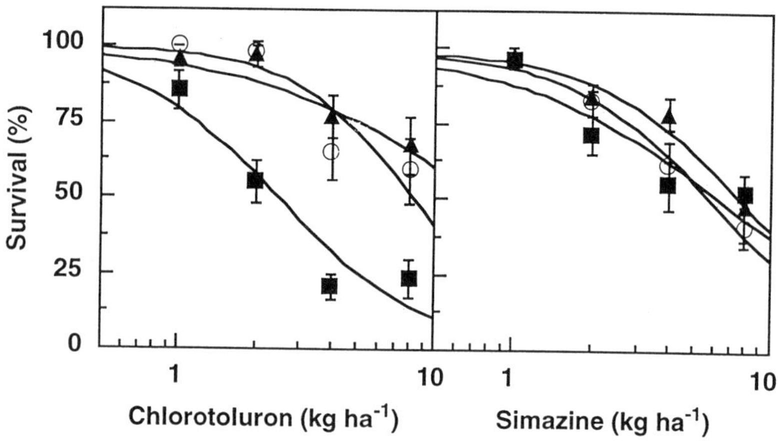

Figure 1. Susceptibility of the resistant biotype (VLR 69) to chlorotoluron and simazine without addition of synergists (O) or in the presence of 2.1 kg ha^{-1} piperonyl butoxide (■) or 1 kg ha^{-1} malathion (▲). Data are the means ± SEM of four replicates. Plotted lines are the fitted logit equation for each data set. For chlorotoluron all lines are not the same ($P > 0.05$), for simazine all lines are the same ($P < 0.05$).

enzymes may be responsible for resistance. Resistance in biotype VLR 69 may be the result of enhanced activity of a single enzyme with broad specificity for numerous herbicides, or it may result from the elevated activity of several enzymes. Plant cytochrome P450 enzymes are highly variable in their substrate specificities. A single enzyme has the ability to metabolize several substrates (*26*), or alternatively, a single substrate may be metabolized by several different enzymes (*27*). Two cytochrome P450 inhibitors, the organophosphate insecticide malathion (*28*) and the insecticide synergist piperonyl butoxide (PBO) (*29*) were applied to soil-grown plants of the resistant biotype. We have previously established that these two compounds can enter soil-grown plants and synergise the action of herbicides (*10,25*). The inhibitors were applied 30 min to 1 h prior to application of the herbicide using a laboratory cabinet sprayer delivering 108 L ha^{-1} at 1 m s^{-1} and a pressure of 250 kPa. Plants were sprayed at the 2-3 leaf stage and were returned outdoors following spraying. Three weeks after spraying plant mortality was assessed. In addition, metabolism of these herbicides in the resistant biotype was determined in the presence of these two cytochrome P450 inhibitors as well as the cytochrome P450 suicide substrate 1-aminobenzotriazole (ABT) (*30*) and the plant growth regulator tetcyclacis (*31*). The inhibitors were applied to the roots of plants growing in nutrient solution 24 h prior to application of ^{14}C-labeled herbicide. The amount of herbicide metabolized was determined 6 h after treatment (chlorsulfuron and tralkoxydim), 24 h after treatment (simazine and chlorotoluron), or 48 h after treatment (diclofop-methyl). Herbicides and metabolites were extracted from leaf tissue by 80 % methanol and separated by HPLC.

Interactive Effects of Cytochrome P450 Inhibitors with PS II-inhibiting Herbicides. Application of PBO to soil-grown plants in combination with chlorotoluron resulted in an interactive effect leading to increased mortality of the resistant biotype (Figure 1). In contrast, malathion applied in combination with chlorotoluron resulted in no interactive effect. In this experiment, the LD_{50} for chlorotoluron decreased from about 8.3 kg ha^{-1} to about 2.5 kg ha^{-1} in the presence of PBO. No interaction was observed between simazine and either malathion or PBO as the dose response curves are nearly identical. PBO, ABT and tetcyclacis all inhibited metabolism of chlorotoluron (Table VII) when applied to the roots of plants growing in nutrient solution. Of these inhibitors, tetcyclacis was clearly the most effective. Similarly, simazine metabolism was inhibited by PBO, ABT and tetcyclacis, (Table VII), with tetcyclacis again the most effective inhibitor. Malathion marginally inhibited metabolism of both herbicides in the resistant biotype. The cytochrome P450 inhibitors had similar effects on metabolism of the two PS II-inhibiting herbicides suggesting that the same enzyme(s) may be responsible for metabolism of these two herbicides in the resistant biotype.

Table VII. Effects of Synergists on the Metabolism of Chlorotoluron and Simazine by the Resistant Biotype of *L. rigidum*

Treatment[a]	*^{14}C Extracted as Parent Herbicide, (%)*	
	Chlorotoluron	Simazine
Control	45 ± 7[b]	36 ± 6
Piperonyl butoxide	63 ± 5	63 ± 9
Malathion	54 ± 6	48 ± 9
1-Aminobenzotriazole	64 ± 5	61 ± 8
Tetcyclacis	82 ± 3	89 ± 1

[a]Metabolism in 24 h of exposure to ^{14}C-labeled herbicides following pre-treatment for 24 h with piperonyl butoxide, malathion, 1-aminobenzotriazole or tetcyclacis each at 70 μM. [b]Data are means ± SEM of four replicate experiments

Interactive Effects of Cytochrome P450 Inhibitors with ALS-inhibiting Herbicides. Malathion has previously been shown to be an excellent synergist for chlorsulfuron in *L. rigidum* due to its capacity to greatly inhibit chlorsulfuron detoxification (*25*). Here we observed a similar result with the resistant biotype in that chlorsulfuron in combination with malathion is considerably more phytotoxic than chlorsulfuron applied alone (Figure 2). The LD_{50} for chlorsulfuron decreased from about 500 g ha^{-1} in the absence of malathion to about 100 g ha^{-1} with the addition of malathion. In contrast, PBO was unable to synergize chlorsulfuron in this biotype. Malathion also dramatically inhibited metabolism of chlorsulfuron in the resistant biotype (Table VIII). PBO did not inhibit metabolism of chlorsulfuron, however, ABT and tetcyclacis had slight inhibitory effects on metabolism of this herbicide. This pattern of inhibition of chlorsulfuron metabolism by the synergists is clearly different from that for simazine and chlorotoluron. This suggests that different isozymes are responsible for metabolism of these herbicides within the resistant population.

Table VIII. Effects of Synergists on Metabolism of Chlorsulfuron by the Resistant Biotype of *L. rigidum*

Treatment[a]	*^{14}C Extracted as Chlorsulfuron (%)*
Control	78 ± 3[b]
Piperonyl butoxide	78 ± 5
Malathion	91 ± 3
1-aminobenzotriazole	84 ± 3
Tetcyclacis	83 ± 4

[a]Metabolism in 6 h of exposure to ^{14}C-labeled herbicides following pre-treatment for 24 h with piperonyl butoxide, malathion, 1-aminobenzotriazole or tetcyclacis each at 70 μM. [b]Data are means ± SEM of four replicate experiments

Interactive Effects of Cytochrome P450 Inhibitors with ACCase-inhibiting Herbicides. The resistant biotype is resistant to a number of ACCase-inhibiting herbicides, but is susceptible to sethoxydim (*15*). The resistant biotype contains an ACCase that is resistant to aryloxyphenoxypropanoate herbicides such as diclofop but is sensitive to cyclohexanedione herbicides such as tralkoxydim and sethoxydim (Table V). This biotype also displays enhanced detoxification of these two groups of herbicides (Table VI). As expected, the resistant biotype was very insensitive to diclofop-methyl application and addition of cytochrome P450 inhibitors did not influence plant survival (Figure 3). Where resistance is primarily due to a change in the target site, inhibition of metabolism should not affect plant survival. The situation is somewhat different concerning tralkoxydim, where tralkoxydim applied in combination with PBO resulted in greater mortality than tralkoxydim applied alone or in combination with malathion. In this experiment, PBO was able to decrease the LD_{50} for tralkoxydim from about 150 g ha^{-1} to less than 100 g ha^{-1}.

The effects of cytochrome P450 inhibitors on metabolism of the ACCase-inhibiting herbicides diclofop-methyl and tralkoxydim were examined. About 62 % of the applied diclofop-methyl was detoxified by conversion to metabolites other than diclofop acid in the resistant biotype by 48 h after herbicide application (Table IX). Pre-treatment with malathion did not influence metabolism of diclofop in this biotype. In contrast, ABT pre-treatment reduced diclofop metabolism such that only 48 % of the applied herbicide had been metabolized to products other than diclofop acid in 48 h. Tetcyclacis and PBO had slight inhibitory effects on diclofop-methyl metabolism. Tralkoxydim was metabolized very rapidly by the resistant biotype with less than 25 % of the herbicide remaining 6 h after treatment (Table IX). None of the cytochrome P450

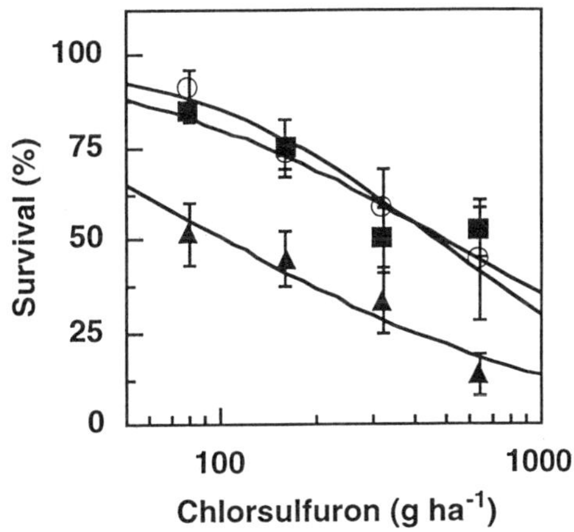

Figure 2. Susceptibility of the resistant biotype (VLR 69) to chlorsulfuron without addition of synergists (O) or in the presence of 2.1 kg ha^{-1} piperonyl butoxide (■) or 1 kg ha^{-1} malathion (▲). Data are the means ± SEM of four replicates. Plotted lines are the fitted logit equation for each data set. All lines are not the same ($P > 0.05$).

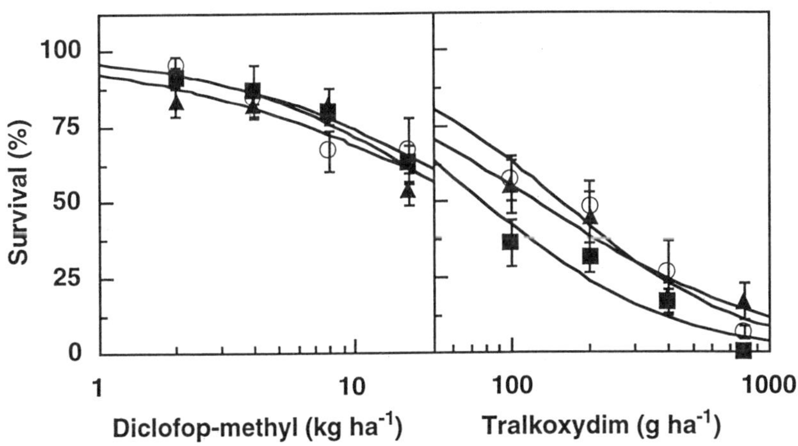

Figure 3. Susceptibility of the resistant biotype (VLR 69) to diclofop-methyl and tralkoxydim without addition of synergists (O) or in the presence of 2.1 kg ha^{-1} piperonyl butoxide (■) or 1 kg ha^{-1} malathion (▲). Data are the means ± SEM of four replicates. Plotted lines are the fitted logit equation for each data set. For diclofop-methyl all lines are the same ($P < 0.05$). For tralkoxydim all lines are not the same ($P > 0.05$).

inhibitors used had any substantial impact on tralkoxydim metabolism in this biotype. The cytochrome P450 inhibitors had different patterns of inhibition on metabolism of diclofop-methyl and tralkoxydim in the resistant biotype. These patterns were also different to those observed for the PS II and ALS-inhibiting herbicides. Overall, there were markedly different patterns of effectiveness of the cytochrome P450 inhibitors in their ability to inhibit metabolism of the various herbicides in the resistant biotype. For example, PBO inhibits metabolism of chlorotoluron and simazine, but not chlorsulfuron, tralkoxydim or diclofop-methyl. These different patterns indicate that enhanced detoxification of herbicides in the resistant biotype is probably not due to a single cytochrome P450 enzyme with broad specificity, but to multiple cytochrome P450 isozymes.

Table IX. Effects of Synergists on Metabolism of Diclofop-methyl and Tralkoxydim by the Resistant Biotype of *L. rigidum*

Treatment[a]	*^{14}C Extracted as Parent Herbicide (5)*	
	Diclofop-methyl and diclofop acid	Tralkoxydim
Control	38 ± 3[b]	23 ± 4
Piperonyl butoxide	44 ± 1	17 ± 2
Malathion	38 ± 3	26 ± 4
1-aminobenzotriazole	52 ± 3	22 ± 6
Tetcyclacis	45 ± 3	18 ± 4

[a]Metabolism following pre-treatment for 24 h with piperonyl butoxide, malathion, 1-aminobenzotriazole or tetcyclacis each at 70 μM. Plants were exposed to ^{14}C-labeled diclofop-methyl for 48 h and to ^{14}C-labeled tralkoxydim for 6 h before harvest. [b]Data are means ± SEM of four replicate experiments

Genetics of Multiple Resistance in *L. rigidum*

Data from biochemical studies show that the resistant biotype contains two resistant herbicide target sites, ALS and ACCase, as well as several different herbicide-degrading enzymes. This obviously makes the genetics of multiple resistance complicated. Data from other studies demonstrated that resistance due to modification of ALS and ACCase are each inherited as a single, nuclear-encoded dominant, or incompletely-dominant, gene (*8,21,24,32,33*).

The inheritance of the genes endowing enhanced herbicide metabolism is not so clear. To date, we have not established the mode of inheritance of such genes, nor have we determined whether these are linked. It is most likely that this enhanced metabolism is the result of an increase in production of specific cytochrome P450 enzymes within the resistant plant. Therefore, changes in the regulation of these enzymes may account for resistance. It is possible that modification of a single regulatory protein may result in enhanced activity of a number of cytochrome P450 enzymes. On the other hand, enhanced activity may be due to mutations within either the 5'-regulatory of the coding regions of the cytochrome P450 gene itself and hence several genes may be involved.

In other resistant *L. rigidum* biotypes, we have observed some common patterns. Resistance to simazine and chlorotoluron due to enhanced metabolism are present in the same biotypes, suggesting that these two mechanisms could be linked (*9,10*). Likewise, two biotypes show enhanced metabolism of both diclofop-methyl and tralkoxydim (Preston, unpublished data). In contrast, non-target site-based resistance to chlorsulfuron can occur concomitant with, or in the absence of, resistance to diclofop-methyl (*22*). Biotype VLR 69 is to date the only population we have documented with enhanced metabolism of all of these herbicides. Therefore, we suggest that the most likely situation is that more than one gene confers multiple resistance due to enhanced metabolism of herbicides in the resistant biotype VLR 69. What is clear from the foregoing observations is that the resistant biotype has multiple resistance to a wide range of herbicides due to the possession of multiple mechanisms of resistance that are encoded by several different genes.

Development of Multiple Resistance and its Implications for Weed Management Strategies

Frequently large populations of *L. rigidum* are treated with herbicides in Australia. In southern Australia, *L. rigidum* can occur at densities exceeding 1000 plants m^{-2} as young seedlings (*34*), and a typical cropping field would exceed 50 ha^{-1}. This means that 500 million *L. rigidum* seedlings might be treated with herbicide in a single, well-infested field. This ensures that at least a few individuals with resistance genes are likely to be present in a population even if the initial gene frequency is low. Herbicides have been heavily relied on to control *L. rigidum* in Australia over the last 15 years, which means the selection pressure is high thereby enriching for resistance genes (*35*). In addition, the short soil seed bank life of this species (*34,36*) ensures that dilution of resistant with susceptible genes in subsequent generations is low. As *L. rigidum* is an obligate outcrossing species, any survivors of herbicide application must cross with another survivor to produce offspring. This outcrossing means that all mechanisms of herbicide resistance present will tend to be mixed and distributed within the progeny of these individuals. In addition, considerable genetic diversity will remain in the population. This is important because if the selection pressure is changed, the population can rapidly respond to the new selecting agent. For example, a study by Matthews and Powles (unpublished) has shown that selection in the field with diclofop-methyl for three years can result in a population highly resistant to diclofop-methyl, and with concomitant, but lower, resistance to a range of other herbicides.

The fortuitous meeting of a highly-variable, outcrossing weed species in high abundance over large areas with highly-efficient herbicides has enabled the development of multiple resistance in *L. rigidum*. These factors, coupled with the extensive and varied herbicide application history of *L. rigidum* biotype VLR 69, have provided ideal conditions for the accumulation of multiple resistance mechanisms. As a result, there are few selective herbicides available for control of this population. While the example of this biotype is extreme, multiple resistance in *L. rigidum* is not rare and, therefore, the use of herbicides as the sole means of control of this species is not a sustainable practice. It is likely that other species that share some of the biological characteristics and management practices of *L. rigidum* will also develop multiple resistance. This has already been observed on a smaller scale in the case of *Alopecurus myosuroides* in the U.K. (*37-39*) and no doubt will occur elsewhere.

The lessons to be learned from the Australian experience with multiple resistance in *L. rigidum* are important; with current herbicide use patterns, other weed species will inevitably exhibit multiple resistance. Once this has occurred, control of these weeds with herbicides will become difficult, particularly where enhanced cytochrome P450-mediated herbicide metabolism is involved in resistance. The varying patterns of multiple resistance evident in *L. rigidum* populations also make recommendation of reliable alternative herbicides virtually impossible. This ensures that in the future, reliance on chemicals alone for control of *L. rigidum* will not suffice. As effective weed control is essential to agricultural productivity, other weed control techniques need to be implemented to supplement chemical control of species such as *L. rigidum*. The development of such Integrated Weed Management strategies will be a considerable challenge in the years ahead.

Literature Cited

1. Reeves, T. G. *Weed Res.* **1976**, *16*, 57-63.
2. Medd, R. W.; Auld, B. A.; Kemp, D. R.; Murison, R. D. *Aust. J. Agric. Res.* **1985**, *36*, 361-371.
3. Powles, S. B.; Matthews, J. M. In *Resistance '91: Achievements and Developments in Combating Pesticide Resistance*; Denholm, I.; Devonshire, A.; Holloman, D., Eds.; Elsevier: London, 1992; pp. 75-87.
4. Heap, J.; Knight, R. J. *Aust. Inst. Agric. Sci.* **1982**, *48*, 156-157.
5. Tardif, F. J.; Preston, C.; Powles, S. B. In *Proceedings of International Symposium on Weed and Crop Resistance to Herbicides*, De Prado, R., Ed.; Kluwer Academic Publishers, Dordrecht, The Netherlands, 1995; in press.
6. Tardif, F. J.; Holtum, J. A. M.; Powles, S. B. *Planta* **1993**, *190*, 176-181.
7. Tardif, F. J.; Powles, S. B. *Physiol. Plant.* **1994**, *91*, 488-494.
8. Tardif, F. J.; Preston, C., Holtum, J. A. M.; Powles, S. B. Aust. J. Plant Physiol. **1996**, *23*, in press.
9. Burnet, M. W. M.; Loveys, B. R.; Holtum, J. A. M.; Powles, S. B. *Pestic. Biochem. Physiol.* **1993**, *46*, 207-218.
10. Burnet, M. W. M.; Loveys, B. R.; Holtum, J. A. M.; Powles, S. B. *Planta* **1993**, *190*, 182-187.
11. Holtum, J. A. M.; Matthews, J. M.; Häusler, R. E.; Liljegren, D. R.; Powles, S. B. *Plant Physiol.* **1991**, *97*, 1026-1034.
12. Christopher, J. T.; Powles, S. B.; Liljegren, D. R.; Holtum, J. A. M. *Plant Physiol.* **1991**, *95*, 1036-1043.
13. Christopher, J. T.; Powles, S. B.; Holtum, J. A. M. *Plant Physiol.* **1993**, *100*, 1909-1913.
14. Burnet, M. W. M.; Christopher, J. T.; Holtum, J. A. M.; Powles, S. B. *Weed Sci.* **1994**, *42*, 468-473.
15. Burnet, M. W. M.; Hart, Q.; Holtum, J. A. M.; Powles, S. B. *Weed Sci.* **1994**, *42*, 369-377.
16. Russell, R. M.; Robertson, J. L.; Savin, N. E. *Bull. Entomol. Soc. Amer.* **1977**, *23*, 209-213.
17. Gronwald, J. In *Herbicide Resistance in Plants: Biology and Biochemistry*; Powles, S. B.; Holtum, J. A. M. Eds.; Lewis Publishers: Boca Raton, Fl., 1994; pp. 27-60.
18. Burnet, M. W. M. Ph.D. Thesis; University of Adelaide 1992.
19. Anderson, J. J.; Swain, R. S. *J. Agric. Food Chem.* **1992**, *40*, 2279-2283.
20. Brown, M. A.; Chui, T. Y.; Miller, P. *Pestic. Biochem. Physiol.* **1987**, *27*, 24-27.
21. Saari, L. L.; Cotterman, J. C.; Thill, D.C. In *Herbicide Resistance in Plants: Biology and Biochemistry*; Powles, S. B.; Holtum, J. A. M. Eds.; Lewis Publishers: Boca Raton, Fl., 1994; pp. 83-139.
22. Christopher, J. T. Ph.D. Thesis; University of Adelaide 1993.
23. Gill, G. S. *Aust. J. Exp. Agric.* **1995**, *35*, 67-72.
24. Devine, M. D.; Shimabukuro, R. H. In *Herbicide Resistance in Plants: Biology and Biochemistry*; Powles, S. B.; Holtum, J. A. M. Eds.; Lewis Publishers: Boca Raton, Fl., 1994; pp. 141-169.
25. Christopher, J. T.; Preston, C.; Powles, S. B. *Pestic. Biochem. Physiol.* **1994**, *49*, 172-182.
26. Pierrel, M. A.; Batard, Y.; Kazmaier, M.; Mignotte-Vieux, C.; Durst, F.; Werck-Reichhart, D. *Eur. J. Biochem.* **1994**, *224*, 835-844.
27. Batard, Y.; Zimmerlin, A.; Le Ret, M.; Durst, F.; Werck-Reichhart, D. *Plant Cell Environ.* **1995**, *18*, 523-533.
28. Kreutz, K.; Fonné-Pfister, R. *Pestic. Biochem. Physiol.* **1992**, *43*, 232-240.

29. Bernard, C. B-; Philogène, B. J. R. *J. Toxicol. Environ. Health* **1993**, *38*, 199-223.
30. Ortiz de Montellano, P. R.; Reich, N. O. In *Cytochrome-P450*; Ortiz de Montellano, P. R., Ed.; Plenum Press, New York, 1986; pp. 273-314.
31. Rademacher, W; Fritsch, H.; Greabe, J. E.; Sauter, H.; Jung, J. *Pestic. Sci.* **1987**, *21*, 241-252.
32. Betts, K. J.; Ehlke, N. J.; Wyse, D. L.; Gronwald, J. W.; Somers, D. A. *Weed Sci.* **1992**, *40*, 184-189.
33. Darmency, H. In *Herbicide Resistance in Plants: Biology and Biochemistry*; Powles, S. B.; Holtum, J. A. M. Eds.; Lewis Publishers: Boca Raton, Fl., 1994; pp. 263-297.
34. Heap, I. M. Ph.D. Thesis; University of Adelaide 1988.
35. Powles, S. B. In *Proceedings, International Symposium, Indian Society of Weed Science*; 1993, Vol. 1; pp. 189-194.
36. Gramshaw, D. *Aust. J. Agric. Res.* **1972**, *23*, 779-787.
37. Kemp, M. S.; Moss, S. R.; Thomas, T. H. In *Managing Resistance to Agrochemicals: From Fundamental Research to Practical Strategies*; Green, M. B.; LeBaron, H. M.; Moberg, W. K., Eds.; American Chemical Society: Washington, DC, 1990; pp. 376-393.
38. Hall, L. M.; Holtum, J. A. M.; Powles, S. B. In *Herbicide Resistance in Plants: Biology and Biochemistry; Powles*, S. B.; Holtum, J. A. M. Eds.; Lewis Publishers: Boca Raton, Fl., 1994; pp. 243-261.
39. Moss, S. R. *Weed Sci.* **1990**, *38*, 492-496.

Chapter 14

Resistance to *Bacillus thuringiensis* in *Plutella xylostella*

The Moth Heard Round the World

Bruce E. Tabashnik[1], Francis R. Groeters[1], Naomi Finson[1], Yong-Biao Liu[1], Marshall W. Johnson[1], David G. Heckel[2], Ke Luo[3], and Michael J. Adang[3]

[1]Department of Entomology, University of Hawaii, Honolulu, HI 96822
[2]Department of Biological Sciences, Clemson University, Clemson, SC 29634
[3]Department of Entomology, University of Georgia, Athens, GA 30602

Insecticidal crystal proteins from *Bacillus thuringiensis* (Bt) offer safe, specific, and effective insect control. Evolution of resistance by pests is the greatest threat to the long-term effectiveness of Bt. The first insect to evolve resistance to Bt in open field populations is the diamondback moth (*Plutella xylostella*), a global pest of vegetables. Resistance to Bt has been found in some populations of diamondback moth from Hawaii, the continental United States, and several Asian nations. Resistance to Bt in diamondback moth is inherited as an autosomal, recessive trait and is associated with reduced binding of toxin to sites on the larval midgut membrane. Resistance to CryIA toxins from Bt subsp. *kurstaki* caused cross-resistance to CryIF, but not to CryIB or CryIC. In laboratory populations containing susceptible and resistant individuals, fitness costs associated with resistance promoted reversal of resistance in the absence of exposure to Bt. In an isofemale line derived from an extensively selected strain, >5,000-fold resistance persisted for >20 generations without selection, which shows that at least one resistant genotype is not inherently unstable. Field-evolved resistance to Bt in diamondback moth has provided a warning and an opportunity to gain knowledge essential for sustainable use of Bt.

Because of their broad spectrum of toxicity, many conventional insecticides raise concerns about human safety and the environment. Further, evolution of resistance to insecticides has occurred in more than 500 species of insects (*1*). These problems with conventional insecticides are spurring the search for alternatives.

Insecticides derived from the bacterium *Bacillus thuringiensis* (Bt) are an attractive alternative for pest control (*2*). Bt is a common soil bacterium that produces proteins (called delta-endotoxins or toxins) that can kill insects. Thousands of strains of Bt have been isolated; each has a characteristic set of toxins. Bt toxins are extraordinarily lethal to certain pests. For example, based on amounts used in agricultural applications, molecules of Bt toxin are 80,000 times more potent than organophosphates and 300 times more potent than pyrethroids (*3*).

In contrast to many conventional insecticides, Bt toxins have a narrow spectrum of toxicity and pose little or no risk to people or wildlife. Bt is generally not toxic to the

0097–6156/96/0645–0130$15.00/0

predators and parasites that attack insect pests. Thus, unlike many conventional insecticides, Bt is compatible with biological control by natural enemies in integrated pest management. Although Bt is the most widely used microbial insecticide, with annual sales of greater than $100 million, it still accounts for only about 1% of all insecticide sales. However, Bt is rapidly becoming a cornerstone of ecologically-sound crop protection. Use of Bt is increasing dramatically because of its safety and recently enhanced efficacy (*2*, *3*).

The most profound advance is the transfer of bacterial genes encoding Bt toxins to plant genomes. Transgenic varieties that express Bt proteins have been created from many crops including broccoli, canola, rice, soybean, tobacco, and tomato. Recent approvals by the U.S. Environmental Protection Agency open the way for large commercial plantings of Bt-expressing varieties of corn, cotton, and potato.

Greater use of Bt, however, increases the risk that pests will evolve resistance (*4*, *5*, *6*). Until recently, some had thought that this risk was low. Sprays of commercial formulations of Bt had been used widely for decades, yet no cases of resistance had been documented in field populations. Important warnings about resistance to Bt came first from laboratory selection experiments and grain bin populations of the Indianmeal moth, *Plodia interpunctella*. Exposure to Bt in the laboratory quickly produced high levels of resistance in the Indianmeal moth (*5*) and the tobacco budworm, *Heliothis virescens* (*6*). This evidence captured the attention of scientists in industry, government, and academia. Yet, some people still asked, "Is it possible for insects to evolve resistance to Bt in open field populations?".

We now know that the answer to that question is "Yes." The insect that provided the answer is the diamondback moth, *Plutella xylostella* (*7*). This pest is found throughout the world wherever one finds crucifers, its host plants. The diamondback moth costs growers of cabbage and related crops about $1 billion each year (*8*). The caterpillars (larvae) eat foliage and can destroy plants when they reach high densities. When susceptible larvae ingest Bt, they are killed by it.

Several recent reviews provide broad perspectives of insect resistance to Bt (*7*, *9-11*). This paper focuses on evolution of resistance to Bt by diamondback moth in Hawaii. The following topics are considered: documentation of resistance, genetics and mechanisms of resistance, cross-resistance, and stability of resistance in the absence of exposure to Bt.

Documentation

Resistance is an evolutionary change (i.e., a change in gene frequencies) that increases the average fitness of a population in an environment where a toxin is present (*7*). The most direct way to demonstrate evolution of resistance is to show that repeated exposure to a particular toxin caused a decrease in susceptibility through time within a population. In most cases of insecticide resistance, documentation has been indirect. The typical evidence is a correlation across populations between the history of exposure to a toxin and the ability to survive exposure to that toxin. Whether the approach is direct or indirect, insects must be raised and tested in a consistent environment to permit the inference that observed differences are genetically based.

In our studies of diamondback moth resistance to Bt, we used both direct and indirect methods of documentation. We had been evaluating resistance to pyrethroids when a grower's suspicions led us to examine responses to Bt. We sampled individuals from various field populations in Hawaii during 1986-1987 and 1989-1990 (*12*, *13*).

We brought field-collected individuals into the laboratory and reared their offspring. Offspring from the first, second, or third laboratory-reared generations were tested for susceptibility to Bt using leaf residue bioassays. We put groups of larvae on cabbage leaf disks that had been dipped in various concentrations of Bt. We first tested responses to Dipel, a popular commercial formulation of the HD-1 strain of Bt subsp. *kurstaki* that contains spores and crystalline toxins called CryIA(a), CryIA(b), CryIA(c), CryIIA, and CryIIB (*14*). We recorded mortality and used the data to estimate the concentration needed to kill 50% of the larvae (LC_{50}). To summarize our results, we calculated a resistance ratio for each field population, which is the LC_{50} of that population divided by the LC_{50} of a standard susceptible laboratory strain called "LAB-P." Although our initial studies were based on mortality of third instar larvae 48 hours after exposure, subsequent work showed that estimates of resistance are correlated across larval instars (*15*) and bioassay time intervals from one to five days (*16*).

Resistance ratios from the 1986-1987 survey showed some variation (range = 0.9 to 6.8) in susceptibility, but provided no strong evidence of resistance in the six field populations tested. During the next three years, two of the original populations were treated frequently with Bt by growers and two were not treated with Bt. Compared with the 1986-1987 data, the 1989-1990 results showed significant decreases in susceptibility in the two heavily treated populations and no such changes in the two untreated populations. The "before and after" data for the two heavily treated sites provide direct evidence of evolution of resistance in the field.

The highest level of resistance occurred in a population (called "NO") from a watercress farm on the island of Oahu that had been treated repeatedly with Bt. The resistance ratio for this intensely treated population was 36. Because the NO site was not included in the 1986-1987 survey, the evidence for resistance in this population is indirect.

Just as we observed with synthetic insecticides (*17*), variation in diamondback moth susceptibility to Bt was related to treatment history; also, variation was greater within islands than between islands. In contrast to the idea that repeated applications "over a wide geographic area" would be required to produce pest resistance in the field (*6*), resistance to Bt in Hawaii was highly localized (*12*). Resistant populations ocurred at some heavily treated small farms; neighboring farms that had not been treated intensively harbored susceptible populations.

In each bioassay, we included leaf disks that were dipped in a concentration of Bt that was comparable to the concentration recommended on the product label. Although laboratory results may not extrapolate reliably to the field, responses to this concentration yield a crude indication of potential efficacy. The recommended concentration killed 90-100% of susceptible larvae, but only 34-35% of larvae from resistant populations. If a field application killed only 35% of larvae treated, a grower would have serious problems controlling a population. Thus, the results imply that the levels of resistance in the field were sufficient to thwart control.

Resistance to Bt is not limited to Hawaii or to diamondback moth. Resistance to Bt has been documented in field and greenhouse populations of diamondback moth from Florida, New York, and at least five countries in Asia (*7, 18*). Although the first well-documented examples from the field involve diamondback moth, laboratory selection has produced resistance to Bt in at least nine other species of insects (*7*). Most of the pests selected for resistance to Bt are moths, but beetles and mosquitoes have also evolved resistance to Bt in the laboratory (*19, 20*). The ability to evolve resistance to Bt appears to be widespread among insects.

Mechanisms of Resistance

To kill insects, Bt proteins must be ingested, solubilized, and proteolytically cleaved from protoxins to active fragments in the midgut (*19*). The next step, which appears to be critical in determining the specificity of Bt, is binding of active toxin to sites on brush border membranes of the midgut epithelium. After binding occurs, insertion of toxin produces pores in the midgut membrane, which disrupts osmotic balance and kills the insect. In principle, toxicity-blocking alterations in any of these steps could confer resistance (*58*).

Initial experiments provided no evidence of altered proteolytic activity in resistant strains of Indianmeal moth (*21*) or diamondback moth (*22*), but some recent data suggest that activation of protoxins is reduced in one resistant strain of Indianmeal moth (*23*). The only well-characterized mechanism of insect resistance to Bt is reduced binding of toxin to midgut membranes (*11*). In Indianmeal moth, binding of CryIA(b) toxin to brush border membrane vesicles was lower in a resistant strain than a susceptible strain (*24*). In diamondback moth, this mechanism was first associated with resistance in a strain from the Philippines (*25*). Binding studies with radioactively-labeled toxin (*25*) and immunohistochemical techniques (*26*) showed little or no binding of CryIA(b) in the resistant Philippines strain compared to an unrelated susceptible strain.

Similar studies with diamondback moth from Hawaii and radioactively-labeled CryIA(c) toxin showed greatly decreased binding to brush border membrane vesicles from a resistant strain relative to two susceptible strains (*27*). However, immunohistochemical tests revealed binding of CryIA(a), CryIA(b), and CryIA(c) in five resistant strains as well as in the susceptible LAB-P strain (*28*). Further, surface plasmon resonance experiments detected no significant difference between resistant and susceptible Hawaiian strains in the kinetics of CryIA(c) binding (*29*). The surface plasmon results suggest that the concentration of binding sites was 3-fold lower in resistant larvae compared with susceptible larvae.

The seemingly paradoxical results with diamondback moth have not yet been resolved. However, these data reinforce the idea that binding is necessary, but not sufficient, for toxicity. Binding is not always correlated with toxicity across toxins for a particular insect strain (*30*) or across resistant and susceptible strains for a particular toxin (*31-33*).

Certain Bt toxins may bind to many sites in the insect midgut, only some of which confer toxicity (*33, 34*). Given that toxin binding alone does not consistently explain patterns of resistance, comparison of post-binding events in susceptible and resistant strains is needed. Analysis of toxin insertion into the midgut membrane, which has been measured as irreversible binding in susceptible insects (*35, 36*), and toxin-induced pore formation may be helpful.

Inheritance of Resistance

Resistance to Bt in diamondback moth is inherited as a recessive, autosomal trait (*37 39*). Analysis of crosses and backcrosses with Hawaiian strains suggests that one or a few major loci are of primary importance (*39*). In selected laboratory strains, resistance continued to increase after repeated exposure to concentrations high enough to kill 100% of putative heterozygotes. These results imply that resistance is not controlled solely by one locus with two alleles (*40*). More conclusive understanding of the genetic basis of resistance in diamondback moth may be facilitated by molecular genetic markers (*41*).

The Spectrum of Resistance and Cross-resistance

Compared with susceptible diamondback moth larvae, resistant larvae from Hawaii were extremely resistant to three CryIA toxins (CryIA(a), CryIA(b), and CryIA(c)) and moderately resistant to CryIIA (*42*). Each of the four aforementioned toxins occurs in Dipel and other commonly used formulations of Bt subsp. *kurstaki* to which resistant strains had been exposed.

Resistant larvae also were strongly cross-resistant to CryIF (*43*). As far as we know, the resistant strains had not been exposed to CryIF, yet the larvae withstood concentrations of CryIF hundreds of times greater than those that killed susceptible larvae (*43*). Like resistant larvae from the Philippines (*25, 44*), those from Hawaii were still susceptible to CryIB and CryIC (*42, 43*). The pattern of cross-resistance correlates with amino acid sequence similarity among toxins. Similarity to CryIA toxins is greater for CryIF than for CryIB or CryIC.

There is no evidence of negative cross-resistance among Bt toxins in diamondback moth. Although negative cross-resistance between toxins CryIA(b) and CryIC against Indianmeal moth was reported initially (*24*), this pattern was not observed in subsequent analyses (*45, 46*). Negative cross-resistance between pyrethroids and Bt has been observed in the sheep body louse (*47*).

Stability of Resistance and What Makes People Sneeze

Does diamondback moth resistance to Bt decline when exposure to Bt stops? Not long ago, we would have answered that question with an emphatic "yes." More recent evidence suggests that "sometimes" is a better answer. The following sections describe initial results that showed declines in resistance in untreated laboratory strains, proximate and ultimate causes of these declines, and cases in which resistance remained stable when treatment with Bt stopped. Finally, we discuss the implications of these results and their relationship to sneezing -- read on, there really is a connection.

Declines in Resistance. To determine how resistant strains respond to lack of exposure to Bt, we collected individuals from the resistant NO population and reared their progeny in the laboratory (*12, 27, 48*). The original NO colony was maintained for three generations without exposure to Bt, then split into four strains. To increase resistance in three of the strains, larvae were fed Dipel-treated cabbage leaves during each of five to nine laboratory generations. One strain (NO-U) was reared only on untreated foliage.

In all four strains, resistance declined when exposure to Bt stopped (*27, 48*). The most spectacular drop occurred in strain NO-Q. Through a combination of field and laboratory selection, NO-Q had achieved a resistance ratio of 2800. After 14 generations without exposure to Bt, however, the LC_{50} for NO-Q was no longer greater than that of the susceptible LAB-P strain. The proportional rate of decline in LC_{50} was similar for the three extremely resistant strains, but resistance decreased more slowly in the untreated NO-U strain. For NO-U, the resistance ratio declined from about 30 to 4 after 15 generations. After 35 generations, the LC_{50} for NO-U did not differ significantly from that of the susceptible LAB-P strain. Similar decreases in resistance occurred in three diamondback moth strains from Japan (*37*) and one strain from Florida (*49*) when they were reared in the laboratory without exposure to Bt.

Proximate Cause of Reversal of Resistance. To better understand the patterns described above, we focused on three laboratory strains: LAB-P, our standard susceptible strain; NO-QA, a reselected resistant strain; and NO-Q, a revertant strain (*27*). When these experiments were started, the LC_{50} of NO-Q had returned to the same level as LAB-P, after having been several thousand-fold higher (as described above). NO-QA was derived from NO-Q and reselected with Dipel for three generations. This reselection produced >1,000-fold resistance to Dipel in NO-QA. The advantage of comparing NO-Q versus NO-QA is that these two strains had a similar genetic background, yet they differed greatly in susceptibility to Bt. This increases the chances that differences between these strains were related directly to resistance.

We honed in on responses to two toxins: CryIA(c) (Figure 1), which occurs in Dipel, and CryIC, which does not (Figure 2). We conducted leaf residue bioassays at the University of Hawaii in parallel with binding studies at the University of Georgia.

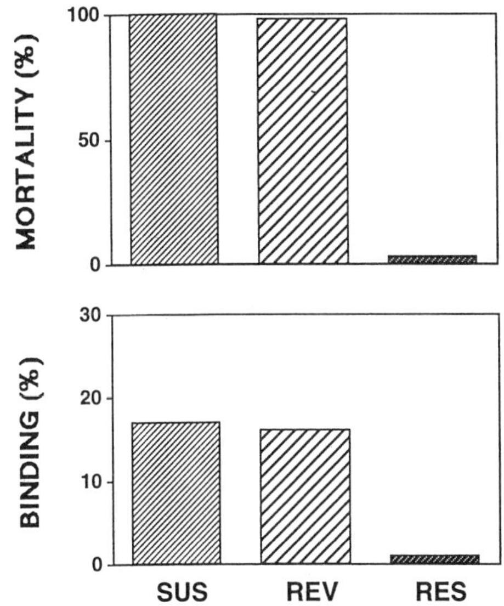

Figure 1. Toxicity and binding of CryIA(c) to susceptible (SUS), revertant (REV), and resistant (RES) larvae of diamondback moth (adapted from *27*).

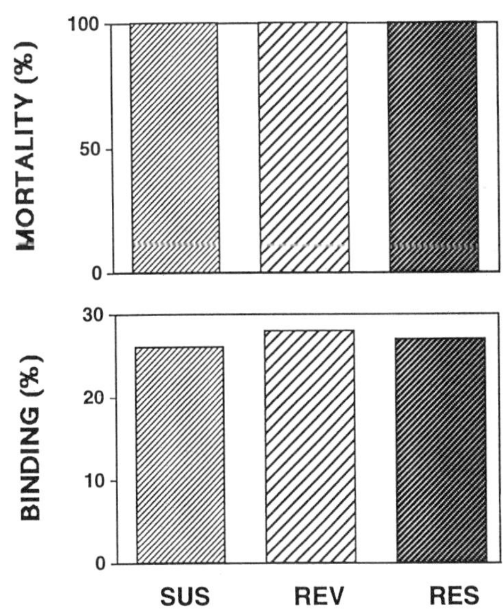

Figure 2. Toxicity and binding of CryIC to susceptible (SUS), revertant (REV), and resistant (RES) larvae of diamondback moth (adapted from *27*).

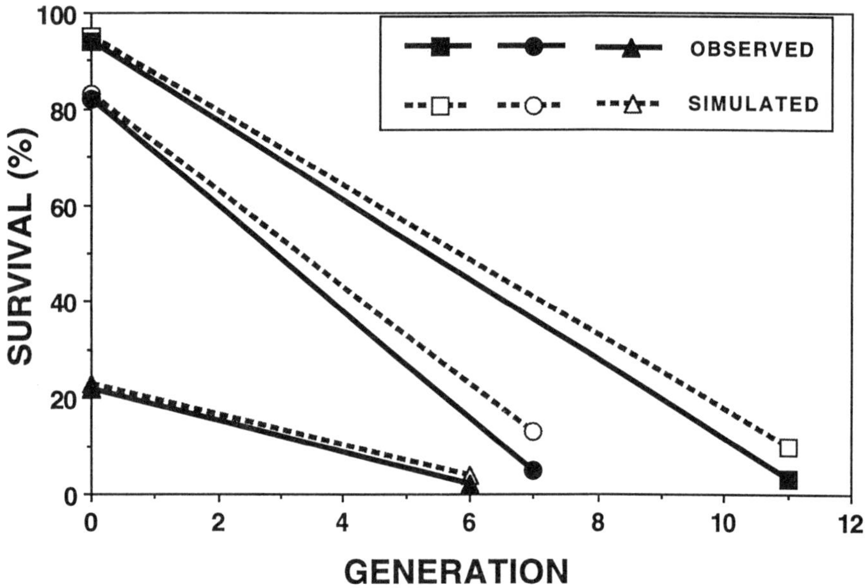

Figure 3. Observed versus simulated rates of decline in resistance to Dipel in three laboratory strains of diamondback moth reared without exposure to Bt (adapted from *27*).

A high concentration of CryIA(c) (10 mg protoxin per liter) killed all of the susceptible LAB-P larvae and 97% of the revertant NO-Q larvae (Figure 1). Brush border membrane vesicles prepared from midguts of thousands of frozen larvae from each strain were tested for binding with radioactively-labeled CryIA(c). Binding of CryIA(c) was similar in the susceptible and revertant larvae, but greatly reduced in resistant larvae (Figure 1). CryIC served as a positive control (Figure 2) : CryIC at 10 mg protoxin per liter killed 100% of the larvae from each of the three strains. Binding of radioactively-labeled CryIC was equally strong across the three strains.

From these results, we conclude that the proximate cause of decreased resistance was restoration of toxin binding. The results suggest that this was a specific change that affected binding and toxicity of CryIA(c), but not CryIC.

Ultimate Cause of Reversal of Resistance. What evolutionary factors favor restoration of binding and toxicity? Is there a fitness cost associated with resistance so that susceptible insects have a competitive advantage when Bt is absent? To answer these questions, we compared fitness parameters between strains NO-Q and NO-QA (*27, 50, 51*). When we conducted the fitness experiments, NO-Q and NO-QA had been separated for only five generations; the LC_{50} to Dipel was 3500-fold greater for NO-QA than NO-Q.

When Bt was absent, some fitness parameters (survival, egg hatch, fecundity, and mating success) were lower for resistant insects than for susceptible insects. The finite rate of increase per individual was 0.59 for the resistant NO-QA strain relative to NO-Q (=1) (*27*). To determine if this fitness cost alone was sufficient to cause the observed declines in resistance, we used computer simulations.

In the simulation model, we assumed that resistance was controlled by a diallelic locus. Although this assumption is probably an oversimplification (*40*), it is a reasonable starting point. We assumed that the fitnesses were 0.59 for resistant homoygotes and 1 for heterozygotes and susceptible homozygotes. This approach is conservative because any cost of resistance in heterozygotes would accelerate declines in resistance. Survivors of a high concentration of Dipel were assumed to be resistant homozygotes. For each of three resistant laboratory strains, we compared the actual declines in the frequency of resistant survivors with those expected from the simulations (Figure 3). Reasonably good correspondence between observed and expected outcomes suggests that the measured fitness cost was sufficient to account for most, if not all, of the observed rates of decline of resistance in the laboratory.

Stable Resistance. If declines in resistance are due solely to reduced fitness in resistant insects relative to susceptible insects, resistance should be stable in lines that are initially 100% homozygous resistant. Such lines can be created by mating a single homozygous resistant male to a single homozygous resistant female. To test the hypothesis, we selected one strain (NO-Y) intensely with Dipel. From this strain, we reared six isofemale lines (each started with a single-pair mating) for many generations without exposure to Bt (*40*). The results suggest that one of the six isofemale lines was fixed (100% homozygous) for resistance. This line (NO-YA) had a resistance ratio of 5,800 after 22 generations of rearing without exposure to Bt. These results support the idea that a fitness cost was the primary cause of declines in resistance.

Just when we seemed to have neatly explained the instability of resistance, new evidence complicated the story. In conjunction with a study of resistance to the *aizawai* subspecies of Bt (Liu, Y. B. and Tabashnik, B. E.; unpublished data), we sampled the NO population in 1993 and tested the stability of resistance to Dipel in a newly established laboratory strain (NO-93). To our surprise, the resistance ratio of NO-93 did not decline from its initial value of 29 after 10 generations without exposure to Bt. The response to selection in a subcolony derived from NO-93 showed that resistance to

Dipel was not fixed in NO-93. These results suggest that continued field use of Bt between 1989 and 1993 selected for amelioration of fitness costs associated with resistance in the NO population. Such reduction of fitness costs might occur by substitution of alternate alleles for resistance at the original resistance locus or loci; or by selection for fitness modifiers at loci not directly related to resistance (*52*).

Conclusions and Implications. Field populations of diamondback moth have evolved resistance to Bt in response to repeated sprays. In some heterogeneous populations, fitness costs associated with resistance favor restoration of susceptibility when exposure to Bt stops. Stabilization of resistance can occur, despite fitness costs, in populations that are homogeneously resistant. Selection for competitive fitness among various resistance-conferring mutations or for fitness modifiers may stabilize resistance in genetically heterogeneous populations. An important lesson from this work is that the stability of resistance in the absence of exposure to a pesticide is not necessarily a fixed trait for a particular pesticide-pest combination. Indeed, stability of resistance to a particular pesticide can change through time for a single pest population.

Receptors for Bt toxins are found in many insects. Presumably, they are not there just so the insects can be killed by Bt. This train of thought shifted from vague speculation to specific hypotheses when two groups reported that the digestive enzyme, aminopeptidase N, is a receptor for the Bt toxin CryIA(c) in the moth, *Manduca sexta* (*53-55*). Binding of aminopeptidase N to CryIA(c) has also been observed in the tobacco budworm (*56*) and diamondback moth (Luo, K.; Tabashnik, B. E.; and Adang M. J.; unpublished data). Some mutations that reduce binding of CryIA(c) might also interfere with the normal functioning of aminopeptidase N. If so, this might explain the fitness costs associated with resistance to Bt in some strains of diamondback moth.

Aminopeptidase N is a glycosolated metalloprotease that removes amino-terminal residues to complete digestion of short peptides in the gut (*54, 57*). This enzyme has some regions that are highly conserved in humans and bacteria, as well as in insects (*54, 55*). In people, aminopeptidase N occurs on intestinal, lung, and kidney epithelial cells. It is a receptor for human coronavirus 229E, which is an important cause of upper respiratory tract infections (*57*). So, the next time you sneeze because you have a cold, think about it: the receptor for the virus that afflicts you may be a close relative of a receptor for Bt toxin in insects.

Acknowledgments

We thank Charles Chilcutt, Juan Ferré, Fred Gould and an anonymous reviewer for their thoughtful comments. This work was supported by USDA Grant HAW00947H, USDA/CSRS Special Grant in Tropical and Subtropical Agriculture 95-34135-7314, and USDA Western Regional IPM Grant 93-34103-8603. This is paper 4142 of the Hawaii Institute of Tropical Agriculture and Human Resources Journal Series, University of Hawaii, Honolulu.

Literature Cited

1. Georghiou, G. P.; Lagunes-Tejeda, A. *The Occurrence of Resistance to Pesticides in Arthropods*; Food and Agriculture Organization of the United Nations: Rome, 1991.
2. Bacillus thuringiensis, *An Environmental Biopesticide: Theory and Practice*; Entwistle, P.; Bailey, M. J.; Cory, J.; Higgs, S.; Eds., Wiley and Sons: New York, 1993.

3. Feitelson, J. S.; Payne, J.; Kim, L. *Bio/Technology* **1992,** *10,* 271-275.
4. Gould, F. *Trends Ecol. Evol.* **1988,** *3,* 515-518.
5. McGaughey, W. H. *Science* **1985,** *229,* 193-195.
6. Stone, T. B.; Sims, S. R.; Marrone, P. G. *J. Invert. Pathol.* **1989,** *53,* 228-234.
7. Tabashnik, B. E. *Annu. Rev. of Entomol.* **1994,** *39,* 47-79.
8. Talekar, N. S.; Shelton, A. M. *Annu. Rev. of Entomol.* **1993,** *38,* 275-301.
9. McGaughey, W. H.; Whalon, M. E. *Science* **1992,** *258,* 1451-1455.
10. Marrone, P. G.; MacIntosh, S. C. *Resistance to* Bacillus thuringiensis *and Resistance Management*; Wiley and Sons: New York, 1993; pp 272-282.
11. Ferré. J.; Esriche, B.; Bel, Y.; Van Rie, J. *FEMS Microbiol. Letters* **1995,** *132,* 1-7.
12. Tabashnik, B. E.; Cushing, N. L.; Finson, N.; Johnson, M. W. *J. Econ. Entomol.* **1990,** *83,* 1671-1676.
13. Tabashnik, B. E.; Finson, N.; Schwartz, J. M.; Caprio, M. A.; Johnson, M. W. In *Proceedings of the Second International Workshop on the Management of Diamondback Moth and other Crucifer Pests*; Talekar, N. S., Ed.; Tainan, Taiwan, 1992, pp. 175-183.
14. Abbott Laboratories. 1992. B.t. products manual. Abbott Laboratories, North Chicago, IL.
15. Liu, Y. B.; Tabashnik, B. E.; Johnson, M. W. *J. Econ. Entomol.* **1995,** *88,* 788-792.
16. Tabashnik, B. E.; Finson, N.; Chilcutt, C. F.; Cushing, N. L.; Johnson, M. W. *J. Econ. Entomol.* **1993,** *86,* 635-644.
17. Tabashnik, B. E.; Cushing, N. L.; Johnson, M. W. *J. Econ. Entomol.* **1987,** *80,* 1091-1099.
18. Shelton, A. M.; Robertson, J. L.; Tang, J. D.; Perez, C.; Eigenbrode, S. D.; Preisler, H. K.; Wilsey, W. T.; Cooley, R. J. *J. Econ. Entomol.* **1993,** *86,* 697-705.
19. Gill, S. S.; Cowles, E. A.; Pietrantonio, P. V. *Annu. Rev. of Entomol.* **1992,** *37,* 615-636.
20. Whalon, M. E.; Miller, D. L.; Hollingworth, R. M.; Grafius, E. J.; Miller, J. R. *J. Econ. Entomol.* **1993,** *86,* 226-233.
21. Johnson, D. E.; Brookhart, G. L.; Kramer, K. J.; Barnett, B. D.; McGaughey, W. H. *J. Invertebr. Pathol.* **1990,** *55,* 235-243.
22. Tabashnik, B. E.; Finson, N.; Johnson, M. W. *J. Econ. Entomol.* **1992,** *85,* 2082-2087.
23. Oppert, B.; Kramer, K. J.; Johnson, D. E.; MacIntosh, S. C.; McGaughey, W. H. *Biochem. Biophys. Res. Comm.* **1994,** *198,* 940-947.
24. Van Rie, J.; McGaughey, W. H.; Johnson, D. E.; Barnett B. D.; Van Mellaert, H. *Science* **1990,** *247,* 72-74.
25. Ferré, J.; Real, M. D.; Van Rie, J.; Jansens, S; Peferoen, M. *Proc. Nat'l. Acad. Sci. USA* **1991,** *88,* 5119-5123.
26. Bravo, A.; Hendrikx, K.; Jansens, S.; Peferoen, M. *J. Invertebr. Pathol.* **1992,** *60,* 237-246.
27. Tabashnik, B. E.; Finson, N.; Groeters, F. R.; Moar, W. J.; Johnson, M. W.; Luo, K.; Adang, M. J. *Proc. Nat'l. Acad. Sci. USA.* **1994,** *91,* 4120-4124.
28. Escriche, B.; Tabashnik, B. E.; Finson, N; Ferré, J. *Biochem. Biophys. Res. Comm.* **1995,** *212,* 388-395.
29. Masson, L.; Mazza, A.; Brousseau, R.; Tabashnik, B. E. *J. Biol. Chem.* **1995,** *270,* 11,887-11,896.
30. Wolfersberger, M. G. *Experientia* **1990,** *46,* 475-477.
31. MacIntosh, S. C.; Stone, T. B.; Jokerst, R. S.; Fuchs, R. L. *Proc. Nat'l. Acad. Sci. USA* **1991,** *88,* 8930-8933.

32. Gould, F.; Martinez-Ramirez, A.; Anderson, A.; Ferré, J.; Silva, F. J.; Moar, W. J. *Proc. Nat'l. Acad. Sci. USA* **1992,** *89,* 7986-7988.
33. Lee, M. K.; Rajamohan, F.; Gould, F.; Dean, D. H. *Appl. Environ. Microbiol.* **1995,** *61,* 3836-3842.
34. Masson, L.; Lu, Y. J.; Mazza, A.; Brousseau, R.; Adang M. J. *J. Biol. Chem.* **1995,** *270,* 20,309-20,315.
35. Ihara, H. E.; Kudora, E.; Wadana, A.; Himeno, M. *BioSci. Biotech. Biochem.* **1993,** *57,* 200-204.
36. Liang, Y.; Patel, S. S.; Dean, D. H. *J. Biol. Chem.* **1995,** *42,* 42,719-42,724.
37. Hama, H.; Suzuki, K.; Tanaka, H. *Appl. Entomol. Zool.* **1992,** *27,* 355-362.
38. Martinez-Ramirez, A. C.; Escriche, B.; Real, M. D.; Silva, F. J.; Ferré, J. *Pestic. Sci.* **1995,** *43,* 115-120.
39. Tabashnik, B. E.; Schwartz, J. M.; Finson, N.; Johnson, M. W. *J. Econ. Entomol.* **1992,** *85,* 1046-1055.
40. Tabashnik, B. E.; Finson, N.; Johnson, M. W.; Heckel, D. G. *J. Econ. Entomol.* **1995,** *88,* 219-224.
41. Heckel, D. G.; Gahan, L.; Tabashnik, B. E.; Johnson, M. W. *Annals Entomol. Soc. Amer.* **1995,** *88,* 531-537.
42. Tabashnik, B. E.; Finson, N.; Johnson, M. W.; Moar, W. J. *Appl. Environ. Microbiol.* **1993,** *59,* 1332-1335.
43. Tabashnik, B. E.; Finson, N.; Johnson, M. W.; Heckel, D. G. *Appl . Environ. Microbiol.* **1994,** *60,* 4627-4629.
44. Ballester, V.; Escriche, B.; Mensua, J.; Riethmacher, G. W.; Ferré, J. *Biocontrol Sci. Technol.* **1994,** *4,* 437-443.
45. Aronson, A. I.; Han, E. S.; McGaughey, W.; Johnson, D. *Appl. Environ. Microbiol.* **1991,** *57,* 981-986.
46. McGaughey, W. H.; Johnson, D. E. *J. Econ. Entomol.* **1994,** *87,* 535-540.
47. Drummond, J.; Kotze, A. C.; Levot, G. W.; Pinnock, D. E. *J, Econ. Entomol.* **1995,** *88,* 1607-1610.
48. Tabashnik, B. E.; Finson, N.; Johnson, M. W. *J. Econ. Entomol.* **1991,** *84,* 49-55.
49. Perez, C. J.; Shelton, A. M.; Derksen, R. C. *J. Econ. Entomol.* **1995,** *88,* 1113-1119.
50. Groeters, F. R.; Tabashnik, B. E.; Finson, N.; Johnson, M. W. *J. Econ. Entomol.* **1993,** *86,* 1035-1039.
51. Groeters, F. R.; Tabashnik, B. E.; Finson, N.; Johnson, M. W. *Evolution* **1994,** *48,* 197-201.
52. Lenski, R. E. *Evolution* **1988,** *42,* 425-432.
53. Sangadala, S.; Walters, F. S.; English, L. H.; Adang, M. J. *J. Biol. Chem.* **1994,** *269,* 10,088-10,092.
54. Knight, P. J.; Crickmore, K. N.; Ellar, D. J. *Mol. Microbiol.* **1994,** *11,* 429-436.
55. Knight, P. J.; Knowles, B. H.; Ellar, D. J. *J. Biol. Chem.* **1995,** *270,* 17,765-17,770.
56. Gill, S. S.; Cowles, E. A.; Francis, V. *J. Biol. Chem.* **1995,** *270,* 27,277-27,282.
57. Yeager, C. L.; Ashmun, R. A.; Williams, R. K.; Cardellichia, C. B.; Shapiro, L. H.; Look, A. T.; Holmes, K. V. *Nature* **1992,** *357,* 420-422.
58. Heckel, D. G. *Biocontrol Sci. Tech.* **1994,** *4,* 405-417.

Chapter 15

High Genetic Variability in *Drosophila melanogaster* for Susceptibility to Lufenuron, an Insecticide That Inhibits Chitin Synthesis

Thomas G. Wilson[1] and Jason R. Cryan[2]

[1]Department of Biology, Colorado State University, Fort Collins, CO 80523
[2]Department of Entomology, North Carolina State University, Raleigh, NC 27695

Chitin synthesis inhibitors are potent insecticides against certain insects, including *Drosophila melanogaster*. The benzoylphenyl urea compound lufenuron was tested against both laboratory and natural population strains of this insect. Laboratory strains were highly susceptible to this compound, while natural population strains from two widely separated locations in the U.S. exhibited considerable (as much as 100-fold) resistance. This insecticide has not seen significant field use—indeed, this entire class of insecticides represents only minor usage—and the resistance may be due to cross-resistance to a previously used insecticide. Since *D. melanogaster* is not under direct insecticide selection pressure, these results suggest that strong selection for insecticide resistance can occur in insects indirectly affected by an insecticide.

The insect growth regulator (IGR) class of insecticides includes chitin synthesis inhibitor compounds that act to interrupt insect development by interfering with chitin synthesis and/or deposition. Chitin is abundant in many invertebrates where it forms the structure of the exoskeleton. Since chitin is not found in appreciable amounts in vertebrates, the biosynthesis of this molecule during insect development presents a particularly inviting target for pesticide action that is not plagued with the disadvantage of vertebrate toxicity. In addition, insects undergo chitin synthesis at frequent intervals during preadult development, typically before each molt, thus presenting this target of opportunity repeatedly during the life cycle of an insect. As well as being larvicidal, these compounds can interfere with embryonic development within oviposited eggs, thus becoming potent ovicides when ingested by adult females of a variety of species (1). Therefore, while chitin synthesis inhibitor insecticides are not directly toxic to adult insects, they have considerable promise for controlling pest insect populations by acting as larvicidal or ovicidal agents or both.

The best known chitin-synthesis inhibitors are the benzoylphenyl urea compounds, of which diflubenzuron has been the most successful (2). Diflubenzuron affects a variety of insect pests, especially those in the orders Lepidoptera, Diptera, and Coleoptera; Grosscurt and colleague (1,3) have summarized the work describing the spectrum of insects affected by this compound. Chitin-synthesis inhibitors appear to act by blocking the polymerization stage in chitin biosynthesis (4), possibly by

0097–6156/96/0645–0141$15.00/0

blocking the transport of chitin precursor molecules (5). The exact mode of action is still uncertain, however, and identification of the molecular target would aid our understanding of the biology of chitin deposition as well as the future development of these insecticides.

Pest resistance to diflubenzuron has been documented for various insects (reviewed in 6), but in general the occurrence is low. Documented instances of resistance include field-selected resistance, laboratory-selected resistance, and cross-resistance to certain other insecticides (6). The mechanism of resistance is unknown, although in certain cases the involvement of an oxidase(s) was inferred from synergist studies (6). Cross resistance to diflubenzuron has been seen in several insects, including the house fly, *Musca domestica* (7,8).

A previous study (9) examined the effect of a putative chitin-synthesis inhibitor on *Drosophila melanogaster*, but resistance was not addressed. Although this insect is only a minor pest, *D. melanogaster* is an attractive model insect due to the enormous body of genetic information gathered on this insect (10). Using *D. melanogaster* for these studies, we are interested in examining the genetic and molecular basis of resistance to benzoylphenyl urea insecticides. The appropriateness of *D. melanogaster* as a model insect for insecticide resistance studies has been previously discussed (11,12).

D. melanogaster are Susceptible to Lufenuron

CIBA-GEIGY Corporation, Greensboro, NC, supplied two benzoylphenyl compounds, and each of these was tested for toxicity to Oregon-RC, a strain of *D. melanogaster* that has been maintained in the laboratory since 1938 (13). Oregon-RC was found to be susceptible to both of these compounds. However, the compound lufenuron(N-[2,5-dichloro-4-(1,1,2,3,3,3-hexafluoropropoxy)-phenylaminocarbonyl]-2,6-difluorobenzamide, formerly termed CGA-184699, possessed about 10-fold greater toxicity than did CGA-112913 (N-[4-chloro-5-trifluoromethyl-2-pyridinyl-oxy)-3,5-dichloro-phenyl-aminocarbonyl]-2,6-difluorobenzamide) toward this strain of *D. melanogaster*. All of the subsequent studies were therefore carried out with lufenuron. Although this compound has been recently marketed in the U.S., it has seen almost no field application (D. Allemann, pers. comm.).

Lufenuron was dissolved in 95% ethanol at 0.5 mg/mL and diluted in the same solvent to obtain the desired concentrations. Stock solutions were stable at 4 °C for at least several months. Since benzoylphenyl insecticides are most effective when ingested by insects (1), lufenuron was incorporated into Drosophila Instant Food (Carolina Biological Supply Co.) by mixing 0.10 mL of the desired concentration of insecticide dissolved in ethanol with 3.9 mL of distilled H_20 and adding 1.0 g of the dried Instant Food in a 22 X 95 mm plastic vial (Sarstedt Co.).

Females were allowed to oviposit on standard medium to obtain an egg collection. From these a total of 35 newly hatched larvae were gently picked from the food and transferred to lufenuron-bearing Instant Food. Control Instant Food had 0.10 mL of ethanol added. Larvae were raised at 25 +/- 1 °C on a 12h:12h L:D cycle. Survival was determined and expressed as adults surviving at least one day following eclosion.

Lufenuron was found to affect all preadult stages of development in *D. melanogaster* (Wilson and Cryan, unpublished). At higher doses this insecticide results in larval death and at lower doses results in pupal or early adult death. Females feeding on lufenuron-bearing Instant Food laid eggs that failed to hatch; upon examination, these eggs contained embryos that died late in embryonic development. These effects are typical of those seen in treatments of other insects with benzoylphenyl urea insecticides (3).

D. melanogaster Strains have High Genetic Variability for Lufenuron Toxicity

We were interested to determine the amount of genetic variability in different *D. melanogaster* strains for susceptibility to lufenuron. Since natural populations of *D. simulans* have been found to frequently possess elevated resistance to organophosphate (14) and *D. melanogaster* to methoprene (15) and dieldrin (16), it might appear that this species is affected by indirect selection pressure by pesticides to a greater extent than was hypothesized (11).

Several wild-type strains were obtained from the Bowling Green *Drosophila* Stock Center, Bowling Green, Ohio. These strains (Table I) were recovered from natural populations more than 50 years ago and have been maintained as inbred lines in the laboratory for many years (13).

In addition, flies were collected from natural populations by baiting with ripe fruit in Vermont during late summer of 1991. Lines were established from single fertilized females and were additionally inbred by a single-pair mating two generations later to minimize heterozygosity. Flies were tested for lufenuron susceptibility 4-6 generations after the second bottleneck.

Results from testing these strains are shown in Table I, and a full dose-response curve is shown in Figure 1 for WC2, the most resistant strain. Among lab strains there is a modest level of variability for lufenuron susceptibility, typically no more than about 5-fold. However, resistance rises dramatically in the natural population strains to a level as much as 100-fold that of the most susceptible lab strain.

Table I. Toxicity of Lufenuron to Various Laboratory and Natural Population Strains of *D. melanogaster*

Strain	Origin	LC_{50} (SD)[a]
Oregon-RC	Laboratory	0.026 (0.014)
Lausanne-S	Laboratory	0.171 (0.237)
Urbana-S	Laboratory	0.072 (0.080)
Swedish-C	Laboratory	0.028 (0.035)
Waterbury	Vermont	0.417 (0.104)
WC1	Vermont	1.37 (0.51)
WC2	Vermont	2.76 (0.71)
WC4	Vermont	2.01 (0.36)

[a] LC50 values are expressed in ppm lufenuron incorporated into Instant Food (Standard Deviation). Each value is derived from triplicate determinations for 6 concentrations of lufenuron.

One possibility for the high resistance seen in the Vermont populations is a local population effect, similar to the high OP resistance in local Florida populations of *D. simulans* that presumably results from increased usage of malathion for mosquito control (14). To determine if the resistance seen in the Vermont populations is more widespread, strains were derived from natural population *D. melanogaster* captured in Colorado during late summer of 1994 and subjected to the same isofemale selection as were the Vermont strains. These strains were tested with high diagnostic doses of lufenuron, and the results are shown in Table II for 12 strains.

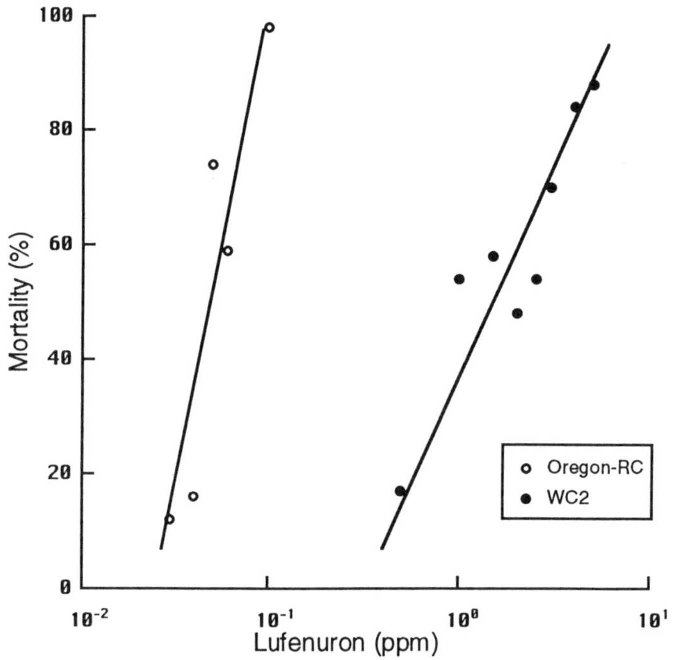

Figure 1. Dose response curve for toxicity of lufenuron to a lab strain and a natural population strain of *D. melanogaster*. Each point represents the mean of three separate determinations.

Table II. Survival upon Exposure to High Doses of Lufenuron in *D. melanogaster* Strains Isolated from Natural Populations in Colorado

Strain	Survival (%)[a] 0 ppm	1 ppm	2 ppm
Oregon-RC	92	0	0
WCC15	96	48	32
WCC16	60	44	20
WCC18	68	8	12
WCC19	56	32	24
WCC20	48	36	20
WCC21	64	16	16
WCC22	60	32	20
WCC23	68	4	0
WCC24	64	20	8
WCC26	52	24	0
WCC27	56	28	24
WCC28	80	56	24

[a] Each value is the average of two determinations

It is clear from examination of the Colorado strains that the resistance is widespread and is not a one-season phenomenon. Moreover, since nearly all of both the Vermont and Colorado strains showed much higher resistance than the lab strains, the allele frequency of the resistance gene(s) appears high. Note that many of the survival values of the WCC strains on 0 ppm lufenuron (control survival) are lower than one might expect. This lowered survival probably results from the inbreeding of each strain, but it is also possible that there is a fitness cost to the flies for maintenance of the resistance gene. It was not due to some toxic component of the Instant Food since reduced survival was also observed in several of the strains raised on a standard agar-molasses-cornmeal cooked medium.

Resistance to lufenuron in these strains appears to be stable. It is not a transient resistance due to resistance induction by an allelochemical compound in a food source in the wild, as has been found for swallowtail butterfly resistance to furanocoumarin-containing plants (17). We have maintained one of the Vermont strains, WC2, in the laboratory on standard laboratory food without insecticide selection since 1991, and we find no diminution of the resistance to lufenuron upon periodic retesting.

Cross-resistance of Insecticide-resistant *D. melanogaster* to Lufenuron

We are interested to determine the mechanism of this resistance found in flies recently derived from natural populations. Efforts are ongoing to map the resistance to determine if resistance corresponds to a known insecticide-resistance locus. A separate approach to elucidating a possible resistance mechanism is to examine other *D. melanogaster* strains of known insecticide resistance (e.g., an elevated level of one of more cytochrome P450 proteins) for cross resistance to lufenuron. If a *D. melanogaster* strain that is resistant to one or more insecticides by a known mechanism is also cross-resistant to lufenuron, then this result would suggest the involvement of that mechanism in resistance to lufenuron. While such a result would not prove that mechanism as the one acting in natural population strains, it would suggest that the mechanism may be important in resistance to lufenuron, at least in *D. melanogaster*.

Therefore, several insecticide-resistant *D. melanogaster* strains that have been characterized for resistance to a particular insecticide were examined for cross-resistance to lufenuron. We were particularly interested in strains resistant to other insect growth regulators such as methoprene and cyromazine. The results are shown in Table III.

Table III. Cross-resistance of Various *D. melanogaster* Strains to Lufenuron

Strain	Insecticide	Mechanism	Lufenuron LC_{50}[a]
91-R	DDT	Cytochrome P450	1.47 (0.43)
91-C	DDT nonselected strain[b]	NA	0.89 (0.26)
Hikone	several	Cytochrome P450	1.11 (0.18)
Met	methoprene	insensitive target protein	0.47 (0.15)
Cyr I	cyromazine	unknown	0.83 (0.23)
Cyr II	cyromazine	unknown	0.94 (0.30)

[a] Same footnote as Table I.
[b] This strain was taken from the same population as 91-R but has not been maintained under selection pressure with DDT (18).

Examination of these results shows that the highest levels of resistance were found in the two strains, Hikone and 91-R. Both of these strains have elevated levels of at least one cytochrome P_{450} (19,20), although the resistance of these strains has not been genetically mapped to the locus responsible for the elevated cytochrome P_{450}. Somewhat lower cross-resistance was found in the two strains resistant to cyromazine (21), but the *Met* strain (22) was not especially cross-resistant. Cross-resistance to diflubenzuron and cyromazine has been reported in *M. domestica* (23), but a population of *Culex pipiens pipiens* resistant to methoprene was not cross-resistant to diflubenzuron (24). Although the cross-resistance is not high, these results suggest that one or more cytochrome P450 enzymes may be involved in lufenuron resistance, which is hardly a surprising result considering their involvement in detoxification of other insecticides (25). The mechanism underlying the smaller level of cross-resistance to cyromazine remains to be determined.

Conclusions

There are two important results from this work. First, resistance to lufenuron (and possibly related compounds) appears already to be present in the field, at least in *D. melanogaster* populations. Since both populations show resistance, it is unlikely to be localized. In both sites examined, the allele frequency of the presumed resistance allele(s) appears to be high, judging from the high percentage of strains that bear resistance. Since lufenuron has not been used to any significant degree in the field, the simplest interpretation of the resistance is that it is due to resistance established to an older insecticide (perhaps a carbamate) that was widely used in the past and is perhaps still in use, and the resistance gene(s) has been maintained in these populations. This resistance may also be due to a recently acquired allelochemical in the diet of natural population flies.

It is also possible that the susceptibility difference seen between lab strains and natural population strains is due to hypersusceptibility of *D. melanogaster* strains that have been maintained in the laboratory for a long period of time. If true, we can presently draw two conclusions: (1) The hypersusceptibility does not extend to all insecticides. Oregon-R, a lab strain similar to Oregon-RC, was found to have a level of resistance to malathion similar to that of each of two natural population strains of *D. melanogaster* from malathion-unexposed areas in Florida (14). (2) The development of hypersusceptibility has a time requirement longer than several years. We retested in 1995 the Vermont strain WC2 that was collected in 1991 and found a resistance level almost unchanged after four years of culture in the laboratory in the absence of any insecticidal selection pressure. Since we have not found a highly susceptible *D. melanogaster* strain recently derived from a natural population, we have been limited to using laboratory strains as susceptible strains in this work. We plan future work examining world-wide strains of *D. melanogaster* that have been maintained in the laboratory for varying (1-50 years) periods of time. It may be possible to determine the temporal appearance of the lufenuron resistance in this manner as well as settle the question of lab strain hypersusceptibility.

Second, these results demonstrate that indirect selection pressure with a pesticide can result in significant resistance. The fact that *D. melanogaster* is seldom under direct insecticidal selection pressure was hypothesized to be a disadvantage of this insect as a model insect for insecticide resistance studies (11), but results from resistance to cyclodienes (16), malathion (26), and lufenuron in the present study demonstrates that high resistance can evolve. We must conclude that other insects may also be "innocent bystanders" of pesticide usage and may respond by evolving resistance. Although this result may be desirable for some economically important predatory insects, it also means that pesticide-susceptible alleles in insect populations will be diminished in both target insects and in nontarget insects.

Acknowledgments. We wish to thank Dr. Don Allemann of CIBA-GEIGY for supplying the insecticides. We thank Joe Todd for his assistance with some of the experiments and Charles Bartlett for helpful suggestions on the manuscript.

References

1. Grosscurt, A.C. *Pestic. Sci.* **1978**, *9*, 373-386.
2. *Chitin and Benzoylphenyl Ureas*, Wright, J.E.; Retnakaran, A., Eds.; Dr W. Junk Publishers, Dordrecht, Netherlands, 1987.
3. Grosscurt, A.C.; Jongsma, B. In *Chitin and Benzoylphenyl Ureas*, Wright, J.E.; Retnakaran, A., Eds.; Dr W. Junk Publishers, Dordrecht, Netherlands, 1987; Chapter 4, pp 75-99.
4. Hajjar, N.P.; Casida, J.E. *Science* **1978**, *200*, 1499-1500.
5. Nakagawa, Y.; Matsumura, F. *Insect Biochem. Molec. Biol.* **1994**, 1009-1015.
6. Grosscurt, A.C.; Stoker, A. *Med. Fac. Landbouww. Rijksuniv. Gent* **1991**, *56*, 1151-1159.
7. Cerf, D.C.; Georghiou, G.P. *J. Agric. Food Chem.* **1974**, *22*, 1145-1146.

8. Oppenoorth, F.J.; Van Der Pas, L.J.T. *Entomol. exp. & appl.* **1977**, *21*, 217-228.
9. Mitchell, M.J.; Smith, S.L. *Experientia* **1988**, *44*, 990-991.
10. Lindsley, D.L.; Zimm, G.G. *The Genome of Drosophila melanogaster*, Academic Press, New York, 1992.
11. Wilson, T.G. *J. Econ. Entomol.* **1988**, *81*, 22-27.
12. ffrench-Constant, R.E.; Roush, R.T.; Carino, F.A. In *Molecular Approaches to Pure and Applied Biology*; Springer Series in Experimental Entomology, Springer Verlag, Berlin, 1992; pp. 1-37.
13. Lindsley, D.L.; Grell, E.H. *Genetic Variations of Drosophila melanogaster.* Canegie Institution Washington Publ. 627, Carnegie Instit., Washington, DC, 1968.
14. Windelspecht, M.; Richmond, R.C.; Cochrane, B.J. *J. Econ. Entomol.* **1995**, *88*, 1138-1143.
15. Wilson,T.G.; Thurston, J. *J. Insect Physiol.* **1988**, *34*, 305-308.
16. ffrench-Constant, R.E.; Roush, R.T.; Mortlock, D.; Dively, G.P. *J. Econ. Entomol.* **1990**, *83*, 1733-1737.
17. Berenbaum, M.R.; Cohen, M.B.; Schuler, M.A. In *Molecular Mechanisms of Insecticide Resistance: Diversity among Insects*, Mullin, C.A.; Scott, J.G. Eds.; ACS Symp. Series 505, Washington, DC, 1992; Chapter 9, pp 114-124.
18. Merrell, D.J.; Underhill, J.C. *J. Econ. Entomol.* **1956**, *49*, 300-306.
19. Hallstrom, I.; Blanck, A.; Atuma, S. *Biochem. Pharm.* **1984**, *33*, 13-20.
20. Waters, L.C.; Zelhof, A.C.; Shaw, B.J.; Ch'ang, L-Y. *Proc. Natl. Acad. Sci. USA* **1992**, *89*, 4855-4859.
21. Adcock, G.J.; Batterham, P.; Kelly, L.E.; McKenzie, J.A. *J. Econ. Entomol.* **1993**, *86*, 1001-1008.
22. Wilson, T.G.; Fabian, J. *Dev. Biol.* **1986**, *118*, 190-201.
23. Shen, J.; Plapp, F.W., Jr. *J. Econ. Entomol.* **1990**, *83*, 1689-1697.
24, Brown, T.M.; DeVries, D.H.; Brown, A.W.A. *J. Econ. Entomol.* **1978**, *71*, 223-229.
25. Hodgson, E.; Rose, R.L.; Goh, D.K.S.; Rock, G.C.; Roe, R.M. *Biochem. Soc. Trans.* **1993**, *21*, 1060-1065.
26. Pralavorio, M.; Fournier, D. *Biochem. Genet.* **1992**, *30*, 77-83.

Chapter 16

Surveillance of Resistant Acetylcholinesterase in *Heliothis virescens*

Thomas M. Brown[1], Patricia K. Bryson[1], Fran Arnette[1], Mitchell Roof[1], James L. B. Mallett[2], Jerry B. Graves[3], and Stanley J. Nemec[4]

[1]**Department of Entomology, Clemson University, Clemson, SC 29634–0365**
[2]**Department of Genetics and Biometry, University College London, London NW1 2HE, England**
[3]**Department of Entomology, Louisiana State University, Baton Rouge, LA 70803**
[4]**Nemec Agriservices, P.O. Box 467, Snook, TX 77878**

AceIn, a gene controlling acetylcholinesterase inhibition and dominantly-inherited resistance to methyl parathion, was monitored in *Heliothis virescens* across the southeastern USA. The resistant allele, *AceIn-R,* was detected in 13 collections from 1989 to 1993 with an average frequency of 14.1±3.9%. This resistance mechanism was found in pheromone-trapped moths from Texas, Mississippi, Georgia and South Carolina. Genotype frequencies agreed with Hardy-Weinberg predictions. This resistance allele occurs at a low frequency in apparent equilibrium despite the widespread substitution of pyrethroid insecticides for methyl parathion. Laboratory colonies collected from South Carolina in previous years when there was more extensive use of methyl parathion exhibited higher frequencies. Results are discussed regarding the evolution of resistance and in relation to resistance management strategies.

Insecticide resistance evolves in populations under selection by insecticides as documented for hundreds of species (*1,2*); however, the role of specific genes in the development of resistance in the field has been studied in only a few cases (*3*). Examples of monitoring for specific genes using enzyme or nucleic acid analyses in microtiter plate readers has demonstrated the correlation to resistance of carboxylester hydrolase activity in aphids (*4-6*) and mosquitoes (*7-9*) and the correlation to resistance of insensitive acetylcholinesterase in planthoppers (*10,11*), house flies (*12*), whiteflies (*13*) and mosquitoes (*14,15*).

In pests with several potential mechanisms for resistance, it will be most important to monitor susceptibility as well as several resistance genes in order to understand the population genetics of resistance development. The tobacco budworm, *Heliothis virescens* (L.) (Lepidoptera: Noctuidae), possesses several mechanisms giving resistance to methyl parathion (*16-18*); therefore, in order to diagnose resistance in a population of this pest, it will be necessary to assay individuals for several potential resistance genes. One step in this direction is to

0097–6156/96/0645–0149$15.00/0

determine the genotype frequencies of current populations of this pest as a baseline for monitoring and as a prerequisite for observing genetic responses to future selection.

Because *AceIn* (Acetylcholinesterase Inhibition) behaves as a single gene in *H. virescens,* and individuals can be assigned one of three genotypes based on enzyme sensitivity to inhibitors (*19,20*), we have initiated monitoring of this gene in populations sampled from agricultural fields. We have also explored the application of this technique to lepidopterous pests of other crops including the diamondback moth, *Plutella xylostella* (L.) (Lepidoptera: Plutellidae). With increasing resistance to pyrethroid insecticides, there are few alternative insecticides for cotton. For this reason, it is important to understand the current state of resistance to methyl parathion and to develop strategies by which it could be reintroduced into cotton IPM in an efficacious manner.

Materials and Methods

H. virescens were collected as adult males from pheromone traps located near cotton fields, except when noted otherwise. Specimens were frozen and shipped on cold packs to Clemson University and then stored at -20°C until analysis. Burleson, TX 1989 was collected by John Goodenough from a cotton field at north Highway 50, 11.7km west of College Station, TX on 28 June; Tift County, GA 1989, was collected by Sammy D. Pair from the Gibbs farm on 2-3 July; Leflore County, MS was collected by J. L. B. M. on 22-23 July and Monroe County, MS was collected by J. L. B. M. on 26, 28-29 July; Snook, TX 1990 was collected by S. J. N. on 10 September; Florence, SC 1990 was collected by F. A. from the Baxley and Woodard farms; Florence, SC 1991, was collected by F. A. from the Rogers farm; Florence, SC 1992, was collected by F. A. from the McSwain, Lowder, Stokes, Hinson, Woodard, and McDaniel farms (10 of each analyzed); Snook, TX 1993 was collected by S. J. N.

Laboratory strain Woodrow, SC 1983 was collected as larvae from cotton plants and strain Florence-tobacco, SC 1987 was collected as larvae from tobacco plants in May 1987. Bossier Parish, LA 1991 (RRRS-DE) and Snook, TX 1991 were laboratory strains obtained from J. B. G. and tested as F_3 adults.

The homogenate of each adult head was assayed for acetylcholinesterase activity in three wells of a microtiter plate containing these treatments (a) control buffer with acetone carrier, (b) 680 μM propoxur, or (c) 680 μM monocrotophos. Complete details of the assay have been published (*20*). The increase in optical density (corrected for reagent blank) at 405 nm in 30 m in the treatment with inhibitor was divided by the increase in the control to obtain the proportion of inhibition. Genotypes were determined from scatterplots comparing the inhibitors (*19,20*). Individuals resistant to inhibition by propoxur were resistant also to methyl paraoxon, paraoxon, and fenitrooxon, while those resistant to monocrotophos were resistant also to 4-nitrophenyl di-2-thienylphosphinate (*20*). *Note: Monocrotophos and propoxur are toxic to mammals. Monocrotophos is mutagenic and its registration as an insecticide has been cancelled in the USA.*

Results

H. virescens collected in pheromone traps possessed acetylcholinesterase activities characteristic of genotypes SS, RR and RS in clusters on scatterplots as illustrated for four locations collected in 1989 (Figure 1), Snook, TX collections in 1990, 1991 and 1993 (Figure 2), and Florence, SC collections for 1990 through 1992 (Figure 3).

The *AceIn-R* allele was found in all 13 uncolonized samples with an average frequency of 14.1±3.9% (Table I). This resistance mechanism was found in moths captured in Texas, Mississippi, Georgia and South Carolina.The three field samples having the lowest proportion of *AceIn-R* alleles were from Texas.

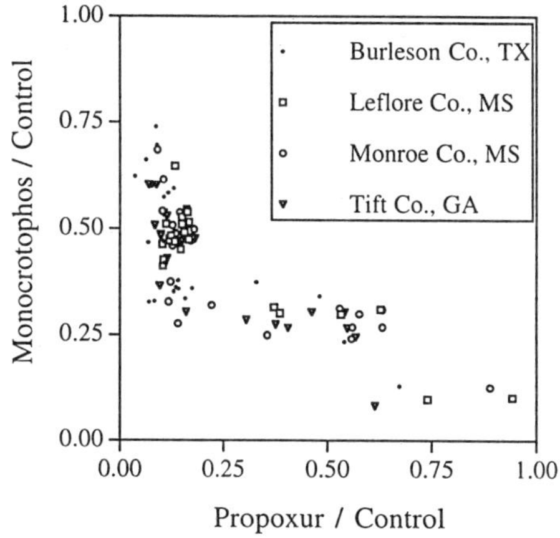

Figure 1. Determination of *AceIn* genotypes in individual *Heliothis virescens* adults captured in pheromone traps in 1989. The cluster to the left was scored *AceIn* genotype SS, the cluster to the lower right was scored RR, and the cluster in the middle was scored RS.

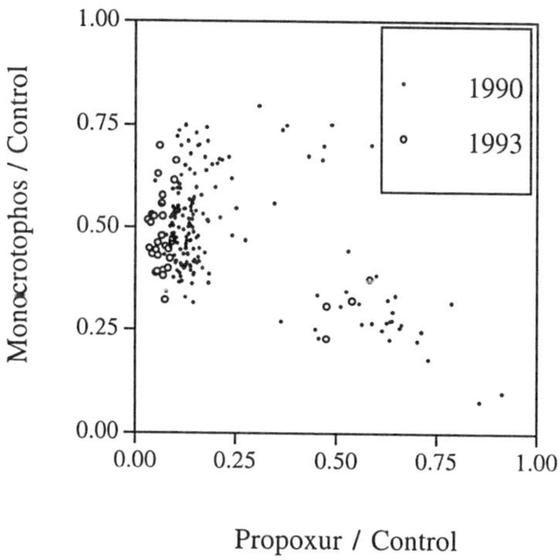

Figure 2. Determination of *AceIn* genotypes in individual *Heliothis virescens* adults captured in pheromone traps in Snook, TX. Clusters were scored as in Figure 1.

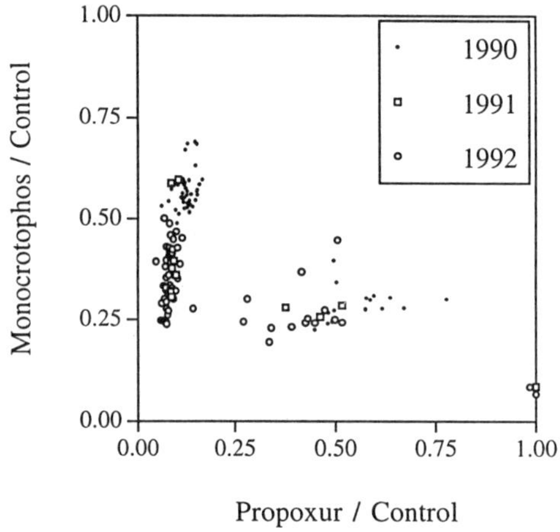

Figure 3. Determination of *AceIn* genotypes in individual *Heliothis virescens* adults captured in pheromone traps in Florence, SC. Clusters were scored as in Figure 1.

Table I. Frequencies of *AceIn* Alleles in *Heliothis virescens* Reared in Laboratory or Trapped in the Field

Collection Site	*Year*	*Generations in Laboratory*	*SS*	*RR*	*RS*	*%AceIn-R*
Woodrow, SC	1983	126	0	34	13	86.2
Florence, SC	1987	18	81	9	64	26.6
Tift County, GA	1989	na[a]	16	1	7	18.8
Leflore County, MS	1989	na[a]	18	2	4	16.7
Monroe County, MS	1989	na[a]	16	1	7	18.8
Burleson, TX	1989	na[a]	20	1	3	10.4
Florence, SC	1990	na[a]	43	0	15	12.9
Snook, TX	1990	na[a]	138	2	34	10.9
Florence, SC	1991	na[a]	10	1	3	17.9
Bossier Parish, LA	1991	3	32	0	0	0
Snook, TX	1991	3	21	0	0	0
Florence, SC	1992	na[a]	45	2	13	14.2
Florence, SC	1992	3	31	0	0	0
Snook, TX	1993	na[a]	28	0	4	6.2

[a] Not applicable; collected in pheromone traps and frozen until analyzed.

The *AceIn-R* allele was found in highest frequency, 86.2%, in the Woodrow, SC colony of *H. virescens* collected from cotton in 1983 (Table I). This colony was maintained in the laboratory without insecticide selection for one year prior to testing, but it retained high resistance. This is in contrast to three more recently founded laboratory colonies (Snook, TX 1991, Bossier Parish, LA 1991 and Florence, SC 1992), all analyzed as F_3 adults, which possessed no *AceIn-R* alleles.

The proportion of SS, RR and RS genotypes in pheromone-trapped and frozen samples did not differ significantly from Hardy-Weinberg predicted values (Table II). A total of 434 moths were analyzed from field trapping without laboratory colonization; of these 100 (23%) carried at least one *AceIn-R* allele.

Table II. Hardy-Weinberg Statistics for *AceIn* in *Heliothis virescens* Collected in Pheromone Traps

	Observed			*Predicted*[a]				
Year	*SS*	*RR*	*RS*	*SS*	*RR*	*RS*	χ^2	*%AceIn-R*
1989	70	5	21	67.5	2.50	26.0	3.54	16.2
1990	181	2	49	182.0	3.03	46.9	0.44	11.4
1991	10	1	3	9.45	0.45	4.11	1.02	17.9
1992	45	2	13	44.2	1.20	14.6	0.71	14.2
1993	28	0	4	28.1	0.12	3.75	0.14	6.2

[a] Predicted values calculated according to Hartl (*21*)

Discussion

It appears that *AceIn-R* was selected to high frequency coincidently with the widespread rise in methyl parathion resistance from 1963 to 1983, declined rapidly from 1983 to 1989, and then reached an equilibrium (Table I). These observations are consistent with the historical use of methyl parathion for cotton insect control. Methyl parathion use was initiated in the 1950's and resistant evolved between 1967 and 1972 as reviewed (*22*). After the introduction of pyrethroids in 1978, methyl parathion use declined; however, it continued to be used for several years in South Carolina in a mixture with EPN which was synergistic against resistant *H. virescens* (*23*) and which was more effective than pyrethroids for control of the boll weevil. The initiation of the boll weevil eradication program in the middle 1980's resulted in nearly total replacement of methyl parathion with pyrethroids for control of *H. virescens*.

Evolution of Resistance. It is likely that *AceIn-R* was extremely rare during the introduction of methyl parathion, that it arose via mutation and was selected to a high frequency during the observed increase in resistance. Although there are few data prior to 1983 during the evolution of resistance, most models assume that alleles for resistance are present initially at a very low frequency, e. g. 10^{-6} (Gressel *et al.,* this volume). Resistant adults from College Station, TX in 1981 possessed acetylcholinesterase activity which was 1.5-fold resistant to sumioxon (oxon of fenitrothion) (*24*); however, homogenates of *AceIn-R* homozygotes exhibited 10.4-fold resistance to this inhibitor (*20*). This may indicate that in 1981 the College Station population had a low frequency of *AceIn-R*, or alternatively, that *AceIn-R* was preceeded by a less potent allele. This mechanism appeared to be completely lacking in the strain NC-86 from North Carolina which was resistant to methyl parathion due to two mechanisms affecting biotransformation of the insecticide (*16*). Unfortunately, we are not aware of any earlier data on the frequency of this mechanism in individuals of this species.

Decline of *AceIn-R*. Resistance by this mechanism is completely dominant in inheritance; i.e. heterozygous individuals are fully resistant (*25*). This means that susceptible alleles can be carried by heterozygous survivors and that it would be unlikely for *AceIn-R* to reach fixation. This mode of inheritance would also slow the rate of contamination of a reservoir population which was not directly selected.

The higher frequencies of *AceIn-R* observed in laboratory strains collected in 1983 and 1987 and the lower values found thereafter (Table I) suggest a decline in this allele which was correlated to the pattern of use of methyl parathion. Rapid loss of the allele from 1983 to 1989 would require that *AceIn-R* (a) was diluted by immigrants with a higher frequency of *AceIn-S,* (b) conferred a fitness deficit, or (c) was selected against by another insecticide favoring *AceIn-S.* Because evidence for (b) or (c) is equivocal as will be discussed below, it is assumed that the high frequencies observed were characteristic of once highly selected local populations subsequently diluted by the more susceptible population at large to produce a state of equilibrium.

Fitness. Three samples from different states all lacked the *AceIn-R* allele when tested after rearing through 3 generations in the laboratory (Table I). This may indicate that this allele is deleterious in some way to insects upon colonization in the laboratory. On the other hand, our Woodrow, SC 1983 colony possessed 86.2% *AceIn-R* allele after prolonged colonization. There was no decline in this allele when mixed with the *AceIn-S* and reared for seven generations (*25*). Also, there is no deficit in the biochemical activity of the resistant acetylcholinesterase (*20*). The widepread presence of *AceIn-R* ten years after the almost complete withdrawl of methyl parathion is evidence that there is no fitness deficit in the field.

Negative Selection. *AceIn-R* confers resistance to inhibition by methyl parathion, fenitrothion, parathion, propoxur and eserine; however, there is concominant hypersusceptibility to inhibition by monocrotophos, dicrotophos, 4-nitrophenyl di-2-thienylphosphinate, and the *N*-propyl analog of carbaryl (*20*). Some insecticides, such as methomyl are approximately equipotent to the S and R forms. While this appears to be the case for the thiopropyl organophosphorus insecticides, profenofos and sulprofos, they are very poor inhibitors *in vitro* and may require bioactivation; therefore, it is unclear whether or not methomyl, thiodicarb, profenofos and sulprofos, the current anticholinesterase insecticides for *H. virescens* control in cotton, have selected against *AceIn-R*. The efficacy of profenofos and sulprofos against methyl parathion resistant populations suggests that these insecticides were unaffected by *AceIn-R*. Registration of monocrotophos for cotton in the USA was cancelled in 1987.

Equilibrium. Observed frequencies from 1989 to 1993 indicate that *AceIn-R* resides at equilibrium (Table II). This hypothesis assumes that there is now no significant selection for or against this allele and that it is not strongly deleterious so that it is maintained in the population at a level of about 14%. At this equilibrium frequency, selection by methyl parathion would be likely to result in very rapid resistance and failure of the insecticide.

Number of Alleles. Scatterplots indicate the presence of *AceIn-R* homozygotes resistant to propoxur (or methyl paraoxon) and also susceptible to monocrotophos (or 4-nitrophenyl di-2-thienylphosphinate). There was no allele giving resistance to both propoxur and monocrotophos in the same individual among 10 RR individuals from the field and 45 RR individuals in laboratory strains. From these observations, there appears to be only one *AceIn-R* allele in *H. virescens.*

Scatterplots indicate that there may be more than one *AceIn-S* allele because there appears to be a tight cluster at about 50% inhibition with additional clustering above and below this grouping (Figure 1). The three clusters could be explained by two *AceIn-S* alleles heterozygous in individuals forming the larger central cluster and homozygous in those forming upper and lower clusters. While there may be two or three *AceIn-S* alleles, they are all equally susceptible to propoxur, but differ in the level of resistance to monocrotophos. In our laboratory strain, there was less variability of response and no clear indication of multiple *AceIn-S* alleles (*25*).

Antiresistance Strategy. The lack of a double-resistant type of allele in *H. virescens* could present an opportunity to select against methyl parathion resistant homozygotes with an appropriate negatively correlated insecticide; however, the only registered compound in the known antiresistant group is dicrotophos. The possibility that heterozygotes will be resistant to both methyl parathion and the antiresistant compound might require that a mixture be applied to kill the heterozygotes. Future application of methyl parathion combined with surveillance of *AceIn-R* would provide an opportunity to observe the affect of selection on a specific gene. Several additional mechanisms are known to confer resistance to organophosphorus and carbamate insecticides; therefore, other genes would have to be monitored concurrently in order to fully understand the process.

Other Pest Species. When this assay was applied to collections of the diamondback moth from Japan, several resistant phenotypes were observed including an activity resistant to both propoxur and 4-nitrophenyl di-2-thienylphosphinate (Miyata and Brown, unpublished results). Codling moths, *Cydia pomonella*, with resistance to azinphosmethyl possessed acetylcholinesterase which was inhibited by propoxur (Dunley and Brown, unpublished results). This indicated the absence of an homologous allele to *AceIn-R*, but it did not eliminate the possibility of another type of allele giving resistance to inhibition by the oxon product of azinphosmethyl (which was not tested).

In whitefly, there are two alleles for resistance and parathion; one found in Sudan being sensitive to azamethiphos and the second found in Nicaragua being resistant to both parathion and azamethiphos (*13*). The allele for parathion susceptibility also gives susceptibility to azamethiphos.

Molecular genetics. We have amplified a portion of the *H. virescens* acetylcholinesterase gene by polymerase chain reaction with the objective of identifying the mutations involved (Evans and Brown, unpublished results). The current state of molecular genetics of insect acetylcholinesterase has been reviewed (Fournier, this volume) and it appears that several mutations may accumulate in the structural gene to produce increasingly resistant forms. Our biochemical surveillance suggests the hypothesis that there is one *AceIn-R* allele as characterized by its responses to inhibitors. This allele may have evolved once and spread through several states or it may have evolved several times in different locations. Analysis of the sequences of this gene from several locations is needed to test this hypothesis.

Conclusions

We have developed a simple method for surveillance of the allele conferring a resistant acetylcholinesterase in methyl parathion-resistant*H. virescens.* The allele, *AceIn-R,* can be monitored to determine the selection exerted by future insecticide applications. The assay, when combined with diagnostic tests for other major genes for resistance, will be useful for understanding the population dynamics of resistance in a major agricultural pest. This method may be adaptable to the diamondback moth and other pestiferous lepidopterans.

It might have been expected that this resistant allele would have disappeared during the last decade in the nearly total absence of methyl parathion. Clearly our results are a demonstration that resistance may linger in the population for years and that the return to older chemistry might be impractical without active management against specific resistance genes.

Acknowledgments

This is Technical Contribution No. 4181 of the South Carolina Agricultural Experiment Station. This research was supported in part by Agreement 92-754SC from Cotton Incorporated.

Literature Cited

1. Brattsten, L. B.; Holyoke, C. W.; Leeper, J. R.; Raffa, K. *Science* **1986,** *231*, 1255-1260.
2. Georghiou, G. P. In *Pesticide Resistance*; Glass, E. H., Ed.; National Academy Press: Washington, D. C., 1986; pp 14-43.
3. Brown, T. M.; Brogdon, W. G. *Ann. Rev. Entomol.* **1987,** *32*, 145-62.
4. Devonshire, A. L.; Moores, G. D. *Pestic. Biochem. Physiol.* **1982,** *18*, 235-246.
5. ffrench-Constant, R. H.; Devonshire, A. L.; White, R. P. *Pestic. Biochem. Physiol.* **1988,** *30*, 1-10.
6. Field, L. M.; Devonshire, A. L.; ffrench-constant, R. H.; Forde, B. G. *Pestic. Biochem. Physiol.* **1989,** *34*, 174-178.
7. Mouches, C.; Pasteur, N.; J. B. Berge, *et al. Science* **1986,** *233*, 778-780.
8. Brogdon, W. G.; Hobbs, J. H.; Jean, Y. S.; Jacques, J. R.; Charles, L. B. *J. Amer. Mosquito Contr. Assoc.* **1988,** *4*, 152-158.
9. Raymond, M.; Callaghan, A.; Fort, P.; Pasteur, N. *Nature* **1991,** *350*, 151-53.

10. Hama, H. *Appl. Ent. Zool.* **1981**, *16*, 52-54.
11. Miyata, T. In *Pest Resistance to Pesticides*; G. P. Georghiou and T. Saito, Ed.; Plenum Press: New York, 1983.
12. Moores, G. D.; Devonshire, A. L.; Denholm, I. *Bull. Ent. Res.* **1988**, *78*, 537-544.
13. Byrne, F. J.; Devonshire, A. L. *Pestic. Biochem. Physiol.* **1993**, *45*, 34-42.
14. Brogdon, W. G. *Bull. World Health Org.* **1988**, *66*, 339-346.
15. ffrench-Constant, R. H.; Bonning, B. C. **1989**, *3*, 9-16.
16. Konno, T.; Hodgson, E.; Dauterman, W. C. *Pestic. Biochem. Physiol.* **1989**, *33*, 189-99.
17. Konno, T.; Kasai, Y.; Rose, R. L.; Hodgson, E.; Dauterman, W. C. *Pestic. Biochem. Physiol.* **1990**, *36*, 1-13.
18. Brown, T. M. In *Managing Resistance to Agrochemicals*; M. B. Green, H. M. LeBaron and W. K. Moberg, Ed.; American Chemical Society: Washington, D. C., 1990; Vol. 421; pp 61-76.
19. Brown, T. M. *Rev. Pestic. Toxicol.* **1991**, *1*, 185-195.
20. Brown, T. M.; Bryson, P. K. *Pestic. Biochem. Physiol.* **1992**, *44*, 155-164.
21. Hartl, D. L.; Clark, A. G. *Principles of Population Genetics*; Sinauer Associates: Sunderland, Massachusetts, 1989.
22. Sparks, T. C. *Bull. Entomol. Soc. Am.* **1981**, *27*, 186-92.
23. Payne, G. T.; Brown, T. M. *J. Econ. Entomol.* **1984**, *77*, 294-97.
24. Bull, D. L. *Bull. Ent. Soc. Am.* **1981**, *27*, 193-197.
25. Gilbert, R. D.; Bryson, P. K.; Brown, T. M. *Biochem. Genet.* **1996** (in press).

Resistance Management Strategies

Chapter 17

The Need for Adaptation to Change in Insecticide Resistance Management Strategies: The Australian Experience

Neil W. Forrester and Lisa J. Bird

New South Wales Agriculture, Australian Cotton Research Institute, Narrabri, New South Wales 2390, Australia

The Australian IRM strategy was first implemented in the 1983/84 season and was based on the rotation of unrelated chemical groups on a generation basis, alongwith with a reliance on ovicide/larvicide mixtures. The pyrethroid window, targeted in the middle of the season, was set at 42 days (one generation in the field). Problems encountered which will be discussed include: cultivation of overwintering pupae; the impact of adult selection; the loss of control options due to environmental considerations and the need to target sprays on smaller larvae. Research on pyrethroid resistance mechanisms indicated the possibility of using monooxygenase synergists such as piperonyl butoxide. However, within a few seasons, there were clear indications of a declining effectiveness of piperonyl butoxide due to the development of alternative resistance mechanisms. Research on new synergists was commenced and alternative larvicide mixing partners were sought. The most successful of these has proven to be Bt which is now used extensively in mixtures with both endosulfan and pyrethroids. However, the main problem with this approach has been the inability to determine when resistant *Helicoverpa armigera* are present in significant enough numbers to warrant the more expensive mixture approach. This problem has now been solved with the recent commercialization of a monoclonal antibody test. The basis of the strategy is now technological rather than biological as previously and relies on the precise targeting of appropriate mixtures. The strategy now encompasses acaricides as well as insecticides and higher early season *Helicoverpa* spray thresholds to reduce overall selection pressure.

There is no one Insecticide Resistance Management (IRM) strategy applicable to all resistance problems. Each resistance situation has its own idiosyncrasies which require a unique solution and indeed this solution can often change markedly with time. The ability to adapt to a changing resistance situation, as indicated by an effective and sensitive resistance monitoring program, is a necessary requisite for a successful IRM strategy (*1*). The changing face of the Australian IRM strategy

0097–6156/96/0645–0160$15.00/0

over the past twelve years will be used as an example of an IRM strategy responding to a dynamic resistance situation.

Early History of the IRM Strategy for *Helicoverpa armigera*

Original Strategy with Pyrethroid Window. The Australian IRM strategy was first implemented in the 1983/84 season and was based on the rotation of unrelated chemical groups on a per generation basis, along with with a reliance on ovicide/larvicide mixtures under high pressure situations (*2*). The pyrethroid window was targeted in the middle of the season to preserve natural enemies as long as possible and to prevent premature flaring of secondary pests such as mites and whitefly. The duration of the window was set at 42 days, the time needed to complete one *Helicoverpa armigera* (Hübner) generation in the field. This strategy kept resistance frequencies in check for a number of years but by the 1988/89 season, resistance frequencies began to escalate markedly, particularly during the late Stage II / early Stage III period (*2*). It was realized that the spraying of the long residual pyrethroids in the last few days of Stage II on crops that were nearing 'cut out' (that is, at a time of reduced growth dilution), was resulting in double selection of overlapping generations. This necessitated a shortening of the pyrethroid window to 35 days in the following seasons (Figure 1) and this also had a positive delaying effect on pyrethroid resistance (*2*).

Additional Early Recommendations. Over this period, there were also a number of important problems encountered which resulted in additional recommendations to the key strategy guidelines.

"Pupae Busters" Campaign. It was recognized that highly resistant populations survived from the end of one season to the start of the next season primarily as overwintering diapausing pupae. The lack of cultivation at soil to kill diapausing pupae in the winter of 1986 because of depressed cotton prices (*2*), highlighted the important role of this life stage in the ecology of resistance in *H. armigera* and resulted in the adoption of the "pupae busters" campaign (*3*). This campaign promotes the timely and effective cultivation of cotton residues as soon as possible after picking, as a non-chemical control measure to reduce resistance selection pressure. This message has been accepted widely in the industry and its adoption rate has been quite good in most areas but not total (*4*).

Impact of Adult Selection and Repellency of Surviving Moths. As the damaging stage of *Helicoverpa* spp. is the larva, traditionally, little thought has been devoted to the impact of insecticides on the non-damaging adult stage. However, when insecticides are applied to a cotton crop, they act on all stages of the pest present in the crop and this includes the adult. It was found that the moth can express resistance to the synthetic pyrethroids but that it is still highly irritated by the pyrethroid deposit (*2*), resulting in the repellency of these resistant moths onto neighboring unsprayed crops or down lower into the untreated portion of the canopy. This helped explain the field observations of resistance failures in crops in which a pyrethroid had not yet been used (but where a neighboring property or properties had) and the claim that resistant moths preferred to lay on flowers (usually lower in the canopy and free of insecticide deposits, particularly if they had opened after spraying).

Need to Target Sprays on Smaller Larvae, Preferably Neonates. The functionally recessive nature of pyrethroid resistance in *H. armigera* (e.g., genetically resistant small larvae can still be killed by normal field rates of pyrethroids alone) has been known for about a decade (*5*) and has been exploited with the recommendation to spray pyrethroids at egg hatch. This has proved successful, particularly as resistance levels have increased and the mixture approach (discussed later) has become less effective and more expensive. However, under protracted and/or heavy pest pressure, this approach is less acceptable as it results in more frequent spraying, sometimes down to 5-6 day spray intervals.

Later Adaptations to the IRM Strategy

Despite the widescale adoption of all the practices discussed above in the earlier phases of the IRM program, resistance frequencies still continued to climb, albeit slowly (Figures 2 and 3). Various tactics have been investigated to reverse this trend and the IRM strategy has been altered recently because of increasing resistance and also because of other factors such as mite problems and environmental problems.

Alternative Insecticides, Synergists, and Mixtures of Insecticides. Research on pyrethroid resistance mechanisms indicated the possibility of using partial resistance breaking pyrethroids (e.g. bifenthrin) or monooxygenase synergists such as piperonyl butoxide (2) with the latter introduced commercially in the 1990/91 season. Once again, this was a successful delaying tactic but by the 1992/93 season there were clear indications of a declining effectiveness of piperonyl butoxide due to the development of alternative metabolic resistance mechanisms such as a monooxygenases suppressible by propynyl ethers but not by piperonyl butoxide, (N.W. Forrester, unpublished data). This mechanism was observed previously in *Heliothis virescens* in the USA. (*6*). Research on new synergists was commenced (e.g. propargite).

Alternative larvicide mixing partners were sought. The most successful of these has proven to be *Bacillus thuringiensis* (Bt) which is now used extensively in mixtures with both endosulfan and pyrethroids when resistant *H. armigera* are present (*7*). However, the main problem with this approach has been the inability to determine when resistant *H. armigera* are present in significant enough numbers to warrant the more expensive mixture approach. This problem has now been solved with the recent commercialization of a monoclonal antibody test, called Lepton, to determine the species of *Helicoverpa* present, either the fully susceptible *H. punctigera* or the multiresistant *H. armigera*. The more expensive mixtures can now be precisely targeted on resistant *H. armigera* populations resulting in lower costs and less spray failures (*8-10*). This new technology has resulted in a relaxation of the previously tightly controlled 35 day pyrethroid window which was extended firstly to 50 days in the 1993/94 (*11*) season and ultimately deregulated completely by the 1994/95 season (*12*).

The basis of the strategy for pyrethroids and endosulfan is now technological rather than biological as previously and relies on the precise targeting of appropriate mixtures (Figure 4). The concept of a window strategy based on rotation of unrelated chemical groups is still being used for those products where resistance levels are still relatively low (i.e. profenofos and thiodicarb) (*13*).

Incorporation of Acaricides into the Resistance Management Strategy. The strategy has also recently expanded to now take into account resistance management for acaricides against *Tetranychus urticae* (*12*), some of which interact with *Helicoverpa* control measures e.g., bifenthrin for direct control of both *Helicoverpa* and mites and propargite, an acaricide with synergistic properties for pyrethroids on *Helicoverpa*. Acaricide use in Australian cotton is increasing and the generally decreasing effectiveness of the organophosphates has put increasing selection pressure on the alternatives, in particular propargite. The

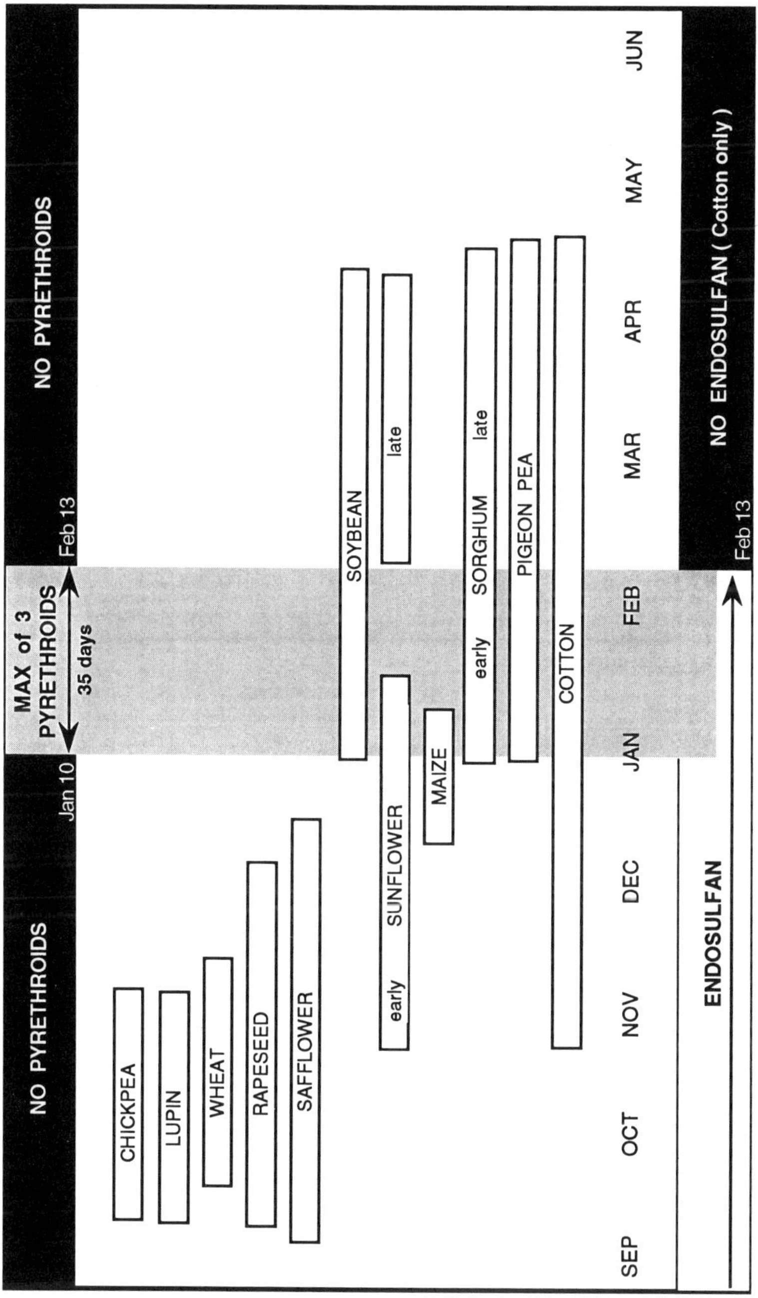

Figure 1. Summer crop Insecticide Resistance Management Strategy for eastern Australia for the 1990/91 season.

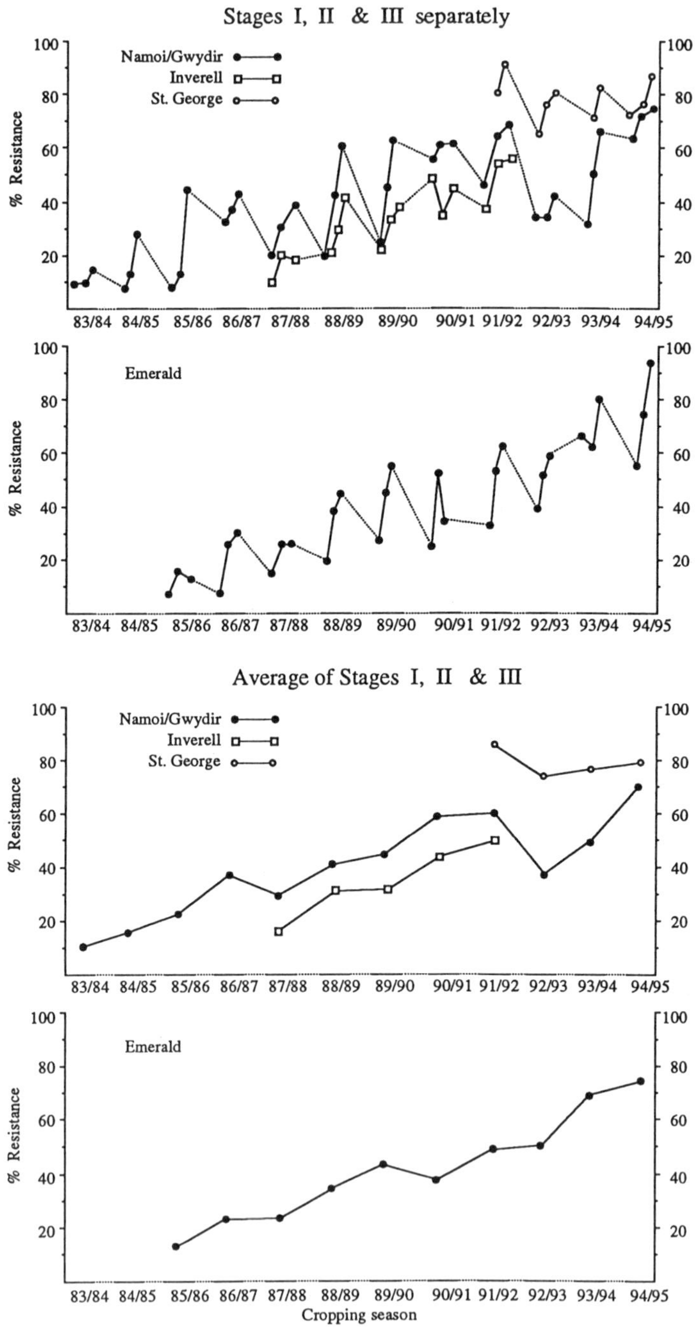

Figure 2. Pyrethroid resistance (% survival at the fenvalerate discriminating dose) in *Helicoverpa armigera* larvae reared from eggs collected from three cotton growing areas, Namoi/Gwydir, St. George and Emerald, and one unsprayed refuge area, Inverell. (*2*).

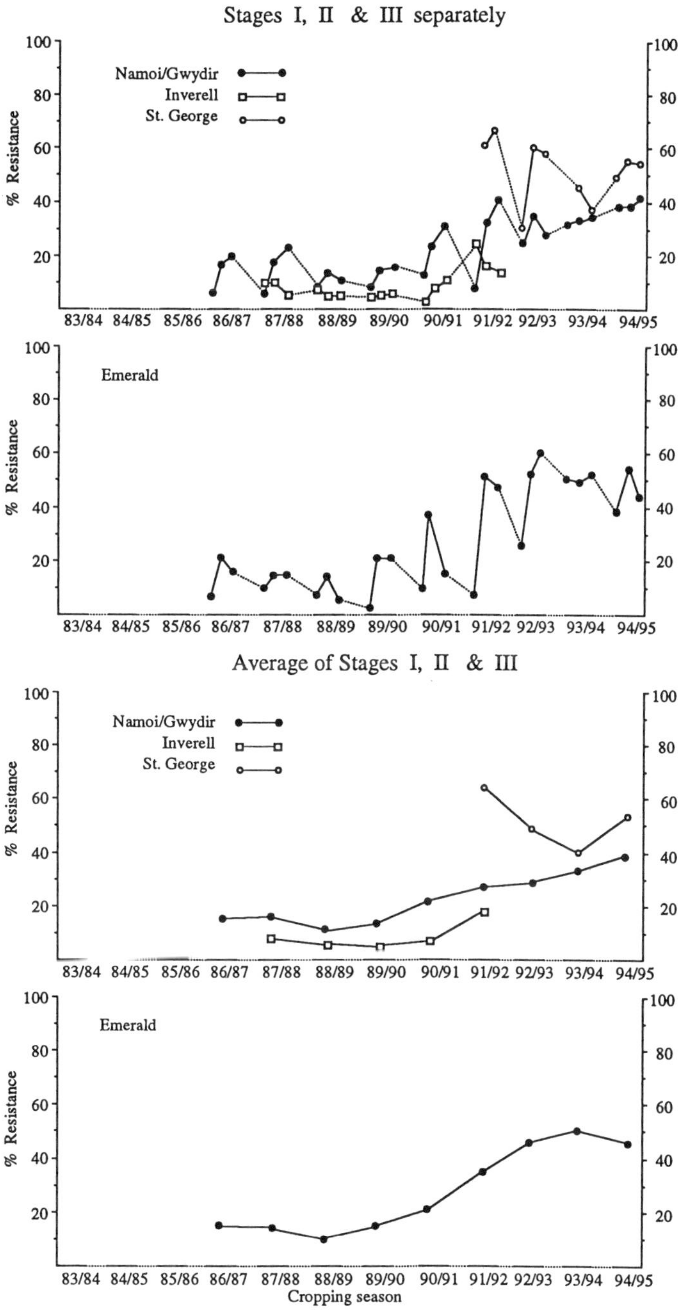

Figure 3. Endosulfan resistance (% survival at the discriminating dose) in *Helicoverpa armigera* larvae reared from eggs collected from three cotton growing areas, Namoi/Gwydir, St. George and Emerald, and one unsprayed refuge area, Inverell. (2).

general approach has been to ground spray the contact acaricides first (e.g. dicofol) and then to use propargite (with its vapour action) when aerial application is required. The pressure on propargite will be relieved somewhat with the registration of alternative acaricides such as abamectin, chlorfenapyr and diafenthiuron.

Expanded Scope of the IRM Program - Higher Early Season Thresholds. The increasing resistance problems over the past few years have resulted in increased use of insecticides (particularly as mixtures of larvicides) with a concomitant increase in the costs of insect control and hence decreased profitability, all classic signs of a deteriorating resistance situation. In an effort to decrease selection pressure, a major campaign was introduced in the 1995/96 season to reduce early season insecticide use by setting a pre-flowering early season threshold of 2 larvae per meter *(13-14)*, instead of the generally used threshold of 1 larva per metre. This was designed particularly to reduce selection pressure on endosulfan and thiodicarb and was generally well accepted, although many growers and consultants were concerned that the threshold was too high and that they would be unable to control the larger larvae. This particular strategy recommendation has been in place for the past season only so it is too early yet to determine its impact.

Impact of Environmental Problems. Environmental problems have also acted to constrain the choice of some products in the IRM strategy *(7)*. For example, the acute fish toxicity of endosulfan and profenofos, the mercaptan odour drift problem of profenofos and the residue problems of endosulfan and chlorfluazuron, have all acted to limit the use of these products (in fact chlorfluazuron was voluntarily withdrawn from use by the Australian cotton industry in the 1995/96 season) and hence put increasing selection pressure onto the few remaining alternatives.

Future

The strategy is also destined to change in the future with the introduction of new synthetic insecticides (e.g. chlorfenapyr) and acaricides (as above), new biological insecticides (e.g. spinosad and new Bt products and viruses, both conventional and genetically engineered), new pests, such as the recently introduced silverleaf whitefly, B type *Bemisia tabaci*, and transgenic cottons. Already, the resistance potential of this last mentioned technology has been recognized and a preventative IRM strategy based on refugia, planting windows and cultivation of overwintering pupae, has been devised and is in the process of implementation.

Summary and Conclusions

In summary, the obvious question that should be asked is : "Has the Australian IRM strategy been a success?" And undoubtedly the answer is yes, as it has clearly bought time (a valuable twelve years or so) to allow the introduction of new technologies such as novel insecticide groups and synergists, the monoclonal antibody based species test kit and most importantly, genetically engineered transgenic cottons. Without this breathing space, the Australian cotton industry would not have been able to survive and prosper during this difficult period and may have declined as in other countries afflicted with the same problem at the same time *(15)*. Also, the successful experience with this conventional IRM strategy has given the local industry confidence that it has the capacity and capability to design and implement an IRM strategy for the forthcoming environmentally desirable Bt transgenic cotton technology.

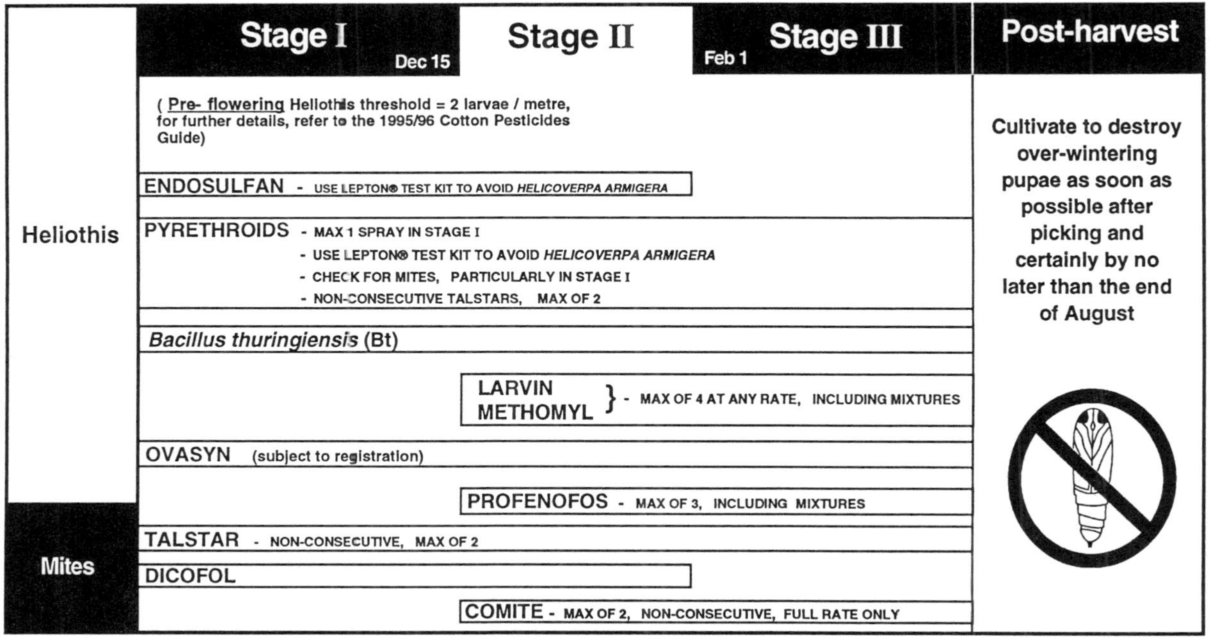

Stage II dates for central Queensland

- irrigated (start December 10th, finish Australia Day = 26th January)
- raingrown (start first flower, finish cut-out)

Figure 4. Cotton Insecticide Resistance Management Strategy for eastern Australia for the 1995/96 season. Ovasyn is amitraz, Talstar is bifenthrin, and Comite is propargite.

Literature Cited

1. Forrester, N.W. *Pestic.Sci.,* **1990**, *28*, 167-179.
2. Forrester, N.W.; Cahill, M.; Bird, L.J.; Layland, J.K. *Bull. Entomol. Res. Special Supplement*, **1993**, *1*, 1-132.
3. Pyke, B. *Aust. Cottongrower* (Australia), **1995**, *16*, 15-16.
4. Pyke, B. *Aust. Cottongrower* (Australia), **1996**, *17*, 28-30.
5. Daly, J.C.; Fisk, J.H.; Forrester, N.W. *J. Econ. Entomol.*, **1988**, *81*, 1000-1007.
6. Brown, T.M.; Bryson, P.K.; Payne, G.T. *Pestic.Sci.*, **1996**, *43*, 323-331.
7. Forrester, N.W. *Agric. Ecosys. and Environ.,* **1994**, *49*, 77-83.
8. Trowell, S.; Garsia, K.; Skerritt, J.; Hill, A.; Forrester, N.; Bird, L.J. In *7th Australian Cotton Conf.*; Australian Cotton Growers' Res. Assoc.: Wee Waa, New South Wales, Australia, 1994, pp 17-20.
9. Forrester, N.W. *Aust. Cottongrower* (Australia), **1993**, *14*, 12-15.
10. Forrester, N.W. *Aust. Cottongrower* (Australia), **1994**, *15*, 18-21.
11. Forrester, N.W.; Anderson, I.; Pyke, B. In *Cotton Pesticides Guide 1993-94*; Shaw, A.J., Ed.; NSW Agriculture: Orange, New South Wales, Australia, 1993, Vol. Agdex 151/680; pp 8-13.
12. Forrester, N.W.; Anderson, I.; Howie, B.; Fitt, G.; Glover, P.; Pyke, B.; Wilson, L. In *Cotton Pesticides Guide 1994-95*; Shaw, A.J., Ed.; NSW Agriculture: Orange, New South Wales, Australia, 1994, Vol. Agdex 151/680; pp 7-15.
13. Forrester, N.W.; Howie, B.; Pyke, B.; Wilson, L. In *Cotton Pesticides Guide 1995-96*; Shaw, A.J., Ed.; NSW Agriculture: Orange, New South Wales, Australia, 1995, Vol. Agdex 151/680; pp 8-17.
14. Gibb, D. In *Newsletter for the Research Extension Education Program*; Cooperative Research Centre for Sustainable Cotton Production: Narrabri, New South Wales, Australia, 1995, Vol. 2, pp 1-4.
15. Cox, P.G.; Forrester, N.W. *J. Econ. Entomol.*, **1992**, *85*, 1539-1550.

Chapter 18

Prevention Versus Remediation in Resistance Management

Jonathan Gressel[1], Shea N. Gardner[2], and Marc Mangel[2]

[1]Department of Plant Genetics, Weizmann Institute of Science, Rehovot, IL 76100, Israel
[2]Section of Evolution and Ecology, University of California, Davis, CA 95616

"After the fact" remedial strategies are often ineffective, especially where resistance is widespread and/or refuges are large. Good pesticides are too often lost. The "it won't happen here" view accounts for the rarity of area-wide management strategies. The successful national example of abolishing agricultural use of DDT in Sri Lanka in favor of its use only in mosquito control precluded resistance until now. Preventive strategies must be immediately cost-effective, as well as useful in delaying resistance, or they will not be implemented. The tendency to cut dose rates is increasing resistances due to multiple-cumulative events (polygenic, amplifications, or sequential mutations within a gene). We have modeled alternating low with intermediate dose rates to delay both major gene and multiple cumulative-resistances as part of IPM. Such novel strategies must be verified with economic and pest control data to convince farmers that they can work.

Burgeoning Resistance - An Ever Increasing Problem

Resistances to pesticides are becoming more widespread, and more resistances are to be expected as more farmers use pesticides. The appearance of pesticide-resistant populations is not a developmental process and has nothing to do with developmental biology, as some specialists imply when they erroneously discuss "development of resistance." Resistant populations do not mysteriously "develop" but evolve according to evolutionary processes. We can try to modulate the rate of evolution, preempting its appearance, or wait and try to selectively rid our crops of resistant pest populations

0097–6156/96/0645–0169$15.00/0

by remedial procedures after resistant populations evolve. This increases the need for both remedial strategies to overcome resistant populations that have already evolved, as well as more cost-effective preemptive strategies to prevent resistance from evolving. This is also termed "managing susceptibility".

The most pernicious cases will be of new resistances to herbicides and not to other pesticide groups for a variety of reasons:

(a) Herbicide usage (whether measured in expenditures, area covered, or weight of active ingredient) is increasing relative to other pesticides.

(b) Herbicide usage is increasing in lower value crops such as wheat and rice, especially in the developing countries. Resistances are following suit (*1-3*). There is a stronger likelihood to repeatedly use single compounds in these typically monoculture crops. Monoculture with one pesticide is a harbinger of resistance.

(c) Earlier single-target herbicides focused on a chloroplast genome coded target (*4*). Mutations to resistance were functionally recessive at field light intensities, as when there are susceptible plastids, they photogenerate large amounts of toxic oxygen radicals that cannot be detoxified. Individuals carrying large populations of resistant plastids are at a very low frequency. This resistance is inherited on a highly conserved gene, and most resistant individuals are rather unfit.

(d) In contrast to (c) there are many current, fashionable, low dose-rate, single-target herbicides where the resistance trait is rather fit, dominantly inherited, and at a high frequency (*5,6*). Thus, resistance is rapidly appearing to these herbicides, due to their large market share.

(e) Resistant populations are beginning to appear to herbicides that have been considered immune to such problems; the phenoxys and chloroacetamides, and other compounds in continuous heavy use for >25 years.

More cases of resistance are also expected to insecticides, as the number of targets of presently available compounds is small, and cross resistances are thus rampant. Still, insecticides are used in higher value crops where farmers have a greater stake in success and can afford a wider variety of pesticides in resistance management schemes. The fewest problems are expected with fungicides, as crop genetics and breeding are more often successful in overcoming pathogens than weeds or arthropods. Pathogens evolve resistance despite sophisticated breeding strategies, so breeders have the same problems as those engaged in rational pesticide use.

We can always expect new problems with pests that were never well controlled filling niches left vacant when successful pesticides eliminate the primary pests. We also have the pests that seem to have the ability to rapidly evolve resistance to every new pesticide developed for their control; e.g. the *Coccidiosis* pathogen of poultry (*7*), the Colorado potato beetle (*8*) and ryegrass species in wheat (*3*), and more recently *Echinochloa* spp in rice (*9*).

Reduced Pesticide Usage - Adding to Resistance Problems. There are increasing economic, regulatory, and consumer-induced pressures to decrease pesticide usage. This can be done in three ways (sometimes with more than one together), each with implications towards the evolution of resistance and its management:

Increased Abstinence. Growers can chose to apply pesticides less often, which can have a variety of effects. If the pesticide was not needed in the first place, as is too often the case, then the effect can clearly be positive as the farmer has met the economic and regulatory goals. The few resistant individual weeds may be successfully suppressed by the crop. All resistant pests will be subject to competition by more fecund susceptibles, if, and only if there is a substantial fitness differential between resistant and susceptible individuals. Abstinence also allows insect populations to be decimated by natural enemies that are suppressed when insecticides are used.

Abstinence can lead to problems when pest populations arise well above their normal levels. Evolution is a numbers game; if other parameters are equal, the more individuals present when a selector such as a pesticide is used, the more likely there will be resistant individuals present per unit area when pesticide treatments begin, assuming a constant mutation frequency. The more resistant individuals present from the start the more rapid the evolution of predominantly-resistant populations. The initial frequency of resistance is a less compelling parameter in pests with high dispersion rates such as insects. When abstinence is to be used, it should be with good scouting, so that pesticides are used when the pests are above thresholds, and abstinence instituted when below.

Using New Low Rate Pesticides. The use of highly potent, low rate pesticides strictly meets the letter of regulatory fiats. Many new pesticides with long-lived residues do not meet the raison d'être for these regulations. Most low dose-rate pesticides have a single target, and it is easier to evolve resistance when only a single target need be mutates, than when many different targets need be mutated. Long residual activity generally increases selection pressure, enhancing the rate of evolution of resistance. Low pesticide rates clearly do not mean low selection intensity for resistance. Conversely, there is a tendency for authorities to demand the deregistration of older, higher dose-rate, often multiple-target pesticides, reducing the flexibility of farmers to mix or rotate them with low dose rate pesticides in well wrought IPM packages. Regulators must be made aware that such deregistrations can be counterproductive; resistance problems will become more rampant, and we will then have to return to the older pesticides to successfully produce crops.

Cutting Pesticide Rates Leads to Further Problems. Recently there have been increasing numbers of cases of resistance evolving where lowered rates of pesticides were used. When high dose rates are used, resistant populations often appear suddenly to the farmer (Fig. 1A), although actually there was a smooth exponential build up of resistant individuals beginning from some low frequency (near the mutation frequency) to resistant populations. When resistant populations appear seemingly without warning, all resistant members of the population are resistant to high dose rates, as seen in Fig. 1B. When a constant lower dose is used, one can select for any one of a plethora of polygenes. As there are many available, the trait is at a much higher frequency than single major genes for resistance. [Polygenes are used loosely herein to cover all multi-cumulative events including accumulating polygenes, as well as multiple mutations within a gene that incrementally confer increasingly greater resistance, as well as gene amplification. A better, but far longer and convoluted descriptor would be "incrementally increasing resistance caused by selection of cumulative, multiple, sequential, mutational events"].

Field data demonstrating such creeping resistance are shown in Fig. 1C. Each dot in Fig. 1C describes the average of a *Lolium* population found in an Australian wheat field. An analysis of such data (Fig. 1D) shows how the average level of resistance in populations is incrementally increasing throughout the population as a function of repeated low dose applications. In North America, where a 3 fold higher dose of the same pesticide was used, target site resistance evolved in three major pest species (*6,12,13*), including *Lolium*, (*6*) the same genus evolving polygenic resistance in Australia.

The nature of polygenic inheritance is such that there are small increments of increase in resistance in such a population (Fig. 1C,D). Perhaps the appearance of a measurable proportion of individuals with the first increment of resistance is delayed (as in Fig. 1C) until the first polygene for resistance has been sufficiently enriched in the population. Formally, at this stage we have a "single gene" resistance, albeit to a very low level of pesticide. After the first increment of resistance appears, some individuals can withstand the evolutionary pressure of higher pesticide doses, enriching more gene doses. Initial models on evolution of quantitative resistance to pesticides were described, but not fully developed (*14*), and "the impact of quantitative trait locus studies on evolution has yet to be felt" (*15*). Presumably this means that while there is considerable circumstantial and epidemiological evidence for polygenic controls, the genetic proofs are rare. There is evidence for polygenic inherited incremental increases in resistance to some fungicides (cf.*16-19*), insecticides (by gene amplification) (cf.*20-21*), herbicides (*22*), and gene amplification resistance to anti-cancer drugs (cf. *23*).

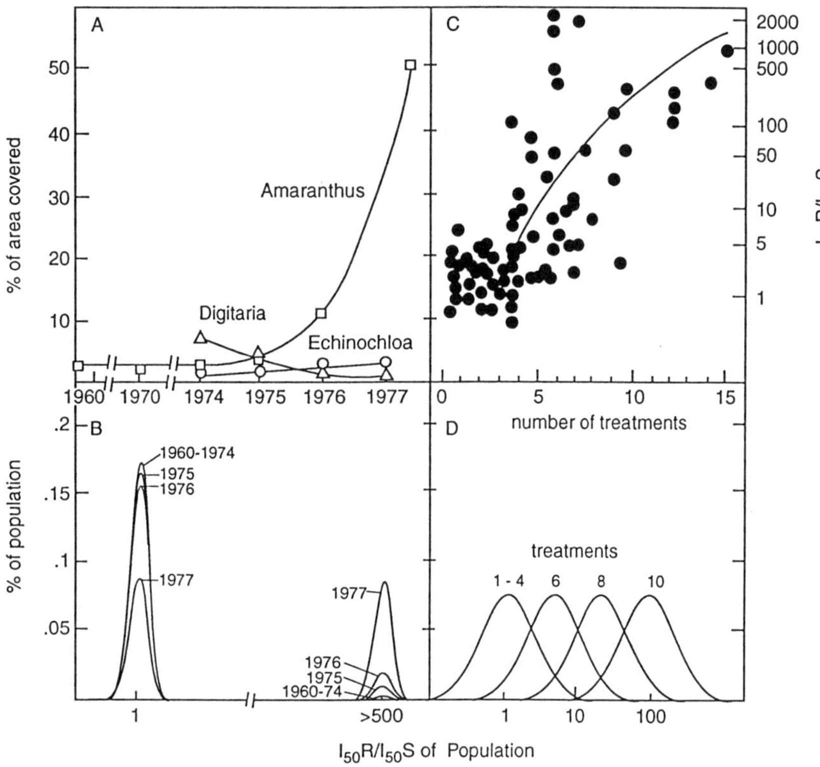

Figure. 1. "Sudden" appearance of major monogene resistance vs. slow incremental creep of polygenic resistance.

A. Actual field data on resistance showing changes in weed populations in a monoculture maize treated annually with atrazine. *Amaranthus retroflexus, Echinochloa crus-galli,* and *Digitaria sanguinalis,* the foremost weeds, were counted. The maize field was treated with atrazine from 1970 onwards. (Data are plotted from Table I in ref.*(10)*.

B. A population distribution description of the same data for *Amaranthus* in A, where the relative dose rates (R/S) on the horizontal axis are arbitrarally plotted.

C. Slow incremental increase in the dose level of resistance in repeatedly treated *Lolium* populations. The line showing how the dose required for control may increase was drawn for demonstration purposes only. *Lolium rigidum* was treated with a typical annual rate of 375g ha^{-1} diclofop-methyl. The relative dose level needed to control resistance in populations is shown as a function of the number of diclofop-methyl treatments. The sensitivity of determination of resistance was lost above a 500-fold increase in relative dose. The populations of seeds were collected in farmer-treated fields and tested by Ian Heap at the Waite Institute, Adelaide, Australia. Modified and redrawn from ref. *(11)*.

D. A population distribution description of the data in C where the dose rates on the horizontal axis are arbitrarally plotted.

In India, there is widespread resistance of *Phalaris minor* (canary grass) to the widely used herbicide isoproturon (over one half million hectares in three years since the first scattered discoveries). The first cases occurred 7-8 years after initial and then continual isoproturon applications to monoculture wheat(*24-25*). A field trip throughout the affected areas, with intensive interviews with farmers showed that resistance evolved first where farmers under-dosed the herbicide *(2)*. In a typical case the farmer initially used only half the recommended rate of isoproturon. This successfully controlled *Phalaris* for three years, but provided inadequate control in the fourth. He then used 0.75 the recommended rate successfully for two years, and unsuccessfully in the third. The full dose rate was then successful for one year but inadequate the next. Fifty percent above the recommended rate worked for a year, but no longer.

This strategy of continually increasing dose rates might be feasible for some insecticides used in high value crops, but is less feasible for fungicides and herbicides where there is far less margin between a utilizable rate and phytotoxicity to the crops. For less valuable crops, economics can also play a role in limiting the rates used for any pesticide. In North America only the manufacturer's recommended application rate of *ca* 1200 g/ha diclofop-methyl was used. In Australia such a rate was not economical in wheat cultivation and the rate of 375 g/ha was chosen because it gives adequate but hardly perfect control of *Lolium.* Thus, only a small increment of resistance was needed to change populations from susceptible to those with a modicum of resistance.

Australian conditions were conducive to rapid evolution, with increases in levels of a polygenic type resistance for the following reasons *(3):* (a) The pest seeds were often at very high initial population densities; *Lolium* is often used as a pasture species prior to planting wheat, leaving behind fields sown with orders of magnitude more *Lolium* seed than wheat. Resistance evolved first when wheat monoculture was begun after *Lolium* pasture, and not in continuous wheat/pasture rotations; (b) *Lolium* is self-incompatible, thus different individuals with different polygenes for resistance are likely to rapidly combine. This process would be slower in a self-compatible species where most plants are pollinated with their own pollen, delaying the combination of different resistant polygenes; (c) *Lolium* produces copious amounts of pollen, facilitating the easy transfer of resistance genes by wind pollination to adjacent, herbicide-treated populations.

At the high field dose rate used in North-America, the single major target-site gene for diclofop-methyl resistance codes for a modified acetyl CoA carboxylase. It is inherited as a semi-dominant trait *(6)*. The trait is functionally dominant at the rates of pesticide used in the field. A higher rate that would require homozygousity for resistance, would also kill wheat.

The Catch 22? It is necessary to understand why high dose rates preclude polygenic resistance and why low dose rates seem to delay major monogenic resistance. Such an understanding of the multiplicity of mechanisms conferring resistance is needed to design evolutionary compatible management strategies to delay or overcome the evolution of all types of resistance.

There has been much discussion by pest management specialists about which dose rates enhance the rate of evolution of resistance to pesticides, antibiotics and anti-cancer drugs. Simultaneously, theoretical geneticists have tried to deal with the first data emerging from use of these biocides, usually showing that resistance was inherited on single major genes. This is contrary to some evolutionary theory that presumed that most evolutionary change is polygenic *(26)*. This conundrum was 'solved' (*27*) with the following explanation: "Empirical data on natural and domesticated populations, and analysis of the models, suggest that strong selection sustained over several generations is usually required for adaptive evolution by a major gene mutation, to overcome deleterious pleiotropic effects generally associated with major mutations. This helps to explain why adaptive evolution by major mutations occurs much more frequently in domesticated and artificially disturbed populations than in natural ones." One might add that the repeated selection is need to eliminate more fit susceptible individuals. Actually, the explanation may be more prosaic; in nature stresses usually appear in a gradual manner, selecting for polygenic mutants, in a similar manner as low pesticide doses, and more rarely in acute manners that are similar to high pesticide levels.

We will try to explain the Catch 22 with *Lolium* as an example, as the same pesticide selected for both types of resistance in different use patterns. In the case of *Lolium* and diclofop-methyl, genetic analysis indicates that the large target-site gene can only be mutated at a few sites under the selection pressure of herbicides, giving various cross-resistances to the two different chemical groups attacking this single target*(6)*. We will assume that 10^{-6} organisms in a pristine population have target site resistance due to one mutation per 10^6 gametes per generation. Such mutations constantly occur in the absence of pesticides. .The presence of such a mutation in a major gene can give rise to target site resistance, and in a minor gene can give rise to an increment of polygenic resistance. The frequency of such mutations do not increase in the population in the absence of selector when there is even the slightest degree of unfitness. This keeps such mutations at a steady state frequency below the mutation frequency. Without pesticide there is no selection for such mutations, precluding their accumulation. On can use as evidence for this the time it took, and number of treatments, etc., to get diclofop-methyl resistance compared to resistance to other herbicides.

The seemingly polygenic resistance to diclofop-methyl (Fig. 1C, D) is probably due to modifications in cytochrome P450s or in their levels. Because of the ubiquity of cytochrome P450 genes in families, the possibility of other mechanisms contributing to resistance(*3,28*), and the variety of cross-resistance spectra (*11, 29*), we can guess that there are at least 20 different polygenes and up to hundreds of genes that can each contribute to resistance. It is assumed that each polygene can independently contribute increments of resistance, and for this analysis that each contributes an average increment of resistance to ca. 50 g/ha. Clearly it is important to have real data from areas where resistance has evolved to replace these assumptions, to allow delineating strategies to delay evolution elsewhere. The possible interdependence among some polygenes, as well as chromosomal linkage groups are presently ignored. Assume here that the frequency of each resistant form of each polygene is also 10^{-6}, but we will also assume that there are potentially 100 resistant polygenes. Thus, at any time $10^{-6}X10^2 = 10^{-4}$ of the pests could bear a single resistant polygene. The likelihood of any individual organism initially having two such resistant polygenes is $10^{-4}X10^{-4}=10^{-8}$, with three resistant polygenes 10^{-12}, etc. Thus, there is a 100 fold greater likelihood of finding any one resistant polygene in a previously untreated population than a resistant major monogene, but a much lower possibility of finding two or more resistant polygenes. This explains how low rates can select preferentially for polygenic-inherited resistance while high rates select only for major monogenes. This has also been found under laboratory simulations for all three major groups of pesticides (*18,30,31*). Recurrent selection at the same rate will continue to enrich for the same type of resistance in the population until resistance predominates.

Selection at a low dose rate could also select for target site resistant alleles - yet the frequency differences between 10^{-6} and 10^{-4} suggest that such an event would be relatively rare, and target-site resistance evolves slowly under low selection pressure. Indeed, years of low dose selections in Australia, resulted in the accumulation of plants containing mixtures of many polygenic, low increment contribution alleles, only recently were populations found that also contained monogenic, target site resistances (*32*).

A pesticide dose response curve generated in the laboratory under ideal conditions is typically linear when plotted using probit techniques. This is not quite the case in the field where it is shallower and sometimes non linear; at higher doses fewer than expected organisms are killed in the field. A sprayer bouncing across a field cannot provide the same uniform pesticide distribution pattern as a laboratory sprayer. In the laboratory pests at highly uniform age are sprayed with a highly uniform spray giving uniform distribution of pesticide and there are no pests hidden in refuges or immigrating after treatment. In the field, weeds germinate at less uniform times and two leaf and four leaf seedlings of the same species often have very different dose response curves. Some seedlings are shielded or shaded from spray by other seedlings or by clods of soil or rocks. The spray pattern is also skewed (Fig. 2). Similarly, there are often large variances in susceptibility among different insect instars, with more advanced instars being less sensitive. In the field, there is often not the synchrony achieved in the lab, and various instars are treated simultaneously. Again a skewed dose-response probit curve will be obtained. Fungi at different stages of development, germination, penetration and establishment are differently affected by fungicides. This would also cause skewing of dose response curves.

Thus, if *Lolium* is 99% controlled by 250-300 g/ha diclofop-methyl in the laboratory, it takes 375 g/ha to get 90-95% control in the field (for the reasons discussed in the previous paragraph), and 1200 g/ha to get the 97-99% achieved in North America. In both cases there are some escapes due to refuges in the field, as well as late germination after the herbicide has dissipated. Presumed doses reaching different plants are depicted in Fig. 2. At 375 g/ha, the typical rate used in Australia for *Lolium,* 5-10% of the plants receive no effective amount of herbicide, and their offspring will be controlled by 375g/ha the following season if they interbreed only with each other. Another 10-20% of the population is subjected to selection for a single polygene (shaded area), because they receive 250-300 g/ha herbicide. Only a small proportion of the individuals receiving 200-250g/ha, (ca. 10^{-4}) have a polygene to allow survival, i.e., those resistant to this dose due to one resistant polygene survive. Those that survive may be severely injured but they recover.

The data in Fig. 1D depict only putatively dead/alive individuals at a fixed time after treatment under controlled conditions and thus "lose" data on sick pests that recover. After a few years of treatment of pristine populations with diclofop-methyl at low rates in Australia, there were often *Lolium* plants that appear very sick or even 'dead' Many such sick plants recovered to produce some seed (Ian Heap, personal communication). These may well be the plants with the first resistant polygene but are not yet classified as "resistant". If they could self pollinate (in *Lolium* they cannot) or are sufficiently close to another plant with the same or different resistant polygene, then 25% of their offspring would have two polygenes, and 50% one polygene. The most likely crosses by the rare individuals that survive the 250-300 g/ha treatment are with the far more ubiquitous healthy plants in the below 250 g/ha class that did not receive an effectual dose. Half the offspring from such crosses will now have one polygene. They will vastly increase the proportion of the population with one polygene the following year, and many more plants receiving 250-300 g/ha will have a modicum of resistance, spreading more pollen, increasing the chances of crosses resulting in two polygenes.

When the high dose rate of 1,200 g/ha is used, it is clear that >97% of the pests are killed (Fig. 2). Most of the survivors were in refuges and received no pesticide at all. An infinitely small proportion of plants received 250-300g, so that the selection for a single resistant polygene would be minimal. Assuming one polygene is required on average for every 50g of pesticide above 200g/ha, then 20 polygenes would be needed to survive 1,200 g/ha. There would theoretically be one plant with 20 polygenes at a frequency of $10^{-4x20}=10^{-80}$ in a pristine population. As we do not know the average increment of resistance provided by each polygene, it is better to use the statistics of polygenic inheritance: If a pristine population has a normal distribution of polygenic resistance centered at 200 g/ha and a standard deviation of 50 g/ha, then the frequency of individuals resistant to 1200 g/ha would be 20 standard deviations above the mean level of pristine resistance i.e., 10^{-88}. Either way, the only likely resistant survivors at 1200 g/ha could be those with a major gene mutant that achieves the needed level of resistance in a single step. In the case of *Lolium,* only a target site resistance seems to be a coded on a major gene. If the field is treated with a

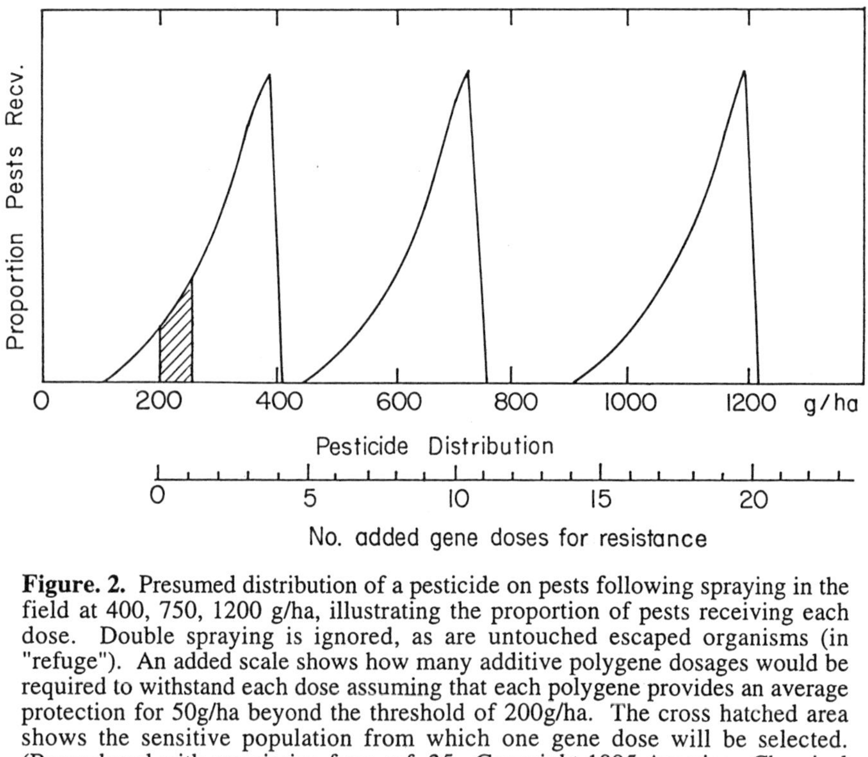

Figure. 2. Presumed distribution of a pesticide on pests following spraying in the field at 400, 750, 1200 g/ha, illustrating the proportion of pests receiving each dose. Double spraying is ignored, as are untouched escaped organisms (in "refuge"). An added scale shows how many additive polygene dosages would be required to withstand each dose assuming that each polygene provides an average protection for 50g/ha beyond the threshold of 200g/ha. The cross hatched area shows the sensitive population from which one gene dose will be selected. (Reproduced with permission from ref. *25*. Copyright 1995 American Chemical Society).

moderate dose (e.g. 700 g/ha in this case), 3-5 % of the plants receive less than a lethal dose, because they are escapees in refuges. Virtually all other plants receive a dose that would still require the combination of many rare resistant polygenes for survival. Probably, for safety sake, an intermediate dose should be chosen to require the presence of 4-8 resistant polygenes for a plant to be resistant.

Remedial Possibilities

Remedial possibilities are too often limited to an abandonment of the pesticide group in favor of others. To often pesticide salespeople have declared: "Don't worry about resistance; we always have alternatives as replacements". Thus, after losing one crop to resistant pests, a grower must often give up an inexpensive pesticide for the salesman's proffered expensive replacement. Too often, this can price the grower out of the market, and the cash crop must be abandoned along with the pesticide. Additionally, there are often cross-resistances that evolved simultaneously that further limit the choice of replacements. This seems especially problematic with the polygenic resistances that seem mostly based on metabolism. This seems to be the case in wheat, both in Australia with diclofop-methyl-resistant *Lolium* (*3,33*) and in India with *Phalaris* and isoproturon (*24*); these are cross-resistances to herbicides acting on a multitude of targets. Farmers are loathe to institute complicated preemptive resistance management schemes, especially if they cost more. Still, the best remedial strategy is to look over one's shoulder and learn from the mistakes of others. When there is resistance somewhere to a pesticide under similar cropping system, it is time to get scared, and not to say "it hasn't happened here, therefore it won't". When the first resistance appears, and it is not spread throughout the population, further enrichment of resistant individuals in the population can be delayed. The delaying tactics that have kept pyrethroid resistance at bay in Australian cotton (*34*) were instituted because it was realized that there was an incipient problem. Thailand, China, and India did not take heed, to the detriment of their cotton industries. Thus, sometimes when resistance is incipient, it is not too late to use the remedial strategy of retroactive preemptive IPM delaying tactics.

Temporary Abandonment. There are some cases where temporary abandonment can later allow for return to the pesticide with resistance. This is basically the preemptive strategy instituted when incipient pyrethroid-resistant insects were found (*34*); the pesticides at risk were used for a short period during each season of multiple sprayings with other compounds. Temporary abandonment is more problematic or even futile once resistance has become fixed throughout a population, as then total eradication of the pest must often be accomplished. Once resistance is fixed, temporary abandonment cannot work when there are internal refuges for the resistant pests to hide, or immigration of resistant individuals. Thus, it was possible, to eradicate small patches of paraquat-resistant wild barley (*35*). This weed species does not have a seed bank as a refuge; it must germinate or die the following year. As resistance was localized, there was no resistant pollen or resistant seed that could immigrate in to the fields to rejuvenate resistance. Still, it took three years to eradicate resistance using far more expensive pesticides than paraquat, and mechanical treatments to prevent seed set (*35*), and there is still a question whether there will be a relapse to resistance in those fields. Too often eradication of resistant populations is attempted after resistances has been fixed in large areas, and refuges had become full. Then it is discovered that resistance is forever.

Models predict that resistance should dissipate from the population as long as there were some susceptible individuals remaining, and there was a large fitness differential (*36*). This was not borne out by field data in one experiment to test the hypothesis (*37*). When neutral pesticides were used (those that controlled the resistant and susceptible individuals to the same extent), the frequency of resistant individuals remained constant for five years, despite the fitness differential. In too many cases resistance is forever, and temporary abandonment is of no avail, leading to permanent abandonment of growers' most cost-effective pesticide.

Selective Abandonment of Some Pesticide Uses. There are cases where the selective abandonment of some uses has saved a pesticide for other more important uses. DDT for mosquito control is performed by coating walls and vegetation near homes with the insecticide. This allows control of mosquitoes that have recently imbibed blood and rest nearby to digest it. Such procedures exert little selection pressure on the whole mosquito population, and do not result in resistance. The widespread agricultural use of DDT surrounding villages inadvertently applied the selection pressure leading to resistance in much of the world. The only regulatory authorities to both appreciate and act on this were those of Sri Lanka, who prohibited agricultural use of DDT, and have successfully saved it for mosquito control. They alone had no problem with DDT resistance *(38)*.

Synergists. Synergists in the context of resistance management are compounds that prevent the degradation of the pesticide (or its toxic products) *(39)*. Such compounds can be of no avail when resistance is due to a mutation in the target-site of the pesticide. In remedial management they can suppress the causes of resistances. Inhibition of cytochrome P-450, NADPH-dependent monooxygenases are of some value already and possibly more so in the future, for managing various resistances where P450s are responsible for pesticide degradation. The P450 inhibitor piperonyl butoxide is used as part of pyrethroid resistance management (*34*), and it partially suppresses diclofop-methyl resistance in *Lolium* (*3*). As these resistances are polygenic, and many P450s are involved, and P450s are known to be differently affected by different inhibitors, cocktails of P450 inhibiting synergists may be needed to fully suppress resistance.

Synergists can also be other pesticides acting synergistically. A recent case is well worthy of mention as the implications go well beyond the initial finding. Propanil-resistant *Echinochloa* has become a scourge in rice in various countries (*9*). An acylamidase degrades propanil both in rice and the weed and it was fortuitously found that piperophos, a rice herbicide that does not effectively control either resistant or susceptible *Echinochloa,* strongly synergizes propanil on both biotypes. Surprisingly, the synergistic mixture does not affect rice. Additionally, the rates used of both in mixture, and the cost of the mixture are far less than either used alone (B. Valverde, Costa Rica, personal communication, 1995). Being less expensive, there is good reason to wonder why it was not discovered earlier, and generally used for more cost-effective pest control,. The mixture would have also been an excellent preemptive tool. Farmers lost a few season's rice crops to resistant weeds before this resistance-managing synergy was found.

The discovery groups of the chemical companies have put little emphasis on finding synergists. As cases of resistance begin to abound at greater frequency, synergists will be found to be a good way to save otherwise useful pesticides from abandonment. This should override the worry that registering a synergist costs the same as registering a new pesticide, as resistance can mean the loss of a registered pesticide. It may well be easier to find new synergists than new pesticides. Wheat should be a case in point. The Australian *Lolium* has cross resistance to all present wheat-selective herbicides, including many that had not been used before in Australia. That is probably because wheat normally uses P450s to degrade herbicides, and closely related *Lolium* seems to have evolved similar resistance as an evolutionary, biochemical mimicry (*33*).

Negative cross resistance. There have been many cases where laboratory studies have shown that some fungicides (*40*) or herbicides *(41)* control resistant biotypes at lower doses than they control the susceptible wild-type, suggesting uses both in prevention and in remediation. The remedial value was demonstrated; pyridate and bentazon selectively depleted triazine-resistant *Solanum* from maize fields *(37)* such that there is a possibility to return (albeit temporarily) to the cheaper triazine herbicides.

Genetic Engineering. In the case of herbicides where it is hardest to find interspecific selectivities, there may be no cost-effective chemical choices, and the only hope may be genetic engineering, if total abandonment is not considered a choice. It is probable that *Lolium* will be resistant to all yet undiscovered wheat herbicides, because wheat and *Lolium* use similar P450 herbicide degradation mechanisms. Wheat needs some new herbicide-resistant genes to which *Lolium* will have a hard time evolving resistance (*33*). Some such genes for herbicide degradation are available from microorganisms, e.g. to glyphosate, to glufosinate, and to dalapon (*42*), but the owners of the first gene have not made it readily available, and the second herbicide is expensive. Similar genetic engineering strategies have been used to control parasitic weeds that are normally naturally resistant to the same herbicides as their crop hosts (*43*).

Preemptive Resistance Delaying Strategies

The best strategy probably always has been to rotate crops and rotate pesticides, and it probably will remain so in the future. The vast majority of cases of resistance comes from monoculture using only one pesticide for a given target pest. Alas, you cannot rotate orchards and many crops are the only successful cash crops in a given agroecosystem. Often there are not a plethora of pesticides to chose from. Resistance management strategies must be elaborated for single pesticide monoculture, as abhorrent as it seems.

Where polygenic inheritance is involved, it has been shown time and again that the initial use of low dose rates facilitate rapid evolution of resistance, and high rates are suggested for prevention. High initial doses have also been proposed as an initial strategy in cancer chemotherapy (*44-46*), because low and then increasing doses have been shown to select for gene amplification (*23*). In the case of anti-cancer drugs, this modeling has suggested that after the first high initial doses are used, the dose can actually be dropped due to an interplay between the remaining cancer cells with the inherent immunological resistance of the patient. This could be extrapolated to agriculture, where the crop has some mechanisms to fend off small infestations of arthropods and pathogens, and can successfully compete with late-germinating weeds.

The strategy often suggested to delay monogenic resistance in monocultural, monopesticidal situations is to lower the dose rate (*36,47*). This decreases selection pressure, as a greater proportion of susceptible individuals remains after treatment, diluting and competing with the infinitesimal proportion of any selected resistant individuals in the population. There are other ways to lower the selection pressure of a pesticide where a single gene target site resistance is expected, depending on the compound and the pest situation. These include using related chemistries with less persistence, or fewer treatments with the same compound. This would allow later influxes of susceptible members oft he same pest species, diluting the proportion of resistant individuals in the population.

Another strategy often proposed is using pesticide mixtures. Too often such proposed mixtures give superior pest control but are counterproductive for resistance management (*48-49*). Indeed, criteria for successful mixtures have been delineated; they include requirements that both components: (a) control the same pest spectrum; (b) have different target sites and modes of degradation; (c) have similar persistence; (d) if possible, exert negative cross resistance or synergism with each other (*49*). The use of synergists (*39*) and negative cross resistance (*40,41*) are probably better preventive delaying tactics than they are for remedial resistance management as such strategies can be less expensive when used and farmers would not have crops lost to evolution of resistant pest populations as quickly as in the past.

Controlled Release - High to Low Dose Due to Pesticide Decay. Another compounding problem, a Catch 22 in its own right, is that when a high dose rate is used, it decays over a period of time. If there is a continuous influx of pests due to immigration of insects, spores, or pollen, or a gradual emergence from refuges (spaced out hatching, or germination of spores or of seeds) then some individuals receive a high dose and some a much lower dose. This can be addressed by repeated spraying, (which growers are now trying to decrease), or by use of controlled release formulations that release a constant pesticide dose, at a rate best attuned to delay the evolution of resistance *(50)*.

Delaying Polygenic and Major Gene Resistances - Modeling

Models for Delaying Polygenic or Monogenic Resistances. We have counted more than 50 models dealing with the evolution and management of resistance in pests; and most modellers seem to believe that the pest group they work with is biologically different from all others; ignoring the rest. Most models for the evolution of resistance and its management have dealt only with major gene effects (e.g.*36,47,51-55*), and only rarely with polygenic resistance (e.g.*56*) and gene amplification *(57)*. None deal with the simultaneous existence of both genetic mechanisms in the same organism, the Catch 22 situation described earlier (58).

Simultaneously Dealing with Mono and Polygenic Resistances. The model is constructed based on the following assumptions:
(A) Polygenic resistance can be delayed by preemptive treatments with low doses of pesticides with synergists, in the rare cases where available, or by treatments with moderate or high levels of pesticide. The moderate or high doses must be applied before too many resistant polygenes have accumulated due to previous treatments with low levels of pesticide; and (B) the rate of evolution of major gene resistance is a direct function of selection pressure, and low and intermediate use rates of pesticide have low selection pressures for major gene resistance (Fig. 3A). We then propose that a rotation of a number of treatments with low doses with a treatment at an intermediate dose will suppress the rate of evolution of both polygenic resistances resulting from low use rates as well as major gene resistance resulting from high use rates (Fig. 3B). The rates can only be chosen after both types of resistance have evolved somewhere, and there is evidence for the different mechanisms. The intermediate treatment is expected to control individuals that have accumulated a few resistant polygenes, setting the situation back to the initial state. The occasional use of intermediate rates might add to the cost of crop production. Still, the alternative pesticides needed when resistant populations become predominant may cost far more, if they exist. The possibility of broad cross resistances with polygenic mechanisms argues that losses will be greater if intermediate dose rates are not occasionally used. We are continually refining our first models (*25*). The modeling is based on standard quantitative population genetics for polygenic resistance and exponentially increasing monogenic resistance, and it disregards fitness differentials in polygenic-inherited resistance (S. N. Gardner, J. Gressel, and M. Mangel, submitted for publication).

This model allows for the use of less pesticide, i.e. levels that do not select only for single gene resistance, with advantages in resistance management, economy, and less environmental impact, all by lowering chemical input. The model is not to be construed as a call for stoppage of rotations of crops, cultural practices, and pesticides, which most feel provide the best possible preemptive resistance management. The model could best be used in situations where alternative crops and pesticides are impractical. Monoculture is imperative in many wheat growing areas, where land, season and/or rainfall, allow only for wheat cultivation as a cash crop, and where evolved polygenic-inherited resistance results in cross resistance to all wheat-selective herbicides.

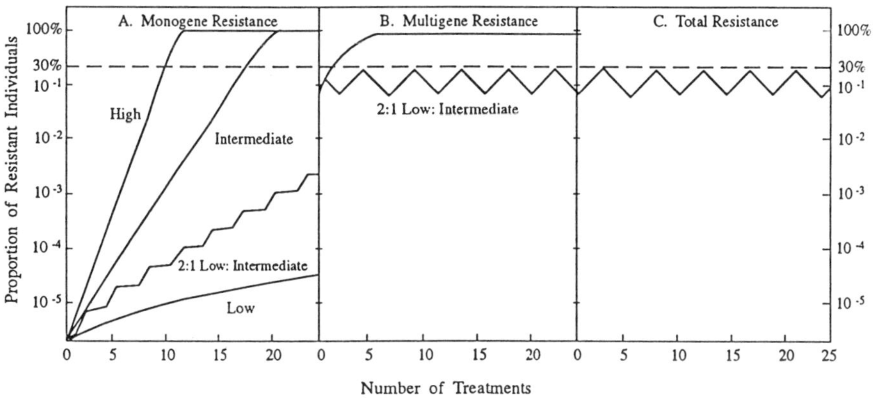

Figure 3. Effects of varying dose rates on the enrichment of different forms of resistance.
A. Enrichment for major gene, usually target site resistance. The effects of dose on the high selection pressure and medium dose with lower selection pressure (straight lines) are plotted from equations similar to those used previously (*36,47*). The calculations for the alternating doses are based on our current models.
B. Enrichment for polygenic resistance by alternating two low and one intermediate dose rates. The calculations for the alternating doses are based on our current models with immigration of susceptibles, and are based on the statistics of population genetic selections. The data are thus expressed as the proportion of the population not controlled at the rate used (see Fig. 2), which is much higher than the actual frequency of genetically-resistant individuals. An intermediate or high dose rate would (theoretically) control such a high proportion of the population that they would plot below the scale.
C. Total enrichment for all types of resistance using the alternating rates in parts A and B based on our current models.

This model has implications beyond preemptive pest resistance management in crops - it could well be considered in management schemes for antibiotic and anti-cancer drug therapies in medicine where resistance, including multi-drug resistance problems are rampant. One of the model scenarios is plotted in Fig. 3. In this scenario, it can be seen that if a high dose were used (Fig. 3A, acute slope) monogenic major gene resistance would quickly appear. If an intermediate dose alone were used, resistance would take far longer (middle straight line). A low dose would hardly select for monogenic resistance (Fig. 3A), but would facilitate evolution of polygenic resistance as shown in Fig. 1C. The application of a few low doses allows for a considerable proportion of the population (but still less than 30%) not controlled, but the intermediate dose represses the level back down (Fig. 3B).

The populaton model does not consider that the intermediate dose leaves about 5% of the population as escapees. These can genetically recombine if proximities are sufficient for breeding. The total effect such a low/intermediate rotation on both monogenic and polygenic resistances is summarized in Fig. 3C. Thus, evolution of resistant populations might occur, but at a slower rate than at the low dose alone, or high dose alone regimes. In addition to modeling, it is imperative to obtain data to further ascertain that this will happen in the field. Once resistance has evolved, e.g. with *Lolium*, one can set out experiments starting with artificially mixed populations with a few percent of resistant individuals to test management strategies, and quickly have resistant populations quickly evolve under the poorer strategies.

Such models must address four types of pest scenarios, and the outcome of resistance management strategies will vary with each. There is some overlap among them.

Type (a) scenario: no immigration and no internal refuges. This scenario fits situations of large scale agriculture where all growers are cultivating the same crop with the same pest management and where the distances are too great for more than a few susceptible pests to fly in as adult insects, pathogen spores, or weed pollen or seeds (as with the Australian *Lolium*). The internal refuges of hidden arthropods or dormant spores or weed seedbanks are very small. The selection will be the most rapid in this scenario, as there will be no influx of diluting susceptible individuals.

Type (b) scenario - where there is immigration of susceptible individuals from the outside. The outside pool is infinitely larger than the pesticide-treated group. The refuges/seedbanks are minuscule. This fits most insect and fungal cases, to a larger or lesser extent, depending on the magnitude of immigration.

Type (c). There is always a large reservoir of susceptible individuals in refuges/ seed banks and there is a constant bi-directional flow between the refuges/seedbanks and the treated population. Initially, in a pristine habitat (never-treated field) the refuge/seedbank population is composed of susceptibles. This situation changes as resistant individuals enter the refuge/seedbanks, such that there is a slow, but delayed, increase in the frequency of resistant individuals emerging from the refuge/seedbanks.

Type (d) scenario - where there is both a sizable immigrant population and a sizable turnover of refuges/seedbanks.

Below we will give further examples from the type (b) situation with varying immigration and describe how resistance depends on both the fraction of immigrating susceptibles and heritability (narrow sense) of the polygenic trait. As the heritability of resistance increases, the immigrant influx required to keep resistance down below the farmer-discernible 30% level varies when a two low dose /one intermediate dose cycle is used (Table I). With the low heritability, only 7% of the treated individuals need be immigrants from the susceptible outside population. With a high heritability, 25% of the treated population must come from outside, which is hard to envisage in many cases.

Immigration has a dual effect of decreasing the rate of enrichment of both polygenic and monogenic resistances. According to the modeled findings, the effect is stronger on monogenic resistance. This allows increasing the level of the intermediate doses to set back the individuals that accumulated a polygene or two for

Table I. Requirement for Influx of Susceptible Individuals to Keep Resistance Below 30% for 50 Cycles of 2 Low and 1 Intermediate Doses

Heritability of Resistance	% Influx Per Treatment Cycle
0.3	7
0.4	10
0.5	13
0.6	17
0.7	25

resistance. In the example shown in Table II a situation is set up whereby populations with >30% resistance are kept at bay for more than fifty cycles of treatment. This includes monogenic as well as polygenic resistance. The modeled scenario in Table II requires a 33% immigration, a situation that can be envisaged with some crops and their pathogens/insects, but is hard to imagine with weeds.

Table II. Threshold Doses to Keep Population with <30% Resistant Individuals (Assuming 33% Influx per Treatment)

	Threshold Doses (g/ha)		
Heritability of Resistance	Two Low Doses	*Then*	One Intermediate
0.3	250		460
0.4	260		480
0.5	280		500
0.6	300		530
0.7	330		562
0.8	380		620

The doses modeled here are based on the system described in Fig. 2; they would have to be modified for other pests based on what is known about the minimal effective dose with adequate control vs. high doses that have selected for monogenic resistance in the past. A similar model can be used for the cases of semi-dominant monogenic resistance, which seems common in insects (*55*). A low dose can be used to control susceptibles and occasional higher doses can be used to obliterate most heterozygotic resistant individuals that may have accumulated, to keep them at a low level.

The frequency of the intermediate dose can be varied, and still preclude resistance for considerable durations (Table III). When the intermediate doses are further apart, the modeled data suggest that the threshold dose must be increased, such that there is little difference in the total amount of pesticide used over a large number of cycles. It is intuitively apparent that one cannot wait too long between intermediate doses, or resistance will be over 30% by the time the grower gets around to using the higher dose. In the conditions shown in Table III, that occurs if the intermediate dose is given after the fifth treatment.

Table III. Threshold Doses to Keep R <30% at Different Dose Frequencies (Assuming 33% Influx; 0.3 Heritability)

	Threshold Dose (g/ha)		Total/30 doses
Frequency of Intermediate	Low	Intermediate	
1:3	250	460	9,600
1:4	280	490	9,825
1:5	320	530	10,500
1:6	breaks down		

The advantage of such models is that there are easy experimental designs for testing them. Experimental verification of such models can be facilitated by mixing pristine wild type material with pests known to contain different levels of resistance. This can ascertain whether the concept of using intermediate doses after a few low doses is more than a theoretical management tool. If it is a valid tool, it must be practiced over wide areas in concert, with all but the most immobile of pests, to prevent mixing of populations allowing for enrichment of genes for resistance. In weeds, there is good evidence in some instances that each case of resistance is due to evolution within a given field, and not due to gene flow (*59*). This is not the case in insects, where flying and human transport allow for easy gene flow (*60*). There is the possibility that gene flow in mosquitoes is not as strong as proposed in (*60*), as DDT resistance has not come in to Sri Lankan populations (*38*), despite their closeness to the mainland.

Polygenic resistance seems to be potentially more dangerous to pest management than monogenic resistance, whether due to genes coding for many overlapping metabolic mechanisms, or to amplifications resulting in multi-drug resistance. This is because resistance can be to a large spectrum of chemicals with different modes of action. Thus, we must weigh the risks of each alternative Catch 22 situation. Many more cases of polygenic resistance should be expected as more farmers cut doses to adhere to strictures to lower pesticide levels. It may well be wise to consider using no pesticide when infestations are low instead of using a low dose, unless the low doses are interspersed with higher doses.

Acknowledgments

Many colleagues have supplied input and ideas and Drs. R.T. Roush, J.A. McKenzie and A. M. Mortimer were kind enough to provide highly constructive criticism. Still, they are not responsible for the conclusions drawn herein. J.G. holds the Gilbert de Botton Chair in Plant Sciences. S.N.G was supported by a NSF Graduate Fellowship and partially by NSF Grant BSR 91-17603. This work was started while M.M. was the Varon Visiting Professor at the Weizmann Institute of Science.

Literature Cited

1. Gressel, J. In *Weed and Crop Resistance to Herbicides*; De Prado, R., Ed.; Kluwer: Amsterdam, 1996, (in press).
2. Gressel, J.; Gianessi, L., Darby, C.P., Saari, L., Seth, A. *Herbicide-resistant weeds: a threat to wheat production in India;* National Center for Food and Agricultural Policy: Washington DC 1994 Discussion Paper PS-94-3, 37 pp.
3. Powles, S. This Volume.
4. Trebst, A. This Volume
5. Guttieri, M.J.; Eberlein, C.V., Mallory-Smith, C.A., Thill, D.C. - This volume
6. Betts, K.J.; Ehlke, N.J., Wyse, D.L., Gronwald, J.W., Sommers, D.A. *Weed Sci.* **1992,** *40*, 184-189.
7. Chapman, H.D. In *Combating Resistance to Xenobiotics* Ford, M.G.; Hollomon, D.W.; Khambay, B.P.S.; Sawicki, R.M. Eds.; Horwood: Chichester,U.K., 1987, pp 54-61.
8. Roush, R.T.; Tingey, W.M. In *Resistance 91 - Achievements and Developments in Combating Pesticide Resistance;* Denholm, I.; Devonshire, A.L.; Hollomon, D.W. Eds.; Elsevier: London, 1992, pp 61-74.
9. Gressel, J.; Baltazar , A. In *Weed Management in Rice;* Kim, K.U., Ed.; 1996 FAO: Rome (in press).
10. Nosticzius, A.; Muller, T., Czimber, G. *Bot. Kozl.* **1979,** *66*, 299-305.
11. Heap, I.M. *Ph.D. Dissertation*, Waite Agric. Inst., Univ. of Adelaide, 1988.

12. Marles, M.A.S.; Devine, M.D., Hall, J.C. *Pestic. Biochem. Physiol.* **1993,** *46,* 7-14.
13. Murray, B.G.; Morrison, I.M., Brûlé-Babel, A.L. *Weed Sci.* **1995,***43,* 233-238.
14. Via, S. In *Pesticide Resistance: Strategies and Tactics for Managment.* National Academy of Science: Washington, DC, 1986, pp 222-235.
15. Cheverud, J.M.; Routman, E. *J. Evol. Biol.* **1993**, *6*, 463-480.
16. Hollomon, D.W.; Butters, J.A., Hargreaves, J.A. In *Managing Resistance to Agrochemicals;* Green, M.B.; LeBaron, H.M.; Moberg, W.K.; Eds.; ACS Symp. Ser.: Washington, DC 1990, pp 199-214.
17. Brent, K.J.; Hollomon, D.W., Shaw, M.W. In *Managing Resistance to Agrochemicals,* Green, M.B.; LeBaron, H.M.; Moberg, W.K., Eds.; ACS Symp. Ser.: Washington DC, 1990, pp 303-319.
18. Peever, T.L.; Milgroom, M.G. *Phytopathology* **1993**, *83,* 1076-1081.
19. De Waard, M.A. In *Resistance '91-Achievements and Developments in Combating Pesticide Resistance,* Denholm, I.; Devonshire, A.L.; Hollomon, D.W. Eds.; Elsevier: London, 1992. pp 48-60.
20. Mouches, C.; Pasteur, N., Berge, J.B., Hyrien, O., Raymond, M., de Saint Vincent, B.R., de Silvestri, M., Georghiou, G.P., *Science* **1986,** *233*, 778-780.
21. Field, L.M., Devonshire, A.L., Forde, B.G. *Biochem. J.* **1988**, *251*, 309-312.
22. Wang, Y.; Jones, J.D., Weller, S.C., Goldsbrough, P.B. *Plant Mol. Biol.* **1991**, *17*, 1127-1138.
23. Schimke, R.T. *Cell* **1984**, *37*, 705-713.
24. Malik, R.K.; Singh, S. *Weed Technology* **1995,** *9*, 419-425.
25. Gressel, J. In *Options 2000-Eighth International Congress of Pesticide Chemistry* Ragsdale, N.N.; Kearney, P.C.; Plimmer, J.R. Eds.; American Chemical Society, Washington, D.C.1995. pp. 330-349.
26. Dove, W.F. *Genetics* **1993,** *134*, 999-1002.
27. Lande, R. *Heredity* **1983,** *50*, 47-65.
28. Powles, S.B. In *Integrated Weed Management for Sustainable Agriculture,* Malik, R.K., Ed.; Ind. Soc. Weed Sci.: Hisar, 1993, Vol. 1 pp 189-194.
29. Powles, S.B.; Matthews, J.M. In *Resistance '91. Achievements and Developments in Combating Pesticide Resistance* Denholm, I.; Devonshire, A.L.; Hollomon, D.W. Eds.; Elsevier: London, 1992 pp 75-87.
30. McKenzie, J.A.; Parker, A.G.; Yen, J.L. *Genetics* **1992,** *130,* 613-620.
31. Mackenzie, R.; Mortimer, A.M.; Putwain, P.D.; Bryan, I.B.; Hawkes, T.R. *Brighton Crop Protection Conf.* pp. 769-774.
32. Tardif, F.J.; Holtum, J.A.M.; Powles, S.B. *Planta,* **1993,** *190,* 176-181.
33. Gressel, J. *Oxford Surv. Plant Mol. Cell Biol.* **1988**, *5,* 195-203.
34. Forrester, N.W.; Bird, L.B. This Volume
35. Powles, S.B.; Tucker, E.S., Morgan, T.R. *Weed Res.,* **1992,** *32,* 207-211.
36. Gressel, J.; Segel, L.A. In *Herbicide Resistance in Plants,* LeBaron, H.M.; Gressel, J. Eds.; Wiley: New York, 1982; pp 325-347.
37. Rotteveel, A.J.W.; Naber, H., Kremer, E. In *Weed and Crop Resistance to Herbicides* ; De Prado, R. Ed.; Kluwer: Amsterdam, 1996, (in press).
38. Lines, J.D. *Parasitol. Today* **1988,** *4,* 517-520.
39. Gressel, J. *Rev. Weed Science* **1990.** *5,* 9-82.
40. Leroux, P. In *Resistance 91 - Achievements and Developments in Combating Pesticide Resistance;* Denholm, I.; Devonshire, A.L.; Hollomon, D.W. Eds.; Elsevier: London, 1992, pp 179-189.
41. Gressel, J.; Segel, L.A. *Z. Naturforsch.* **1990,** *45c,* 470-473.
42. Gressel, J. In *Herbicide Resistant Crops: Agricultural, Economic, Environmental, Regulatory and Technological Aspects,* Duke, S. O., Ed.; Lewis Publishers: Boca Raton FL, **1995.** pp 231-250.
43. Joel, D.M.; Kleifeld, Y., Losner-Goshen, D., Herzlinger, G., Gressel, J. *Nature* **1995** *374,* 220-221.
44. Harnevo, L.E.; Agur, Z. *Math. Biosci..* **1991**, *103*, 115-138.
45. Harnevo, L.E.; Agur, Z. *Canc. Chemoth. Pharma.* **1992**, *30,* 469-476.

46. Agur, Z.; Harnevo, L. In *Biomedical Modelling and Simulation,* Eisenfeld, Witten and Levin Eds.*(in press.)*
47. Gressel, J.; Segel, L.A. *J. Theor. Biol.* **1978**, *75,* 349-371.
48. Roush, R.T. *Pestic. Sci.* **1989,** *26*, 423-441.
49. Wrubel, R.P.; Gressel, J. *Weed Technol.* **1994,** *8,* 635-648.
50. McKenzie, J.A. *Chem. Ind.* **1987,** *8*, 266-269.
51. Mortimer, A.M.; Ulf-Hansen, P.F., Putwain, P.D. In *Achievements and Developments in Combating Pesticide Resistance.;* Denholm, I.;Devonshire, A.L.; Hollomon, D.W., Eds.; Elsevier: London, **1992**, pp. 148-164.
52. Maxwell, B.P.; Roush, M.L., Radosevich, S.R. *Weed Technol.* **1990,***4,* 2 13.
53. Taylor, C.E., In *Pest Resistance to Pesticides;* Georghiou, G.P.; Saito, T., Eds.; Plenum: New York, 1983, pp. 163-173.
54. Roush, R.T.; Daly, J.C. In *Pesticide Resistance in Arthropods,* Roush, R.T.; Tabashnik, B.E., Eds.; Chapman Hall: New York, 1990, pp. 97-152.
55. Georghiou, G.P.; Taylor, C.E. in *Pesticide Resistance: Strategies and Tactics for Management,* National Acad. Press: Washington, 1986**,** pp 157-169.
56. Shaw, M.W. *Plant Pathol.* **1989,** *38*, 44-55.
57. Tabashnik, B.E. *J. Econ. Entomol.,* **1990,** *83,* 1170-1176.
58. Gressel, J.; In *International Crop Science I.* Buxton, D.R.; Shibles, R.; Forsberg, B.L.; Blad, K.H., Asay, K.H., Paulsen, G.M., Wilson, R.F., Eds.; Crop Sci. Soc. America: Madison WI., 1993, pp 121-127.
59. Darmency, H.; Gasquez, J., *New Phytol.* **1981,** *89,* 487-496.
60. Raymond, M. This volume

Chapter 19

The Insecticide Resistance Action Committee: A Continuing Industry Initiative

Gary D. Thompson[1] and Paul K. Leonard[2]

[1]IRAC-U.S., Dow-Elanco, 9330 Zionsville Road, Indianapolis, IN 46268
[2]IRAC Central, Cyanamid International, Chaussee de Tirlemont 105, B-5030 Gembleux, Belgium

Resistance to crop protection agents threatens the economic viability of the crop protection industry. To address resistance in arthropods to crop protection agents industry has formed Insecticide Resistance Action Committees (IRAC) at the global and country level. IRAC committees have three major actives: 1) developing methods to conduct surveys and conducting resistance surveys on the extent and type of resistance, 2) developing sound resistance management guidelines and sponsoring research to confirm their effectiveness, and 3) assisting with educational efforts and implementation of management strategies. An overview of IRAC progress and future plans in these areas is provided.

The crop protection industry has always had a vested interest in preventing the development of resistance to their products. It is an industry dependent on a narrow market segment for survival. The development of resistance by an important pest can have devastating economic effects in farming communities that can rapidly expand to regions and even national levels eliminating an entire customer base. It is also quite evident that there have been very few crop protection agents discovered with different modes of action that are practical to use when safety profiles, economics and efficacy are considered. The time and cost to discover and develop crop protection agents has increased from 3 to 5 years and 10 million dollars to 7 to 10 years and 60 to 100 million dollars or more. It takes industry more than 10 years of commercial success to recover these costs. The failure of a product at any time in its life cycle due to resistance seriously limits the ability of the registrant to recover the investment, pursue advanced technologies, and sustain a viable customer base. The net result is an

0097–6156/96/0645–0187$15.00/0

overwhelming economic incentive for industry to understand and combat resistance and to do it in concert with university researchers, government agencies such as the Cooperative Extension Service, the EPA, and other registrants.

History and Organization

Insecticide Resistance Action Committee (IRAC) was founded in 1984 (1,2) to organize these cooperative industry efforts. It operates as an expert committee of GIFAP, the international voice of the crop protection associations, the American and European Crop Protection Associations (ACPA & ECPA) and SACI, the Japanese Agrochemical Association (Figure 1). IRAC is recognized as an advisory body to the World Health Organization (WHO) of the United Nations and the Food and Agricultural Organization (FAO). Similar committees to fight herbicide (HRAC), fungicide (FRAC) and rodenticide (RRAC) resistance also have been formed. Country and regional IRAC sub-committees have been formed including IRAC U.S., IRAC China, IRAC Pakistan, and IRAC Mexico to improve the effectiveness of the committee. Commodity specific working groups such as Top Fruit and Cotton working groups meet at the global and national level. In the past, working groups such as the Pyrethroid Efficacy Group (PEG) have been formed around modes of action to improve efficiency and to focus on crisis situations. Working groups are formed as needed but are disbanded as soon as a general approach is developed which will satisfy the need (3) for the group.

Membership of IRAC and it's sub-committees is open to any company producing or planning to market insect or mite control products. Each participating company in good standing is allowed one voting member on each working group. There is an annual assessment fee to fund educational and

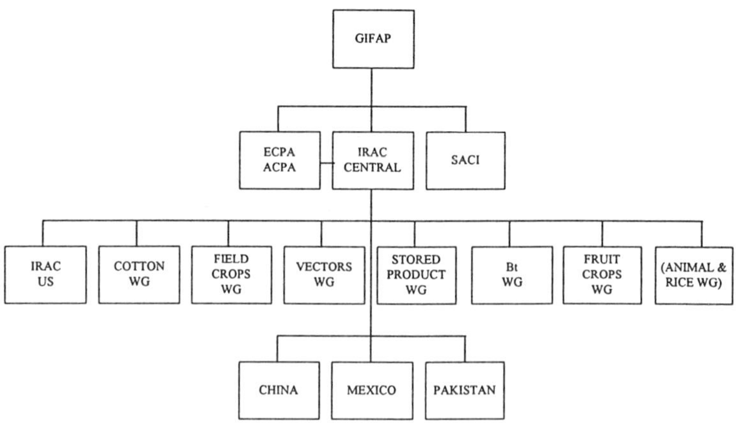

Figure 1. Organizational structure of IRAC committees and working groups during 1995.

research activities. Additionally, and just as importantly, members are able to coordinate their internal resistance related educational and research activities with either IRAC projects or with other member company activities to maximize the benefits to the science of resistance management. *Ad hoc* members also join the committees as needed. IRAC U.S. now has permanent *ad hoc* members from the National Cotton Council, Cotton Incorporated and the WRCC 60.

Functions and Achievements

IRAC activities to date have focused on 1) developing methods to conduct surveys and conducting resistance surveys on the extent and type of resistance, 2) developing sound resistance management guidelines and sponsoring research to confirm their effectiveness, and 3) assisting educational efforts and the implementation of management strategies.

Surveys. The basic method for solving most problems is to first characterize the problem, identify the root causes, develop a solution, and establish measurements which can be monitored to see if progress is or is not being made. The majority of IRAC's efforts during the past ten years have been along these lines. The first effort was an internal survey based on member companies' documented experiences of product failures in field use. This comprehensive survey of known or suspected cases around the world was conducted in 1985 and it has been updated approximately every other year. In recent years inputs from additional researchers and sources have been included. The survey was updated in 1995 and is being prepared for distribution. The 1992-93 version was published in the Pesticide Manual (4) and the output from the survey is also available in a spreadsheet format which can be searched or sorted. The occurrence of resistance in the survey is declared based on the judgment of area specialists and it is not totally complete on a global basis since some areas are not reported. It is, however, the most comprehensive and accurate survey on insecticide resistance available. Guidelines for including a case in the survey are:

1. The product for which resistance is being claimed carries a use recommendation against the particular pest mentioned and has a history of successful performance.
2. Product failure is not a consequence of incorrect storage, dilution or application and is not due to unusual climatic or environmental conditions.
3. The recommended dosages fail to suppress the pest population below the level of economic threshold.
4. Failure to control is due to a heritable change in susceptibility of the pest population to the product.

It should be noted that mention of a particular country/crop in the survey database does not mean that the entire crop area will be affected by the resistant arthropod and the susceptibility and distribution often vary considerably.

An example of the survey database under the category of resistance to organophosphorus insecticides applied to cotton, where chemical control was difficult or impractical during 1992 to 1993, resulted in the following partial list:

Aphis gossypii: Greece, Turkey, Thailand, South Africa, USA, China.
Bemesia tabaci: widespread extending to Ethiopia, Israel, Turkey, Peru, Mexico.
Bucculatrix thurberiella: Peru & Mexico
Heliothis virescens: USA
Helicoverpa armigera: Thailand
Spodoptera exigua: Central America and Mexico
Spodoptera littoralis: Israel, Turkey, & Egypt
Tetranychus cinnabarinus: South Africa and Turkey

Additional categories include cases of resistance which are less important economically but still require careful observation: rice, fruit crops, field crops, vegetables, ornamentals, stored products, public health vectors, and animal health; and the chemical classes of carbamates, chlorinated hydrocarbons, organotins, benzoylurea, ovicides of the clofentezine and hexythiazox type, formamidines, phosphine, and pyrethroids.

Susceptibility Tests. The establishment of measurements to monitor resistance management progress has been primarily conducted by establishing base line susceptibility test methods and databases. IRAC has facilitated this process by collecting existing methods from the WHO, member companies, and elsewhere. IRAC volunteers have simplified the techniques for field uses in remote areas when needed, validating their usefulness and publishing the methods. IRAC and its members have also been very active in the establishment of baseline surveys; an example was the implementation of the Adult Vial Test (AVT) to monitor *Heliothis virescens* resistance to pyrethroids in the U.S.. There was an emerging problem with *H. virescens* in cotton. Dr. F. W. Plapp, Jr. of Texas A&M, perfected a survey technique that used discriminating doses in pretreated glass vials. An IRAC subgroup, PEG US (Pyrethroid Efficacy Group United States), recognized the utility of the technique, further refined it, and produced and distributed tens of thousands of the vials. PEG US coordinated the early testing and data processing throughout the U.S. Cottonbelt. The result was a more timely and extensive database than would have been developed without PEG's assistance. Once the utility of the program was demonstrated, state or regional coordinators were identified through the Land Grant University system to continue the baseline monitoring. This rapidly focused response of expertise and "Seed" research money is one example of the way that IRAC complements the efforts of government and universities.

IRAC has also been very active in encouraging the development of baseline data in other areas of the world and has sponsored numerous training events. The data has proven invaluable in monitoring the progress of programs and in obtaining grower support.

The following methods were published (5) and proposed to be used in baseline monitoring programs:

IRAC Test #	Pest	Suitable Test Substance	General Comment
1	*Myzus persicae*	OP's & Carbamates	leaf dip
2	*Psylla spp*	OP's & amitraz	shoot dip
3	*Tetranychus or Panonychus*	clofentezine, hexy-thiazox or tetradifon	ovicide leaf dip
4	*Tetranychus or Panonychus*	several	adult leaf dip
5	*Nialparvata & Nephotettix*	all insecticides	seedling dip, cages
6	*stored product beetles*	malathion, pirimiphos-methy	filter paper
7	*leaf feeding lepidoptera & coleoptera*	most products	leaf dip
8	*Bemisia tabaci*	amitraz	leaf dip, cage
9	*Leucoptera & Lithocolletis*	benzoylureas	terminal dip on tree

Resistance Management Guidelines - The IRAC committees have worked hard to develop practical resistance management guidelines. A set of general guidelines have been developed as well as more specific ones for country and crop commodity groups. The current general guidelines are:

- Always consult with your local crop advisor/crop protection specialist for guidance and information on resistance management strategies in your area.
- Always include efficient cultural/biological control practices in your pest control program.
- Time the application of insect control products against the most susceptible life stages based on local pest thresholds.
- Do not rely on or treat sequential generations of insects with the same class of products.
- Use insect control products specifically as labeled including rates and spray intervals.

- In the event of a control failure due to resistance, *do not* re-treat with an insect control product of the same class.
- Insure mixtures components of different classes of insecticides are used at rates that provide equivalent control and persistance.

Communication and acceptance of resistance management guidelines is one of the largest challenges that IRAC faces. We are using multiple approaches to communicate these efforts including: the farm press, demonstration projects, educational efforts, promotional literature and labeling. Statements on use labels citing resistance management guidelines were extremely rare five years ago but are very common today. An excellent example is the four METI (mitochondrial electron transport inhibitor) miticides that were recently introduced in Europe (Figure 2). IRAC recognized the potential for over use of the METI's and the

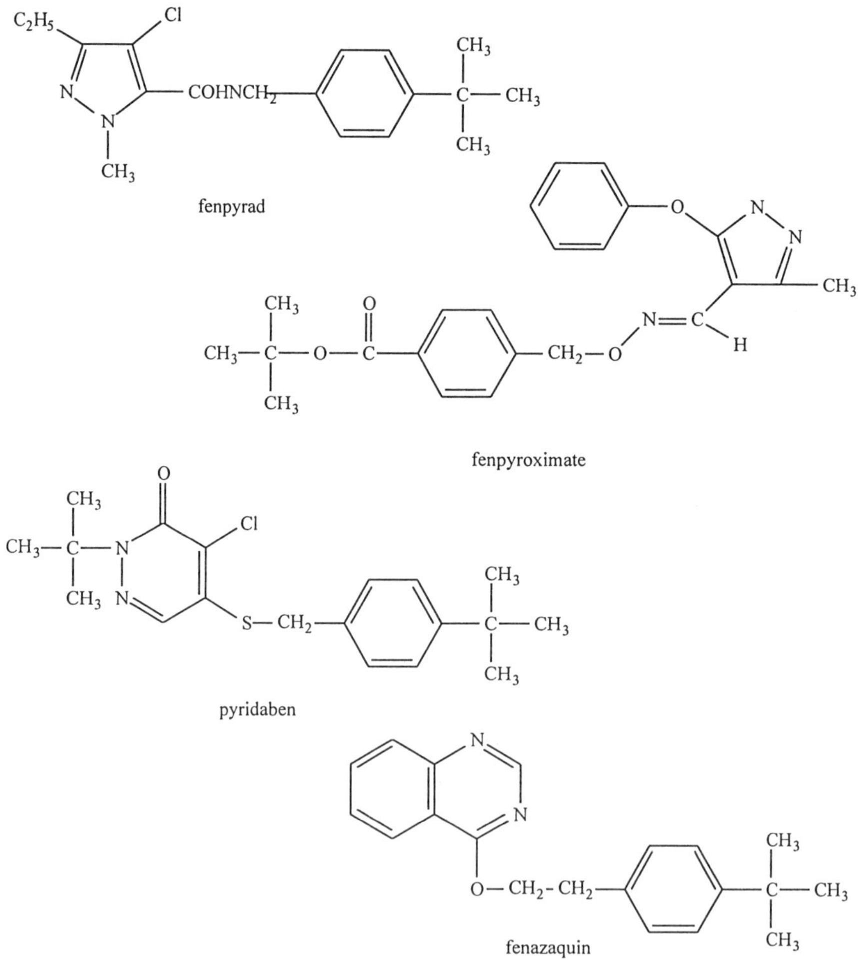

Figure 2. METI acaricides with IRAC resistant management label statements.

consequent risk of resistance development. The companies responsible for distributing these products in Europe, BASF (pyridaben), DowElanco (fenazaquin), Cyanamid (tebufenpyrad) Agrevo and Zeneca (fenpyroximate), worked together as a sub-group of IRAC's Fruit Crops Working Group to address this problem. In this way it was possible to agree on a label statement which restricts applications to one METI per season at any location and prohibits mixing or rotation of METI compounds. This unprecedented agreement came before resistance had been observed or reported in Europe. IRAC is therefore keen to see this type of proactive approach being adopted in other parts of the world or for other product types. IRAC remains committed to the conviction that prevention is far better than cure!

Research Projects - An additional component of IRAC activities is the identification and partial funding of critical research needs. The funding levels are modest by today's standards but the total cumulative investment will soon reach one million dollars. The value of the funding is often enhanced with additional project funding from member companies and with the timeliness of which it can be supplied. An example of IRAC's commitment to research has been the partial funding of the work by Alan McCaffery at the University of Reading and Jim Ottea at Louisiana State University. Their studies over several years have contributed to the understanding of how resistance mechanisms evolve over time in relation to selection pressure and provided IRAC valuable insight into how resistance mechanisms can be modified. Current projects include:

- Malaria mosquito control in Mexico
- Colorado Potato beetle control in Poland
- *Heliothis* control in China
- *Heliothis* and whitefly control in Pakistan
- Diamondback moth control in Taiwan
- Western flower thrips - new monitoring techniques
- Whiteflies control in cotton in the Southwest U.S.
- Monitoring techniques for new acaricides
- Rotation demonstrations for acaricides in California cotton
- Rotation demonstrations for acaricides in Washington apples
- Surveys of resistance levels and management demonstrations of plant bugs in midsouth U.S. cotton

A closer look at the mosquito control project in Mexico reveals that IRAC has created a unique umbrella for cooperation between the manufacturing industry, government, and international institutions:

- First field evaluation of resistance strategies on malaria vectors that claim two million lives annually
- IRAC'S largest project in terms of funding
- Three years at a cost of $352,000
- One third from IRAC, balance from supporting companies and government
- WHO and FAO involved

- IRAC volunteers designed test kits to monitor resistance
- Different management strategies evaluated across entire regions

Education - IRAC's principal tactic for implementing resistance management strategies is through demonstration projects and educational efforts. As the committee has matured it has become evident that these efforts should be equal to or greater than new research activities in priority. Specific educational efforts to date have included the production of the video "The Paradox of Resistance", major assistance with the continued publication of the *Resistant Pest Management Newsletter* by the Pesticide Research Center of Michigan State University, and organizational efforts and funding support for numerous workshops, symposiums and conferences, including this one.

IRAC US and the IRAC Central committee are currently collaborating on the largest educational program to date. A pamphlet targeted at growers and dealers will be mass produced for the U.S., Europe, China, Pakistan, and Mexico. Additional distribution will be made as funding permits. The pamphlet will emphasize the important economics of resistance management to sustainable agriculture production costs. A poster with the message "Stop - Have You Considered Resistance Management in Your Crop Protection Purchase" is also being produced in large quantities for display at points of purchase. Approximately 1000 educational packages that contain reference material and leader guides for county agents and others to conduct grower meetings is planned as well. Fleishman Hillard Inc., the world's largest public relations firm, is producing the educational materials at their cost and Cotton Incorporated and the Cotton Foundation have pledged support as well. IRAC invites others to join them in this most important phase of implementation.

Future Projects - Considerable progress has been made in recent years by organizations and individuals in understanding and communicating resistance management issues. The fact remains, however, that the majority of growers in both developed and developing countries have few economic alternatives for protecting crops, and the threat of resistance is high and increasing in most situations. Industry will continue to provide new technologies and modes of action . Still, it is extremely important that we continue to protect our current technology as well.

The world has taken for granted that a few acres can feed many and that disease transmitting insects can be controlled. However, the fact is that in many instances only one or two modes of action are available for any given pest control situation, this should raise the level of concern for all. Industry is recognizing the importance of the issue and is responding through sensible use patterns, appropriate labeling, promotional literature and training. Industry's support of the Resistance Action Committees has provided a unique focus for additional efforts on arthropod resistance management.

There are numerous additional examples of cooperative efforts between government, academia and industry but there is room for improvement in forming true partnerships in addressing resistance management. The first step in forming partnerships is open communications. IRAC has attempted to increase communications through the publication of its minutes, holding open meetings at commodity production conferences and inviting Environmental Protection Agency and university members to their meetings. We ask that organizers remember to include an industry group or member in their future plans.

Literature Cited

1. Jackson, G. J.; *Pest and Diseases*; Proc. British Crop Protection Conf.; **1986**; Vol 2, pp 943-949.
2. Leonard, P. K., Perrin, R. M. *Pest and Diseases*; Proc. Brighton Crop Protection Conf.; 1994.
3. Wege, P.; Leonard, P.; *Pest and Diseases*; Proc. Brighton Crop Protection Conf., 1994, Vol. 1, pp 427-430.
4. *Pesticide Manual Tenth Edition*; Tomlin C., Ed.; Publisher: Crop Protection Publications, BCPC and Royal Institute of; 1994, pp xxviii-xxxv.
5. *Bulletin OEPP/EPPO*; **1990,** Vol. 20, pp 389-404.

Ecology, Transgenesis, and Regulatory Issues

Chapter 20

Molecular Genetics and Ecology of Transposon-Encoded Streptomycin Resistance in Plant Pathogenic Bacteria

George W. Sundin[1] and Carol L. Bender[2]

[1]Department of Microbiology and Immunology, University of Illinois—Chicago, 835 South Wolcott Avenue, Chicago, IL 60612
[2]Department of Plant Pathology, Oklahoma State University, Stillwater, OK 74078

Plasmid-encoded streptomycin resistance in the plant pathogenic bacteria *Erwinia amylovora*, *Pseudomonas syringae* pvs. papulans and syringae, and *Xanthomonas campestris* pv. vesicatoria is conferred by the *strA-strB* aminoglycoside phosphotransferase genes. These genes are located on a transposable element designated Tn*5393*. The expression of the *strA-strB* genes is directed from a promoter within Tn*5393* in *P. syringae* pv. syringae; the insertion sequence elements IS*1133* and IS*6100* are involved in the expression of *strA-strB* in *E. amylovora* and *X. campestris*, respectively. Epidemiological studies have shown that single plasmid species in *E. amylovora*, *P. syringae* pv. papulans, and *X. campestris* pv. vesicatoria are the main carriers of Tn*5393* within local populations; however, Tn*5393* is distributed among a large group of plasmids in *P. syringae* pv. syringae. The location of Tn*5393* on stable indigenous plasmids in *P. syringae* pv. syringae suggests that this transposon may persist in this species. Streptomycin-resistant nontarget bacteria isolated from plants and soil from several agroecosystems contained Tn*5393* sequences. These bacteria were postulated to be the source of streptomycin resistance in plant pathogenic bacteria. The *strA-strB* genes, but not Tn*5393*, are also widespread among commensal and pathogenic bacteria from humans and animals indicating that these bacteria, and plant pathogenic bacteria, share a common gene pool.

The aminoglycoside antibiotic streptomycin was discovered in 1944 and was one of the first significant antimicrobial compounds to be utilized in clinical medicine. Streptomycin has also been used therapeutically and as a feed additive to farm animals and as an agricultural bactericide to control certain plant diseases. The amount of streptomycin applied to agricultural crops in the United States (estimated at 17,525 kgs. in 1991) is significantly less than that of alternative bactericides such as copper compounds, or of fungicides, herbicides, and insecticides (*1*). However,

0097–6156/96/0645–0198$15.00/0

the application of streptomycin to plant surfaces is of concern due to the development of streptomycin resistance (Sm^r) in bacterial plant pathogens and saprophytes, and the potential relevance of this resistance to human health.

The emergence of antibiotic resistance (Ab^r) has been studied extensively in clinical bacteria; typically, resistance is conferred by enzymatic alteration of the antibiotic, alteration of the target site, or efflux of the antibiotic from the cell (*2*). Antibiotic resistance in clinical bacteria is so widespread throughout the world that the effective control of numerous important clinical pathogens has been seriously compromised. While Ab^r may be conferred by chromosomal mutations, usually, Ab^r genes are located on extrachromosomal genetic elements called plasmids. Indigenous plasmids are found in almost all bacterial genera and tend to encode genes which are necessary for the adaptation of a bacterium to a particular environmental niche (*3,4*). Many plasmids are also self-transmissible, encoding genes necessary for their transfer either to related bacteria or to a broad range of hosts. Gene transfer is recognized as a major component of the rapid evolution of Ab^r in clinical bacteria, and is presumably responsible for the observations of similar or identical Ab^r genes in otherwise unrelated bacterial genera (*5-7*).

Streptomycin resistance in clinical bacteria has been well studied and at least four streptomycin-inactivating enzymatic mechanisms have been characterized; these enzymes inactivate the streptomycin molecule by adenylylation or by phosphorylation at either of two sites (*8*). Streptomycin-resistance genes are widely distributed among clinical genera and can be found on all common bacterial gene transfer vehicles, i.e. plasmids, transposons, and integrons (*8*). Plant pathogenic bacteria with resistance to streptomycin have been isolated on at least four continents (*9*); however, until recently, little was known about the genetics and ecology of plasmid-encoded, transferable streptomycin resistance. In this review, we will describe a transposon, Tn*5393* (*10*), which confers streptomcyin resistance in three important genera of plant pathogens. Tn*5393* encodes the *strA-strB* Sm^r genes which are also widely distributed among bacterial pathogens of humans and animals (*11*). We will discuss the genetics of Tn*5393* and also examine information concerning the dissemination and persistence of this transposon within populations of plant-associated bacteria.

Genetic Characterization of Tn*5393*

Streptomycin-resistance genes have been cloned from the plant pathogens *Erwinia amylovora*, *Pseudomonas syringae* pvs. papulans and syringae, and *Xanthomonas campestris* pv. vesicatoria (*10,12-14*). Using hybridization and restriction analyses, Sundin and Bender showed that the Sm^r determinants from *P. syringae* pvs. papulans and syringae and *X. campestris* pv. vesicatoria were highly similar to the *strA-strB* aminoglycoside phosphotransferase genes from the broad-host-range clinical plasmid RSF1010 (*14*). *strA* and *strB* are two distinct genes arranged in tandem; each gene encodes a phosphotransferase enzyme which acts on different positions on the streptomycin molecule (*8*). While it is unusual for a bacterium to carry two Ab^r genes targeted at a single antibiotic, it has been shown that both *strA* and *strB* are required for cells to exhibit high levels of streptomycin resistance (*14a*). Chiou and

Jones sequenced the Sm^r determinant from *E. amylovora*, showed that it had 99.8% nucleotide identity with the *strA-strB* genes, and also identified by sequence analysis that *strA-strB* were located on a transposable element which was designated Tn*5393* (*10*). Tn*5393* is a 6.7-kb transposable element which is organized similarly to members of the Tn*3* subgroup of the Tn*21* family (*15*). Tn*5393* contains 81-bp inverted repeats, and encodes a transposase (*tnpA*) and resolvase (*tnpR*) genes [*10*; (Figure 1)]. In Tn*3*, the *tnpA* and *tnpR* genes are transcribed divergently from promoter sequences located on opposite DNA strands within a central intergenic region termed the recombination site (*res*) (*15,16*). *res* also contains three binding sites for TnpR (*17*); the multifunctional protein TnpR serves to resolve cointegrates formed during the transposition process and also binds at *res* to repress the transcription of *tnpA* and *tnpR*.

With subsequent DNA sequence and hybridization analysis, it was determined that the *strA-strB* genes in *P. syringae* pv. syringae and *X. campestris* pv. vesicatoria were encoded on transposable elements designated Tn*5393*a and Tn*5393*b, respectively (*18*). In *P. syringae* pv. syringae, the *strA-strB* genes are located downstream of *tnpR* and are transcribed along with *tnpR* as an operon [*18*; (Figure 1)]. The minimal inhibitory concentration of streptomycin in this bacterium is only 75 μg/ml, and promoter fusion studies indicated that the expression of *strA-strB* was repressed by TnpR (*18*). In *X. campestris* pv. vesicatoria, the insertion sequence (IS) element IS*6100* is located within the *tnpR* gene and increases the expression of *strA-strB* [*18*; (Figure 1)]. Many bacterial IS elements have been shown to contain outwardly-directed promoter sequences which function to activate the expression of adjacent genes (*19*). The presence of IS*6100* is also significant in that this IS is distributed among widely divergent bacteria including *Flavobacterium* sp., *Mycobacterium fortuitum*, and *Pseudomonas aeruginosa* (*20, 21*). In *E. amylovora*, the IS element IS*1133* is located upstream of *strA-strB*, and deletion analysis suggested that this IS provided a promoter sequence for the expression of the Sm^r genes [*10*; (Figure 1)]. Thus, Tn*5393* is a versatile transposon which is capable of acquiring IS elements to express the *strA-strB* genes in different plant pathogenic bacterial genera.

Population Dynamics of the *strA-strB* Genes in Plant Pathogenic Bacteria

In Sm^r plant pathogenic bacteria, the *strA-strB* genes are encoded on Tn*5393*, a transposon which is usually borne on conjugative plasmids. The dissemination of conjugative plasmids is an established mechanism for gene transfer within populations of plant pathogenic bacteria. The location of Tn*5393* on plasmids of varying size also suggests that interplasmid mobilization of this transposon occurs within natural populations. Epidemiological studies incorporating frequent and widespread samplings have provided meaningful information concerning the important plasmids involved in the dissemination of the *strA-strB* genes within natural populations of plant pathogenic bacteria (summarized in Table I).

Erwinia amylovora. Plasmid-borne streptomycin resistance in *E. amylovora* was originally detected in 1990 in strains isolated from one apple orchard in Michigan

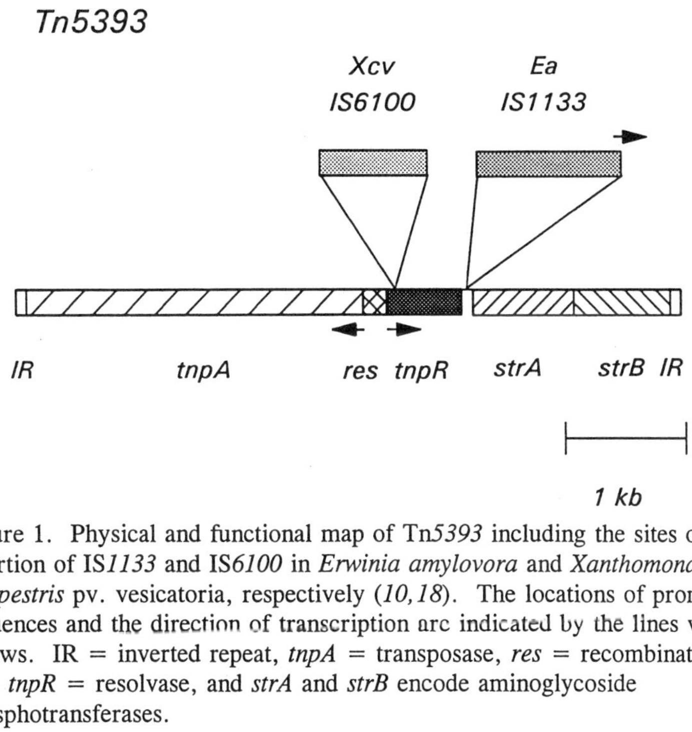

Figure 1. Physical and functional map of Tn*5393* including the sites of insertion of IS*1133* and IS*6100* in *Erwinia amylovora* and *Xanthomonas campestris* pv. vesicatoria, respectively (*10,18*). The locations of promoter sequences and the direction of transcription are indicated by the lines with arrows. IR = inverted repeat, *tnpA* = transposase, *res* = recombination site, *tnpR* = resolvase, and *strA* and *strB* encode aminoglycoside phosphotransferases.

Table I. Summary of results of epidemiological analyses of streptomycin-resistant (Sm^r) plant pathogenic bacteria including the frequencies of *Tn5393*-containing plasmids among the total Sm^r populations.

Pathogen	Host[a]	Loc.[b]	Plasmid	Size[c]	Tra.[d]	Freq.[e]	Reference
Erwinia amylovora	A	MI[f]	pEA29	36	++	1.9	23
			chrom.	---	--	1.4	23
			pEa34	34	++	95.3	23
	A	MI[g]	pEa34	34	++	100.0	23
Pseudomonas syringae							
pv. papulans	A	NY	pCPP501	108	++	100.0	24
pv. syringae	C	OK	pPSR2	190	++	100.0	14
	P	OK	pPSR1	68	++	31.3	14
	P	OK	pPSR5	68	++	68.8	14
	W	OK	pPSR3	220	++	100.0	14
	P	OK	pPSR14	68	++	2.3	27
	P	OK	pPSR15	125	++	15.8	27
	P	OK	pPSR17	190	++	57.1	27
	P	OK	pPSR18	210	++	3.0	27
	P	OK	pPSR19	220	++	16.5	27
	P	OK	pPSR20	105	++	3.0	27
	P	OK	pPSR21	200	--	2.3	27
Xanthomonas campestris							
pv. vesicatoria	T	ARG	[h]	68	--	100.0	12

[a] Host of isolation: A = apple, C = cottonwood, P = ornamental pear, T = tomato, W = willow.
[b] Location of isolation: ARG = Argentina, MI = Michigan, U.S.A., NY = New York, U.S.A., OK = Oklahoma, U.S.A.
[c] Size of individual plasmids in kilobase pairs.
[d] Transferability of plasmids: ++ = self-transmissible by conjugation, -- = nonself-transmissible.
[e] Frequency of isolation of individual plasmids in the studies indicated.
[f] pEA29, chromosome, and pEa34 location of Tn*5393* in strains from Van Buren county, Michigan.
[g] pEa34 location of Tn*5393* in strains from Newaygo county, Michigan.
[h] The 68-kb Sm^r plasmid from *X. campestris* pv. vesicatoria was not given a designation.

(*22*). A single plasmid, designated pEa34, was associated with Sm^r *E. amylovora*, and Tn*5393* was isolated from this plasmid (*10*). Continued intensive usage of streptomycin in Michigan apple orchards was correlated with an increased distribution of Sm^r *E. amylovora* in subsequent years (*23*). Tn*5393* sequences were detected on an additional plasmid, pEA29, and on the chromosome in some strains; however, pEa34 remained the dominant Sm^r plasmid isolated [*23*; (Table I)].

Pseudomonas syringae. In apple orchards in New York, streptomycin resistance was reported to have spread quickly within populations of the blister spot pathogen *P. syringae* pv. papulans (*24*). A 108-kb conjugative plasmid, pCPP501, was detected in all Sm^r strains and was not detected in any Sm^s strains [*24*; (Table I)]. Since this plasmid was transferred at high frequencies by conjugation *in vitro*, it was hypothesized that the dissemination of this plasmid within field populations was responsible for the rapid increase in streptomycin resistance in this pathogen (*24*). Streptomycin-resistant strains of *P. syringae* pv. papulans which encode the *strA-strB* genes also occur within apple orchards in Michigan (*25*). In this case, the *strA-strB* genes were encoded on plasmids of at least five different sizes, none of which were 108 kb (*25*).

In Oklahoma, the intensive usage of streptomycin in ornamental tree nurseries has selected resistance in populations of *P. syringae* pv. syringae, a pathogen which causes dieback and cankers of stems (*14*). Extensive field surveys determined that 36 - 54% of the *P. syringae* pv. syringae strains isolated were resistant to streptomycin in nurseries in which streptomycin was applied [*14, 26, 27*; (Table I)]. In nurseries where streptomycin was not applied, as many as 54% of the *P. syringae* pv. syringae strains were streptomycin resistant [*27*; (Table I)]. A similar situation exists in ornamental tree nurseries in Oregon. Surveys done in 1982/83 and 1992/93 indicated that the percentage of Sm^r *P. syringae* pv. syringae had increased from 15.2% to 62.2% (*28*). Scheck *et al.* have also shown that 93.4% of the 211 Sm^r strains tested hybridized with the *strA-strB* genes (*28*). Sustained usage of streptomycin in nurseries in Oregon was implicated in the increase of the frequency of streptomycin resistance observed between 1982 and 1993.

Streptomycin resistance in *P. syringae* pv. syringae from Oklahoma was conferred by the *strA-strB* genes located on Tn*5393* in all cases, and the transposon was detected on several plasmid species ranging from ca. 68 - 220 kb [*14,27*; (Table I)]. The Sm^r plasmids isolated differed according to the host plant in one study (*14*), and seven Sm^r plasmids were isolated in varying frequencies from ornamental pear in a subsequent study (*27*). Four of the Sm^r plasmids (68 - 125 kb) were grouped based on their hybridization to the *oriV* and *par* loci of plasmid pOSU900, a cryptic plasmid from *P. syringae* pv. syringae (*27*). These essential plasmid replication and stability determinants are widespread within the species *P. syringae* and have been detected in at least 16 other pathovars (*26,29*) which suggests that plasmids containing these sequences may encode additional determinants that are important for the association of *P. syringae* with plant hosts.

Copper is also intensively utilized as a bactericide in nurseries in Oklahoma, and *P. syringae* pv. syringae has developed resistance to copper as well as to streptomycin (*14*). Analyses of plasmid profiles from copper resistant (Cu^r), Sm^r, Cu^r Sm^r, and Cu^s Sm^s strains revealed 18 distinct profiles mostly containing one plasmid per strain; the resistance phenotypes were always plasmid-associated (*27*). The Cu^r determinant was linked with the Sm^r transposon Tn*5393* on the same plasmid in five of the six different plasmid profiles; in one Cu^r Sm^r plasmid profile, the Cu^r determinant and Tn*5393* were located on separate plasmids (*27*). As stated above, the frequencies of specific plasmid profiles differed among the nurseries sampled suggesting that the plasmids and/or their host strains encoded additional traits which were beneficial in particular nursery environments. Genomic fingerprinting of the strains using the arbitrarily-primed polymerase chain reaction (AP-PCR) technique revealed a high level of genetic diversity and indicated that the *P. syringae* pv. syringae strains sampled comprised two distinct subpopulations (*27*). Other studies have shown that the pathovar syringae of *P. syringae* is genetically diverse and consists of subpopulations which are related to the host of isolation (*30,31*). The results of our populations studies implied that the interplasmid movement of Tn*5393* and interstrain plasmid transfer played important roles in the evolution of streptomycin resistance in these *P. syringae* pv. syringae populations. Interestingly, each resistance plasmid isolated was only detected among members of a single subpopulation suggesting that plasmids were compartmentalized within specific genomic backgrounds. It should be noted, however, that Tn*5393*-containing plasmids were effectively disseminated among both chromosomal groups (*27*). The distribution of a resistance plasmid among many host strain backgrounds may ultimately contribute to the persistence of these plasmids within the populations of *P. syringae* studied.

Xanthomonas campestris. The bacterial spot pathogen of pepper and tomato, *X. campestris* pv. vesicatoria, has evolved streptomycin resistance in several independent locations (*12*). Strains from three locations were shown to contain the *strA-strB* genes. Two strains isolated eight years apart in Argentina contained a 68-kb Sm^r plasmid, which was unique to this population of Sm^r strains [12; (Table I). The isolation of this plasmid over an eight-year interval suggests that it is stable and may be indigenous to *X. campestris*. Tn*5393* containing the IS*6100* insertion in *tnpR* was cloned and characterized from the 68-kb Sm^r plasmid (*18*).

Ecological Fitness of Tn*5393*-containing Plasmids

The role of plasmids containing Tn*5393* on the survival of plant pathogenic bacteria on plant hosts sprayed with streptomycin, and the potential longterm persistence of Tn*5393*-containing plasmids within populations are two questions which have received limited attention. Sundin and Bender examined the ability of indigenous Cu^r and Sm^r plasmids to persist in *P. syringae* pv. syringae and determined whether plasmid carriage affected epiphytic fitness (*32*). Studies performed *in vitro* indicated that all plasmids studied were stable for over 200 generations in the host strain *P. syringae* pv. syringae FF5 which was grown in a carbon-limited minimal medium that did not contain copper or streptomycin (*32*). The presence of the Cu^r Sm^r plasmid pPSR1 in *P. syringae* pv. syringae FF5.1 increased the survival and growth of this strain on ornamental pear leaves sprayed with copper and streptomycin while populations of the plasmid-free Cu^s Sm^s strain FF5.1 were significantly reduced (*32*). Epiphytic populations of FF5.1(pPSR1) were similar to those of FF5.1 on unsprayed trees over a 12-week period (*32*). Likewise, the presence of pEa34 enabled populations of *E. amylovora* G11nal^r to increase on apple blossoms sprayed with streptomycin while populations of the Sm^s plasmid-free G11nal^r were reduced (*23*). In growth chamber competition experiments, *P. syringae* pv. syringae FF5 containing either Cu^r pPSR4, Sm^r pPSR5 (Tn*5393*), or the recombinant Cu^r Sm^r pPSR4::Tn*5393* maintained epiphytic populations similar to the plasmid-free FF5 strain on bean leaves (*32*). Regarding *P. syringae* pv. syringae, the results of experiments *in planta*, coupled with plasmid stability data gathered *in vitro*, suggested that the indigenous Cu^r and Sm^r plasmids studied would persist within *P. syringae* pv. syringae populations.

Dissemination of Tn*5393* among Nontarget Phylloplane and Soil Bacteria. The presence of Tn*5393* in diverse plant pathogenic bacteria isolated on two continents raises the question of the origin and selection of this transposon. The results of several studies have shown that the *strA*-*strB* and Tn*5393* sequences are also distributed among nontarget bacterial inhabitants from plants and soil taken from locations where streptomycin was previously applied (*13,33-35*). The frequency of Sm^r bacteria containing *strA*-*strB* appeared to be correlated with the number of applications of streptomycin in some studies. *strA*-*strB* and Tn*5393* sequences have also been detected in isolates from agroecosystems geographically separated from regions where streptomycin had been applied (*35*). These data imply that the *strA*-*strB* genes and Tn*5393* may be indigenous in many environments and perhaps were initially selected in soil bacteria in response to natural streptomycin-producing bacteria in soil. It appears likely that Tn*5393* was introduced independently to the different plant-pathogenic bacterial populations from Sm^r environmental bacteria; however, the close nucleotide sequence similarity of the *strA*-*strB* genes suggests that this element was recently disseminated. The amplification of Ab^r nontarget bacteria followed by the dissemination of Ab^r determinants to pathogenic bacteria is an established route in the evolution of antibiotic resistance in clinical bacterial pathogens (*36*). In these cases, the tremendous selection pressure imposed by antibiotic usage is thought to result in the selection of an obscure Ab^r strain(s), amplification of the Ab^r strain(s), and subsequent transfer of the Ab^r determinant to pathogenic bacteria (*7*).

Dissemination of the *strA*-*strB* Genes among Human and Animal-associated Bacteria. The deployment of antibiotics in clinical medicine has been called the "largest experiment in bacterial populations dynamics on earth" (*37*). The observation of similar Ab^r genes in widely disparate organisms has shown that gene transfer can and does occur between organisms which are evolutionarily unrelated and inhabit differing environmental niches (*38*). The *strA*-*strB* genes were first identifed as important Sm^r determinants in clinical bacteria; they have since been identified in 17 gram-negative genera (*11*; Table II). *strA*-*strB* are only associated with Tn*5393* in environmental bacteria. The genes are mostly found on small broad-host-range plasmids which are ubiquitous in clinical bacteria; however, the right inverted repeat of Tn*5393* is conserved downstream of *strB* in several of the clinical plasmids (*11*). This observation suggests that Tn*5393* was disseminated originally to clinical plasmids, but later the transposition functions were lost.

The presence of *strA-strB* on Tn*5393* and conjugative replicons undoubtedly contributes to the successful transfer of these genes in plant and soil habitats. Transfer presumably would not require the successful establishment of the donor replicon in the recipient strain; the transposon itself could excise following conjugal transfer of the host plasmid and reinsert in the recipient genome. While *strA-strB* are not associated with a transposable element in bacteria of human or animal origin, the location of the genes on broad-host-range plasmids contributes to their successful establishment in novel hosts and abrogates the need for a transposon location. Thus, the *strA-strB* genes are linked with broad-host-range plasmids and transposons in clinical and environmental bacteria, respectively, and have adapted to facilitate the colonization of bacterial hosts from these distinct ecological niches.

Table II. Genera of gram-negative bacteria which are documented to contain the *strA-strB* streptomycin-resistance determinant and the genomic location of *strA-strB* in these organisms[a].

Actinobacillus	P	*Pantoea*	P
Bordetella	P	*Pasteurella*	P
Eikenella	C	*Proteus*	P
Enterobacter	P	*Providencia*	P
Erwinia	C, P	*Pseudomonas*	P
Escherichia	P	*Salmonella*	P
Haemophilus	P	*Shigella*	P
Klebsiella	P	*Xanthomonas*	P
Neisseria	P		

[a] The information from this table is taken from reference 11.
C = chromosome, P = plasmid.

Conclusions

Streptomycin resistance is prevalent among plant pathogenic bacteria in agroecosystems in which the antibiotic is utilized as a bactericide. A transposable element, Tn*5393*, confers streptomycin resistance in three important plant pathogens, *E. amylovora*, *P. syringae*, and *X. campestris*. The expression of the *strA-strB* Sm^r

genes on Tn*5393* in *E. amylovora* and *X. campestris* is positively affected by the upstream insertions of IS*1133* and IS*6100*, respectively.

Because Tn*5393* is encoded typically on conjugative plasmids, gene transfer is implicated in its rapid dissemination within populations. Also, the mobility of Tn*5393* between plasmids and chromosomes increases the chances for this element to become associated with superior genotypes which are stable without streptomycin selection. These associations are postulated to result in the persistence of Tn*5393* within populations.

The widespread distribution of the *strA-strB* and Tn*5393* sequences among nontarget bacteria in agroecosystems is thought to result in the transfer of this Sm^r determinant to plant pathogenic bacteria. The *strA-strB* Sm^r genes are also distributed widely among evolutionarily unrelated human and animal pathogenic bacteria. This indicates that a common bacterial gene pool is available to bacterial inhabitants of humans, animals, and plants.

Acknowledgments

We thank Alan Jones and Heather Scheck for communicating unpublished information. Research done in the Bender laboratory was supported by the Oklahoma Agricultural Experiment Station and National Science Foundation grant EHR-9108771.

Literature Cited

1. Vidaver, A.K. *Phytopathol. News* **1993**, *27*, 6-7.
2. Hayes, J.D., and Wolf, C.R. *Biochem. J.* **1990**, *272*, 281-295.
3. Eberhard, W.G. *Plasmid* **1989**, *21*, 167-174.
4. Tschape, H. *FEMS Microbiol. Ecol.* **1994**, *15*, 23-32.
5. Amabile-Cuevas, C.F., and Chicurel, M.E. *Cell* **1992**, *70*, 189-199.
6. Courvalin, P. *Antimicrob. Agents Chemother.* **1994**, *38*, 1447-1451.
7. Davies, J. *Science* **1994**, *264*, 375-382.
8. Shaw, K.J., Rather, P.N., Hare, R.S., and Miller, G.H. *Microbiol. Rev.* **1993**, *57*, 138-163.
9. Cooksey, D.A. *Annu. Rev. Phytopathol.* **1990**, *28*, 201-219.
10. Chiou, C.-S., and Jones, A.L. *J. Bacteriol.* **1993**, *75*, 732-740.
11. Sundin, G.W., and Bender, C.L. *Mol. Ecol.* **1995**, *4*, in press.
12. Minsavage, G.V., Canteros, B.I., and Stall, R.E. *Phytopathology* **1990**, *80*, 719-723.
13. Norelli, J.L., Burr, T.J., LoCicero, A.M., Gilbert, M.T., and Katz, B.H. *Appl. Environ. Microbiol.* **1991**, *57*, 486-491.
14. Sundin, G.W., and Bender, C.L. *Appl. Environ. Microbiol.* **1993**, *59*, 1018-1024.
14a. Chiou, C.-S., and Jones, A.L. *Gene* **1995**, *152*, 47-51.
15. Grinsted, J., de la Cruz, F., and Schmitt, R. *Plasmid* **1990**, *24*, 163-189.
16. Sherratt, D. In *Mobile DNA*; Berg, D.E., and Howe, M.M., Ed.; American Society for Microbiology: Washington, D.C., 1989, pp. 163-184.

17. Grindley, N.D.F. *Cell* **1983**, *32*, 3-5.
18. Sundin, G.W., and Bender, C.L. *Appl. Environ. Microbiol.* **1995**, *61*, 2891-2897.
19. Galas, D.J., and Chandler, M. In *Mobile DNA*; Berg, D.E., and Howe, M.M., Ed.; American Society for Microbiology: Washington, D.C., 1989, pp. 109-162.
20. Martin, C., Timm, J., Rauzier, J., Gomez-Lus, R., Davies, J., and Gicquel, B. *Nature(London)* **1990**, *345*, 739-743.
21. Kato, K., Ohtsuki, K., Mitsuda, H., Yomo, T., Negoro, S., and Urabe, I. *J. Bacteriol.* **1994**, *176*, 1197-1200.
22. Chiou, C.-S., and Jones, A.L. *Phytopathology* **1991**, *81*, 710-714.
23. McManus, P.S., and Jones, A.L. *Phytopathology* **1994**, *84*, 627-633.
24. Burr, T.J., Norelli, J.L., Katz, B., Wilcox, W.F., and Hoying, S.A. *Phytopathology* **1988**, *78*, 410-413.
25. Jones, A.L., Norelli, J.L., and Ehret, G.R. *Plant Dis.* **1991**, *75*, 529-531.
26. Sundin, G.W. Ph.D. dissertation. 1994, Oklahoma State University.
27. Sundin, G.W., Demezas, D.H., and Bender, C.L. *Appl. Environ. Microbiol.* **1994**, *60*, 4421-4431.
28. Scheck, H., Pscheidt, J., and Moore, L.W. 1995. Personal communication.
29. Murillo, J., and Keen, N.T. *Mol. Microbiol.* **1994**, *12*, 941-950.
30. Denny, T.P., Gilmour, M.N., and Selander, R.K. *J. Gen. Microbiol.* **1988**, *134*, 1949-1960.
31. Legard, D.E., Aquadro, C.F., and Hunter, J.E. *Appl. Environ. Microbiol.* **1993**, *59*, 4180-4188.
32. Sundin, G.W., and Bender, C.L. *Can. J. Microbiol.* **1994**, *40*, 279-285.
33. Burr, T.J., Norelli, J.L., Reid, C.L., Capron, L.K., Nelson, L.S., Aldwinckle, H.S., and Wilcox, W.F. *Plant Dis.* **1993**, *77*, 63-66.
34. Sobiczewski, P., Chiou, C.-S., and Jones, A.L. *Plant Dis.* **1991**, *75*, 1110-1113.
35. Sundin, G.W., Monks, D.E., and Bender, C.L. *Can. J. Microbiol.* **1995**, *41*, 792-799.
36. Levy, S.B. *Banbury Rep.* **1986**, *24*, 17-30.
37. O'Brien, T.F. *et al.* *Rev. Infect. Dis.* **1987**, *9*, S244-S260.
38. Mazodier, P., and Davies, J. *Annu. Rev. Genet.* **1991**, *25*, 147-171.

Chapter 21

Movement of Resistance Genes Among Plants

Henri Darmency

Laboratoire de Malherbologie, Institut National de la Recherche Agronomique, BV 1540, 21034 Dijon, France

Resistance to herbicides, pathogens and insects may occur in wild plant species through mutation and introgression with other species. The movement of genes among plants provides weeds with the opportunity to express new properties and modify their invading potential. Examples of gene flow within and between weed populations, between weed species, and between crops and weeds are reviewed. Special attention is given to herbicide resistance genes and gene flow between genetically engineered crops, that display new highly adaptive genes, and weeds.

In Roman times and for early biologists, species were immutable units. Variations within species represented mere imperfections. It was only since the early-twentieth century that the idea of dynamic variation emerged. Mutation, migration and selection were suggested to create and maintain variation within populations. Migration refers to all mechanisms resulting in the movement of genes from one individual to another, and from one population to another. It generally occurs within a species, but interspecific gene flow is not excluded. In plants, it is realized through seed migration and pollen flow. Fundamentals of gene flow have been reviewed extensively by Slatkin (*1*), and numerous examples of gene flow in plants have been provided in a comprehensive compilation by Levin and Kerster (*2*).

Gene flow provides plants with the opportunity to express new properties. This is why I will focus, in this paper, on plants which have a particular ability to take advantage of human disturbances: weeds. Weeds display variation within species in numerous traits including caryology, morphology, phenology, reproduction, response to insects and pathogens, and molecular markers. Some traits, such as resistance to herbicides, adapt weeds to farmer's cropping practices (*3*). Others, such as resistance to pests, led to increased aggressiveness of weeds (*4*). I address here the question of how resistance genes are disseminated within and among weed populations, between different weed species, and between weed and crop species. This last issue is

0097–6156/96/0645–0209$15.00/0

becoming more important as crops continue to be engineered with genes encoding resistance to herbicides, pathogens and insects that might spread to related weed species through introgression. The majority of examples deal with the movement of herbicide resistance genes, some with marker genes, but unfortunately none with pest or desease resistance genes which have been poorly documented in weeds populations.

Gene Flow Within Populations

For geneticists, a population is a group of plants that interbreed. The actual size of a population, therefore, remains unknown until the study of the movement of a gene is carried out. In contrast, for ecologists, a population is a group of plants that grow in a location submitted for long time to homogeneous selection pressures. In this paper, a weed population is defined as all the individuals, including flowering plants as well as seeds buried in the soil, that are present in a cultivated field.

Mating System and Population Size. The mating system is one of the main factors regulating the movement of genes through pollen. For inbreeding weed species, gene flow is due primarily to seed dispersal which, in turn, depends on both the biology of the species and farmer's tillage and harvesting equipment. For instance, triazine resistance appeared initially as patches of resistant plants in maize fields, then spread slowly within the field in spite of high seed output. Meanwhile, movement of combine harvesters contaminated other fields with resistant seeds (*5*).

In partially and completely outbreeding species, pollen dispersal provides another opportunity of gene flow that is different from that realized by seed. Pollinators and wind may spread pollen over large distances, and outcrossing may result in new genotypes having characteristics different from their parents. Theory and models of population genetics provide the basis for the understanding of the effects of factors such as outbreeding rate, migration-distance of pollen, population size, plant density, gene inheritance, and selective value of a gene, but several conditions prevailing in farmers' fields are still poorly considered, like population disequilibrium due to changing farmers' practices, burried seed, mixture of generations, presence of crop cover and patchy distribution.

Outbreeding probably allowed some allogamous species like rigid ryegrass (*Lolium rigidum* Gaud.) and blackgrass (*Alopecurus myosuroides* Huds.) to develop multiple resistances and cross-resistances (*6*). When resistance is due to enhanced metabolism of herbicides, plants having weak resistance mechanisms have some chance to survive and therefore contribute to the next generation. It is likely that the allogamous mating system, under continuous selection pressure, favors concentration of genes which results in level of resistance high enough to survive herbicide field rates. When resistance is due to strong resistance mechanisms (target site resistance), mechanistic enrichment of resistance genes among survivors may lead to individuals expressing multiple resistance. However, large population size, herbicide selection pressure and rapid turn-over of generations are required for such a process. If the size of a population is too much reduced by farmers' practices, then rare alleles are lost, and even population structures are altered. For instance, calculations showed that at least 5000 interbreeding individuals are necessary to avoid losses of alleles in poppy populations (*Papaver rhoeas* L.), a self-incompatible species that produces huge amount of seeds (*7*). This number of plant is seldom reached in cultivated fields since poppy cannot make large populations in crops other than cereals and anti-dicotyledonous herbicides in cereals kill poppys easily.

Outbreeding and Environment Conditions. A good example of the influence of environment on the movement of genes among plants is found in lambsquarter (*Chenopodium album* L.). In a study to determine mating system, plants marked with known alleles at several isozyme loci were grown at various distances from each other in a crop-free area. When two plants were very close, they produced 50 % hybrid seeds, indicating that outcrossing was exactly proportional to the rate of foreign pollen in the immediate vicinity of a plant. In such a condition, lambsquarter behaved as a complete outbreeder. The rate of hybrid seed decreased linearly with the logarithm of the distance between plants, no more hybrids being found at 2.6 m or more. This result suggested that lambsquarter is only wind pollinated with reduced pollen dispersal even in a crop-free area. In a maize field, hybrid output was reduced by ten, probably because pollen flow by wind was prevented under the cover of maize (*8*).

Thus, crops such as sugar beet or peas, which are small plants, would allow lambsquarter plants to exchange genes if they are at a sufficiently high density. In contrast, in maize, oilseed rape and wheat, lambsquarter flowers below the top of the crop so that movement of lambsquarter genes would seldom occur. In addition, if a farmer keeps plant density as low as possible, the population becomes a mere group of inbred lines. When a mutation for herbicide resistance occurs, founder effect and selection pressure intensify the lack of polymorphism. Paradoxically, herbicide resistance lead to high plant density in infested area, thus creating suitable conditions for gene flow, but there is no more polymorphism within the population. This is why all triazine resistant populations of lambsquarter were completely deprived of polymorphism: the mutated line could not exchange genes with other lines (*9*).

Contrary to a commonly held idea, this lack of polymorphism is not a consequence of a cytoplasmically inherited resistance. In a contrasting environment like a private garden, where high plant density may occur during short time and no herbicide is sprayed, lambsquarter populations showed resistant mutants with various isozymes patterns, indicating either the mutation was frequent and recurrent, or polymorphism originated through gene exchange (*9*). Gene exchange with migrant plants from adjacent populations was shown to led to a polymorphic triazine resistant population of annual bluegrass (*Poa annua* L.), a moderately outcrossing species (*10*). High polymorphism and no difference with adjacent susceptible populations was found in a triazine resistant population of blackgrass, a complete outbreeder, therefore indicating that cytoplasmic inheritance was not a major factor limiting polymorphism (*11*).

Variation in patterns of gene flow may be specially important at the time a new gene first arises in a weed population. Its rarity makes it sensitive to a series of variations which are ignored on the average but which have local significance. Randomness and aggregation of plant distribution give local variation within the field. An analysis of isozyme patterns of half-sib families of wild mustard (*Sinapis arvensis* L.), an obligate outbreeder, showed non-random mating. Differences of allele frequency were observed among pollen clouds fertilizing plants at different places in the field, and genetic distances between sub-populations in areas of low plant density were lower than in areas of high plant density (Lefol, E. *et al.*, *Weed Res.*, 1996, 2, in press). This probably corresponds to pollinator-visitation differences between patches and areas of low density. As the probability of dispersal varies with the degree of patchiness, the frequency of favourable genes, such as those encoding for herbicide resistance, will increase inside small patches rather than at random over all the field. Thus, the overall gene flow will be smaller than expected from panmixia. In musk thistle (*Carduus nutans* L.), however, while the averaged dispersal distance of marker alleles was shown to be low, the distribution of pollen may extend to some distance, resulting in occasional long-distance movement of genes that could broaden the genetic diversity of founding sub-populations (*12*). The foraging behaviour of different pollinator species, such as butterflies or bumblebees in *Senecio* (*13*), also influences the pattern of gene flow and makes it more and more unpredictable.

Gene Flow and Inheritance. The mode of hereditary transmission of a resistance gene is a classical factor involved in modelling gene flow. Most of the herbicide resistances known to date are encoded by monogenic dominant genes (*14*). The only case of a recessive allele was found in green foxtail (*Setaria viridis* (L.) Beauv.), a highly self-fertilized species in which the release of homozygotes is favoured (*15*). Recessive alleles are likely to have few chance to be selected in the field and to spread through crossing compared to dominant alleles. With recessive mutation, heterozygotes are killed by herbicides. However, if homozygotes appear in a population, continuous selection pressure may result in fixation of the recessive allele. In contrast, pollen-mediated gene flow of dominant alleles confers immediately the resistance to the outcrossed progeny, that lead to a quick spread and establishement of the resistance. Dominant alleles always preserve polymorphism as the heterozygotes survive and transmit susceptible alleles to subsequent generations, especially in outcrossed species.

Oligogenic inheritance would be rare as it requires gene exchange to combine favorable alleles into the right genotype to express resistance. Indeed, the only case reported to date, with probably two additive genes, occurred in populations of an outbreeder, blackgrass (*14*). Outbreeding and large population size are necessary to allow the presence of a large number of partial resistance genes in a single generation. For inbreeder, or when field conditions reduce outcrossing, recombination among individuals takes many generations before releasing field resistant plants. For outbreeder, the combination of favorable alleles accumulated in a highly resistant phenotype is segregated each generation, so that the spread of resistance is very slow. However, as herbicide pressure tends to decrease due to environmental and economic concerns, it is likely that conditions for a continuous increase of oligogenic resistances are met.

In the case of cytoplasmically inherited resistance, like for triazine resistance, the transmission of alleles occurs through ovules only. Therefore, the spread of resistance genes is limited to seed dispersal from the mother plant, without segregation of alleles in offspring. In consequence, migrant seeds are immediately resistant, and the frequency of resistant alleles in the population is independent of pollen flow and determined only by the rate of survival. This property has been used to study the quantitative and spatial evolution of a single plant progeny of lambsquarter year after year in a maize monoculture (*5*). However, some exceptional transmission by pollen was demonstrated to occur in up to 1% of crosses in lambsquarter, and to a smaller extent in other species (*14*), but this is likely to have very few consequence on gene flow.

Movement of resistance genes may also be modified according to gene characteristics independent of inheritance. For instance, the possible expression of the resistance gene in the gametophyte would change pollen flow. Resistance to herbicides inhibiting ALS is expressed during pollen growth if the resistance is due to the mutation of the gene encoding for ALS, but not if it is due to detoxification (*16*). This allows selection for the mutated ALS gene if the pollen population is subjected to herbicide: spray or herbicide remaining in the plant. Spraying weeds at flowering was not likely in the past, but this now occurs more frequently in fallowing to prevent seed set. Pleiotropic secondary effects of the resistance gene might also be expressed in gametophytes and allow selective responses to other environmental factors which could change allele frequency during gene flow.

Secondary effects of resistance genes are known to have important effects on the fitness of resistant plants and, consequently, on the behaviour of the realized gene flow. Fitness is a measure of the number of progeny contributed by a genotype to the next generation. Male and female components of the fitness may be different (e.g. cytoplasmic resistance). It is generally difficult to obtain accurate estimates of relative fitness that can be extrapolated to field conditions. Fitness in the absence of herbicide is particularly important: resistance genes may have a biological cost. For instance, triazine resistant plants show a less efficient photosynthesis than susceptible plants. Therefore, herbicide rotations may be effective in slowing the spread of resistance in weed populations. However, secondary effects are not always detrimental on a biological or ecological ground, but farmer may use them in resistant-weed management strategies. Diclofop-methyl resistance in Italian ryegrass (*Lolium multiflorum* L.) seems to be associated to late anthesis. Inundating receptive stigmas of resistant plants with susceptible pollen before resistant pollen was released would reduce the development of herbicide resistance (*17*). Sulfonylurea resistance in kochia (*Kochia scoparia* L.) affects germination at low temperature. Earlier spring emergence of the resistant plants may allow some selective control decreasing the proportion of resistant plants in the population (*18*).

Gene Flow Between Populations

Although quantitatively small, gene flow between populations may have dramatic consequences on the spread of resistances genes on a regional scale. When resistant plants, previously confined within a field plot, become uncontrolled and produce very large amounts of seed, inevitably some seeds and pollen grains will successfully migrate. This movement provides new initial sources of resistance genes in farther fields. The importance of gene flow between populations is often compared to the mutation rate, but this has been examined in a few studies. It is often difficult to discriminate between gene flow and independent mutation events. Distinct isozyme phenotypes were observed in triazine resistant populations of lambsquarter from different regions of France, but the same phenotype was constantly found in populations from the same region (*9*). These data suggest spread of resistance both through gene flow and independent mutations. The existence of multiple resistance alleles in sulfonylurea resistant kochia also indicates that multiple founding events occurred rather than long distance spread (*19*).

Local seed dispersal may certainly play a major role in weed populations, especially for inbreeder species. The resultant migrant plant immediately display resistance. Seeds may spread through cultivation and harvest machines and are generally deposited at the field entry, as evidenced in the study of the distribution of a resistant pale smartweed (*Polygonum lapathyfolium* L.) population over three years (*20*). Roadsides are contaminated in the same way. The use of contaminated manure resulted in some case in new infested area although never treated with herbicides (*21*). Birds may be responsible for nearby spread as for black nightshade (*Solanum nigrum* L.), and wind allows large distance spread of horseweed (*Conyza canadensis* (L.) Croq.). Plants of Russian thistle (*Salsola iberica* Sen. and P.) may move up to several kms according to wind direction and velocity, thus disseminating seed at varying distances from the original plant site and allowing contamination of fencelines and roadside ditches with sulfonylurea resistance genes (*22*).

Gene flow through pollen is often a favourable factor for the resistant population. Due to the high selective value confered by triazine resistance, a resistant population of annual bluegrass developed in an habitat (roadside) selectively colonized by the annual erect ecotype. By chance, the initial resistant plant in this area was a perennial prostrate ecotype. The prostrate growth form provided a marker of the genetic evolution of a population facing pollen migration from populations of erect annual ecotype growing in untreated adjacent open areas. Although the polymorphism of the resistant population was lower than that of adjacent populations, it consisted of a high frequency of heterozygotes and large number of hybrids between the two ecotypes, features which have not been observed naturally elsewhere (*10*).

At the extreme of allogamy, the comparison of chlorotoluron resistant and susceptible populations of blackgrass from wide geographical origins showed very few differences. There was no departure from expected heterozygosity, and very low genetic differentiation among populations. Since there was no cost associated to the resistance, it was possible that the resistance genes would be distributed at random within populations long time before the herbicide selection pressure applied (*11*). In that case, gene flow between populations would be a serious concern for the spread of resistance genes, unless it is counterbalanced by spread of still larger amount of susceptible alleles.

Pollen-mediated gene flow of herbicide resistance genes is a function of distance, pollen characteristics, abundance of pollen, mating system and environmental conditions. Very few studies analysed pattern of pollen flow between weed populations. Indirect evidence was inferred from paternity analysis of progeny of natural populations of wild radish (*Raphanus sativus* L.), a self-incompatible insect-pollinated species. Gene flow rate of isolated populations ranged from 3.2 to 18.0 %, that suggested high potential for rapid spread of herbicide resistance genes (*23*). Outcrossing at long distance was lower in kochia as no more than 0.01 % of progeny of sulfonylurea susceptible plants grown 29 m away resistant plants expressed resistance (*24*). Pollen-mediated gene flow between populations of lambsquarter seems to be impossible since no outcrossing was detected between plants at distance higher than 2.6 m (*8*).

The immigration of susceptibility alleles into a weed population that is resistant can lead to its return to susceptibility, or at least can slow the evolution of resistance. To be effective, gene flow of susceptible alleles must be associated to some fitness advantage as reported above for diclofop-resistant Italian ryegrass (*17*). The source of susceptible alleles must be close to the resistant population as pollen is not likely to spread over large distances. This stategy, that consists in maintaining susceptible populations of weeds, pathogens and insects to oppose selection for resistance is called "the refugee concept". However, one must care that the refugee niche itself cannot be contaminated by migration of resistance genes.

Gene Flow Between Weeds Species

Botanists and agronomists have often described forms intermediate between two species that they interpreted taxonomically as spontaneous interspecific hybrids without providing genetic evidence. Gene flow intrinsically leads to variability and genetic continua, and therefore should be described within the framework of gene pool (*25*). Introgression, which describes how genetic material is occasionally transferred across taxa by accidental hybridization and backcrossing with either of the taxa, is a driving force in the make-up of gene pool (*26*).

Interspecific hybridizations and introgressions among wild species, and between native and invading plant species, have been described repeatedly (*26, 27*). They have led to new ecotypes, subspecies, homo- and allopolyploid species that sometimes survived in peculiar habitats only or also could invade large areas. Interspecific hybrids are generally more vigorous than both parents and probably allow introgression between the two parental species, as shown in thistle, *Carduus acanthoides* L. (2n=22) and *C. nutans* L. (2n=16) (*28*), and violet, *Viola arvensis* Murr. and *V. calaminaria* Lej. (*29*).

Speciation after hybridization has been described in some cases involving alloploid progeny and confirmed through artificial crossing in the laboratory. Deadnettle (*Lamium moluccellifolium* Fries., 2n=36) appears to be obtained from crosses between *L. purpureum* (2n=18) and *L. amplexicaule* L. or *L. bifidum* Syr. (2n=18) (*30*). Similarly, an allohexaploid groundsel species, *Senecio cambresis* Rosser (2n=60) formed after hybridization between *S. vulgaris* L. (2n=40) and *S. squalidus* L. (2n=20) (*31*). In both cases, a synthetic alloploid was derived from colchicine treated hybrids and looked like the wild polyploid type. These hybrid progeny are normally found only at sites where both parents occur.

In some other cases, hybrid descendants have spread to locations beyond the range of its progenitors. Common hempnettle (*Galeopsis tetrahit* L., 2n=28) is thought to have originated from crosses between *G. pubescens* Bess. (2n=14) and *G. speciosa* Mill. (2n=14). Artificial hybrids had low fertility, but one hybrid was triploid (2n=21) and could be backcrossed to *G. pubescens* and provided fertile progeny resembling the natural tetraploid species (*32*). Although not typically a weed, the wild salsify (*Tragopogon miscellus* Own., 2n=24), which originated from crosses between *T. dubius* Scop. (2n=12) and *T. pratensis* L. (2n=12), has broader ecological amplitude and expanded more than one of its parent, *T. pratensis*. It now grows sympatrically with *T. pratensis* in a few locations only (*33*).

Similarly, cordgrass (*Spartina anglica* Hubb., 2n=120-124), an amphiploid between the native *S. maritima* Curt. (2n=60) and the recently introduced *S. alterniflora* Lois. (2n=62), has colonized a much wider area than its progenitors (*34*). Just a century after its origin, it has become a dominant component of salt marshes on both sides of the Channel. Annual bluegrass (*Poa annua* L., 2n=28), which probably originated during the recent quaternary period after hybridization between *P. infirma* Kunth and *P. supina* Schrad, two diploid species (2n=14), spread worldwide. In contrast, its progenitors separated and occupy only small areas on the Mediterranean and temperate Atlantic borders and mountains of Central Europe, respectively (*35*).

However, apart from the studies reported above which were undertaken to demonstrate the origin of some new or invading taxa, no studies have been carried out on the hybridization between related weed species. Morphological, and especially caryological variations within populations and species are normally ascribed to troubles in the regulation of the chromosome number or selection for genetic differentiation, never as the result of introgression. There is here an open field for research on population polymorphism, as many *hybridus* taxa are found in weed species.

Gene Flow Between Crops and Weeds

By domesticating plants and improving crops, man has had a tremendous influence on the diversification of plants and has even created new species which are sometimes genetically well isolated from related species (*36*). Genetic engineering is a new technology which provides traits with high agronomic value. However, if wild organisms acquire the transgenes through pollen exchanges with genetically engineered crops, they may display new adaptive advantages or disadvantages which could change biological equilibria and cause new problems for farmers. For instance, it may enable wild plants to mimic crops, escape weed killers, resist pests, alter their phenology and reproduction.

The likelihood of such an event depends on the coexistence of the crop and wild relatives within a distance pollen may travel, on the simultaneousness of flowering of the various species, on their crossability, on the survival and reproductive ability of hybrids, and on the fate of the genes in wild populations (*37-39*). Weeds belonging to the same gene pool as crops may be more susceptible to such introgression than other wild species which fill niches less affected by man. To estimate frequency and impact of such events, we may learn from past introgressions.

Introgression and Succesful Speciation. Several recent studies have reviewed cases of introgression between crops and weeds which resulted in hybrids that were adapted to new environments. Examples include hybrid sorghum, wild sugar beet, wild radish, giant green foxtail, etc. (*40, 41*, Darmency, H., In *Weed and Crop Resistance to Herbicides*, De Prado, R.; Jorrin, J., Eds.; Kluwer Acad. Press: Dordrecht, NL, 1995; in press.). This confirms that genes can flow over species and be used to enhance the colonizing ability of weeds.

Loss of Species Identity. There are also cases where introgression with a cultivated plant seems to have provided no benefit to the wild plant. Carrots in Europe illustrate this. A single species is present, *Daucus carota* L., which includes two subspecies: *carota*, the wild carrot, and *sativus*, the cultivated carrot. *Carota* is annual in the Netherlands whereas *sativus* is biennial, and several morphological characters distinguish these subspecies. "Wild" populations at the edge of fields have intermediate characters, such as leaf shape, and are biennial.. Introgression has therefore occurred in the narrow contact zones between the two subspecies which are widely pollinated by many insects (*42*). How does this change in biology affect the wild plant? Ecological studies are required, but one can already guess that as these wild populations flower less often, they will change their life history strategy and become less noxious for farmers.

The case of American squashes is similar. The size and edibility of the fruit are visible signs of the gene exchanges going on between the cultivated species *Cucurbita pepo* L. and *C. moshata* Duch. and the related wild species *C. sororia* Duch. and *C. texana* (Scheele) Gray, respectively, with which they cross easily over long distances thanks to insects (*43*). One may wonder whether and how the wild plants benefit from this. This state of affairs amounts to gene contamination which may lead some of the wild populations to an ecological dead end. Introgression makes squashes expend a lot of energy for a type of reproduction which is unadapted without the help of man. However, plants that have not undergone introgression probably still hold a large adaptive advantage in the wild, enabling them to maintain the populations alive and to keep introgression at low level.

In rice, situations may arise which are beneficial to a taxon and others which may contribute to its disappearance. In the 1940's, rice growers in an Indian province started growing a variety of rice with a red pigmentation so that wild rice would show up easily for uprooting (*44*). This weeding strategy was thwarted by introgression: wild rice picked up the pigmentation gene which had an extremely high selective value since it was used as the weeding criterion. The *O. perennis* plants were intermediate in morphology, autogamy rate and seed dormancy (*45*).

On the other hand, it has been suggested that introgression of *O. perennis* by cultivated rice in Taiwan brought about the extinction of the local wild taxon. The frequency of wild characteristics, the fertility of pollen and the amount of seed produced all decreased in a series of samples collected from 1929 to 1976 (*46*). This coincided with an increase of the area where a second crop was grown, which flowers at the same time of year as wild rice does. Ecological investigations reveal that land development has modified the water balance in such a way that areas where wild rice grew opened up to competitive weeds which finally kept the wild rice from developing. A reduction in vegetative propagation and the loss of seed dormancy, both caused by introgression, may have contributed to the extinction of wild rice in Taiwan, although they were not the main factors (*47*).

Genetic Barriers. Apart from the litterature, possible introgression between crops and weeds may be predicted from studies on the nature of the genetic barriers between species and conditions under which they may be overcome. It is likely that gene systems which are lethal to, or weaken and degenerate, interspecific hybrids between wild and cultivated species became selected as a system against loss of species identity. In other terms, the more sympatric species are, the more they benefit from and the more likely they are to have such a barrier or different ploidy levels as a protection against gene homogenisation.

For instance, in maize (*Zea mays* L.), introgression is strongest with ssp *parviglumis* Ilt. & Doeb. which grows in wild habitats, not with the more sympatric subspecies with which pollination in the field is most likely, i.e. the wild mimetic maize ssp *mexicana* Ilt. & Doeb. (*48*). The genome structures of ssp *mexicana* and maize differ considerably from each other, which therefore reduces the likelihood of recombination and creates fertility anomalies protecting ssp *mexicana* from gene introgression from maize (*49*). In rice (*Oryza sativa* L.), either sterility of the hybrids, as with *O. glaberrima* Steud., or a lethal gene system, as with *O. perennis* Moench, give rise to a balance between isolation and introgression (*50*).

The various subspecies of pearl millet, (*Pennisetum americanum* (L.) Leeke), allogamous annual plants with the same number of chromosomes, all interbreed readily. African farmers regularly find hybrid forms in their fields which are very hard to distinguish from cultivated pearl millet until flowering or even ripening. It seems that the wild pearl millet growing in and around fields (*P. violaceum* (Lam.) L. Rich.) is better protected from cultivated pearl millet than the wild pearl millet which grow outside of cultivated areas (*P. mollissimum* Hochst). Cultivated characteristics are less likely to appear among the descendants of hybrids between the crop and the former taxon than among those of hybrids of crop and the latter taxon (*51*). Seed malformation may affect the F_2 generation (*52*). One may wonder whether these related varieties managed to maintain their own characteristics sympatrically because they had the appropriate genetic systems. The differences between wild and cultivated pearl millets are indeed arranged into linkage groups of coherent sets of characters (*51*).

Current Researchs on Introgression. Among the first commercially released herbicide and insect or virus resistant crops, rape is probably the crop for which recent investigations were the most detailed because every year millions of flowers undergo pollination. An indication of eventual hybridization between oilseed rape (*Brassica napus* L.) and a related wild Brassiceae, however rarely it may occur and even if hybrids are sterile, can have serious consequences. Rape itself is an allopolyploid resulting from a cross between *B. oleracea* L. and *B. campestris* L. (*53*). Hybrids between wild radish (*Raphanus raphanistrum* L.), or hoary mustard (*Hirschfeldia incana* Lagr-Foss), and oilseed rape were naturally produced in field experiments at a rate up to 800 hybrid per ha (*54*). The use of male sterile rape varieties increased this yield up to several million hybrid seeds per ha. Very few hybrids between rape and wild mustard (*Sinapis arvensis* L.) were obtained (*39*), but hybrids with *B. campestris* L. have been obtained at high rate (*55*). Hybrids developed well and were more competitive than the wild parent (*56*) but had very low fertility. Hybrids grown among the wild parent for backcrossing purposes produced fewer than two seeds per hybrid, i.e. 0.01% of what the wild parent produces. The odds that the offspring will establish themselves among a normal population are therefore quite low but cannot be ruled out. Population studies to detect past introgression using marker genes could provide an estimate of the normal gene flow. However, new genes obtained by genetic engineering are expected to confer a selective advantage on the hybrids and may change the effective gene flow.

Conclusion

Genes encoding herbicide and pest resistance may occur naturally in weed populations or be transmitted through pollen exchange with herbicide resistant crops. In the first case, their original frequency depends on the mutation rate and the ability of a species to maintain polymorphism within populations. Herbicide treatments and pest infestation directly select for them. Genes for resistance may remain very locally distributed or be widespread. Field conditions affect realized gene flow since they change mating system and pollen flow. Paying more attention to the control of weed patches in fields will reduce the potential of gene flow through pollen and seeds.

In the second case, all weed populations exposed to pollination by a related herbicide-resistant or pest-resistant crop may produce resistant hybrids at rates which are a function of plant genotype and environmental conditions. Waste habitats in which escaped crop plants grow together with weeds are also sources of hybridization. The spread of resistance genes will depend on the fitness of the hybrids and their descendants. Once transferred into the genome of weed species, these resistance genes will evolve in populations as well as naturally occurring genes.

In conclusion, gene flow among plants is a concern of special importance because it could potentially lead to the release of new weed genotypes that combine adaptive traits that are more and more problematic for the farmer. Although herbicide resistance is currently a primary concern for farmers, is easy and safe to study in scientific experiments, and is now ready for commercial release in transgenic cultivars, one must also consider seriously the spread of insect and virus resistance genes in weeds. The latter will strongly interfere with the ecological balance of wild habitats and the ease of resistance management

References

1. Slatkin, M. *Ann. Rev. Ecol. Syst.* **1985**, *16*, 393-430.
2. Levin, D.A.; Kerster, H.W. In *Evolutionary Biology*; Dobzhansky, Th., Hecht, M.K., Steeze, W.C., Ed.; Plenum: New York, USA, 1974, Vol. 7; pp 139-220.
3. Warwick, S.I. *Proc. Brighton Crop Conf. Weeds*; BCPC: Farnham, UK, 1991; Vol. 3, pp 997-1006.
4. Burdon, J.J.; Groves, R.H.; Cullen, J.M. *J. Appl. Ecol.* **1981**, *18*, 957-966.
5. Darmency, H.; Gasquez, J. *Weed Technol.* **1990**, *4*, 173-177.
6. Hall, L.M.; Holtum, J.A.M.; Powles, S.B. In *Herbicide Resistance in Plants: biology and biochemistry*; Powles, S.B.; Holtum, J.A.M., Ed.; Lewis Pub.: Boca Raton, USA, 1994; pp 243-261.
7. Lawrence, M.J.; O'Donnell, S. *Heredity* **1981**, *47*, 53-61.
8. Gasquez, J. *In Genetic Differentiation and Dispersal in Plants*; Jacquard, P. Ed.; NATO ASI Series, G5; Springer-Verlag: Berlin, D, 1985; pp 57-66.
9. Gasquez, J.; Compoint, J.P. *Agroecosystem* **1981**, *7*, 1-10
10. Darmency, H.; Gasquez, J. *New Phytol.* **1983**, *95*, 299-304.
11. Chauvel, B.; Gasquez, J. *Heredity* **1994**, *72*, 336-344.
12. Smyth, C.A.; Hamrick, J.L. *Evolution* **1987**, *41*, 613-619.
13. Schmitt, J. *Evolution* **1980**, *34*, 934-943.
14. Darmency, H. In *Herbicide Resistance in Plants: biology and biochemistry*; Powles, S.B.; Holtum, J.A.M., Ed.; Lewis Pub.: Boca Raton, USA, 1994; pp 263-297.
15. Jasieniuk, M.; Brule-Babel, A.L.; Morrison, I.N. *Weed Sci.* **1994**, *42*, 123-127.
16. Richter, J.; Powles, S.B. *Plant Physiol.* **1993**, *102*, 1037-1041.
17. Ghersa, C.M.; Martinez-Ghersa, M.A.; Brewer, T.G.; Roush, M.L. *Weed Technol.* **1994**, *8*, 139-147.
18. Thompson, C.R.; Thill, D.C.; Shafii, B. *Weed Sci.* **1994**, *42*, 50-56.
19. Guttieri, M.J.; Eberlein, C.V.; Thill, D.C. *Weed Sci.* **1995**, *43*, 175-178.
20. Darmency, H.; Compoint, J.P.; Gasquez, J. *C. R. Acad. Agric. Fr.* **1981**, *67*, 213-238.
21. Eberlein, C.V., Al-Khatib, K., Guttieri, M.J.; Fuerst, E.P. *Weed Sci.* **1992**, *40*, 507-512.

22. Stallings, G.P.; Thill, D.C.; Mallory-Smith, C.A.; Lass, L.M. *Weed Sci.* **1995**, *43*, 63-69.
23. Ellstrand, N.C.; Devlin, B.; Marshall, D.L. *Proc. Natl. Acad. Sci. USA* **1989**, *86*, 9044-9047.
24. Stallings, G.P.; Thill, D.C.; Mallory-Smith, C.A.; Shafii, B. *Weed Sci.* **1995**, *43*, 95-102.
25. Harlan, J.R.; de Wet, J.M.J. *Taxon* **1971**, *20*, 509-517.
26. Heiser, C.B. *Bot. Rev.* **1973**, *39*, 347-366.
27. Anderson, E. *Introgressive hybridization*; J. Wiley & Sons, New York, USA, 1949; pp 109.
28. Warwick, S.I.; Thompson, B.K.; Black, L.D. *Can. J; Bot.* **1990**, *68*, 1675-1679.
29. Kakes, P. *Acta Bot. Neerl.* **1982**, *31*, 371-378.
30. Bernstrom, P. *Hereditas* **1955**, *41*, 1-122.
31. Ashton, P.A.; Abbott, R.J. *Heredity* **1992**, *68*,25-32.
32. Muntzing, A. *Hereditas* **1932**, *16*, 105-154.
33. Novak, S.J.; Soltis, D.E.; Soltis, P.S. *Amer. J. Bot.* **1991**, *78*, 1586-1600.
34. Gray, A.J.; Marshall, D.F.; Raybould, A.F. *Adv. Ecol. Res.* **1991**, *21*, 1-62.
35. Tutin, T.G. *Watsonia* **1957**, *4*, 1-10.
36. Harlan, J.R.; de Wet, J.M.J.; Price, E.G. *Evolution* **1973**, *27*, 311-325.
37. Ellstrand, N.C. *TREE* **1988**, *3*, S30-S32.
38. Manasse, R.B. *Ecol. Appl.* **1992**, *2*, 431-438.
39. Darmency, H. *Mol. Ecol.* **1994**, *3*, 37-40.
40. Gepts, P. *Evol. Biol.* **1993**, *27*, 51-94.
41. Raybould, A.F.; Gray, A.J. *J. Appl. Ecol.* **1993**, *30*, 199-219.
42. Wijnheijmer, E.H.M.; Brandenburg, W.A.; Ter Borg, S.J. *Euphytica* , **1989**, *40*, 147-154.
43. Wilson, H.D. *Biosci.* **1990**, *40*, 449-455.
44. Dave, B.B. *Ind. J. Agric. Sci.*, **1943**, *13*, 46-53.
45. Oka, H.I.; Chang, W.T. *Phyton* **1959**, *13*, 105-117.
46. Kiang, Y.T.; Antonovics, J.; Wu, L. 1979. *J. Asian Ecol.* **1979**, *1*, 1-19.
47. Oka, H.I. *Bot. Bull. Acad. Sinica* **1992**, *33*, 75-84.
48. Doebley J.; Goodman, M.M.; Stuber C.W. *Econ. Bot.* **1987**, *41*, 234-246.
49. Smith, J.S.C.; Goodman, M.M.; Kato, T.A. *Econ. Bot.* **1982**, *36*, 100-112.
50. Chu, Y.E.; Morishima, H.; Oka, H.I. *Japan. J. Genetics* **1969**, *44*, 207-223.
51. Pernès, J. *La vie des Sciences* **1985**, *2*, 429-447.
52. Amoukou, A.I.; Marchais, L. *Euphytica* **1993**, *67*, 19-26.
53. U, N. *Jap. J. Bot.* **1935**, *7*, 389-452.
54. Darmency, H.; Fleury, A.; Lefol, E. *Proc. Brighton Crop Conf. Weeds*; BCPC: Farnham, UK, 1995; Vol. 2, pp 433-438.
55. Jorgensen, R.B.; Andersen, B. *Amer. J. Bot.* **1994**, *81*, 1620-1626.
56. Lefol, E.; Danielou, V.; Darmency, H.; Boucher, F.; Maillet, J.; Renard, M. *J. Appl. Ecol.* **1995**, *32*, 803-808.

Chapter 22

Genetically Altering Insects: Promise, Prospects, and Limitations

K. J. Hughes[1], S. K. Narang[2], R. A. Leopold[1], O. A. Johnson[1], and J. D. DeVault[1]

[1]Bioscience Research Laboratory, Agricultural Research Service, U.S. Department of Agriculture, 1605 Albrecht Boulevard, Fargo, ND 58105

[2]National Program Staff, Agricultural Research Service, U.S. Department of Agriculture, Building 005, BARC-West, Beltsville, MD 20705

Recent advances in molecular genetics have shown the potential of transgenic technology in a number of scientific disciplines. The use of transgenic technology in the field of entomology will revolutionize our understanding of beneficial insect species while providing new avenues towards the control of pesticide resistant species. This review highlights the most common techniques used to introduce desired gene constructs into insects and explores potential uses for this technology.

The control of agriculturally and medically important insect pests over the past forty years has been achieved mostly through the use of chemical insecticides. However, insect resistance to these compounds, increasing concerns about the impact of insecticide use on the environment and human health, and the high cost of insecticide development have resulted in fewer pesticides being available to replace older and less effective compounds. Consequently, the disciplines of horticulture, chemistry, and entomology have been challenged to identify and develop environmentally safe methods to control important insect pests. These methods have included the release of natural predators to control infestations e.g. the release of *Rodolia cardinalis* to control cottony cushion scale, cultivation of plants resistant to either insects or other plant pathogens e.g. the use of resistant rootstock to control *Phylloxera vitifoliae* in France, the mass release of sterilized insects that was used successfully to eradicate the screwworm *Cochliomyia hominivorax* from the Southeastern United States, and the identification of biologically based- insecticides such as *Bacillus thuringensis* (*Bt*) toxin.

Recently, the powerful tools of molecular biology have been used to develop

Published 1996 American Chemical Society

cotton plants that harbor the genes encoding the *Bt* toxin. The expression of the *Bt* toxin genes in these plants confers resistance to most cotton pests. The development of transgenic cotton plants illustrates the enormous potential of molecular genetics in the design of innovative insect control strategies. The application of molecular genetics will enhance current insect control strategies, such as sterile insect technology, by allowing the mass rearing of only male or female insects before sterilization. What is more important, transgenic technology will permit the manipulation of the genomes of both insect pest species and beneficial insect species.

However, the full potential of transgenic technology in the design of effective insect control strategies has not been realized. This is because molecular biology based insect control programs require the integration of several distinct technologies. First, traits that are desirable must be identified and characterized. Next, the gene(s) involved in the expression of the desired trait must be cloned and characterized. Finally, these genes must be suitably modified and successfully introduced into the genome of the target organism so that the desired trait will be transmitted in a Mendelian fashion.

The aim of this chapter is two-fold: 1) provide an overview of current methods used to introduce foreign DNA into insect embryos and stably integrate foreign DNA into the host genome, and 2) discuss ways in which transgenic methodology can be used to design effective insect control programs. We hope these discussions will suggest other research avenues and identify potential problems that need to be addressed to achieve maximum benefit from transgenic technology in insect control.

DNA Delivery Methods

Microinjection. The success of any transgenesis technology depends upon the effectiveness of introducing foreign DNA into an organism. Presently, the most commonly used technique to introduce DNA into insect embryos is microinjection. This technique involves: chemical or mechanical removal of the egg chorion; desiccating the dechorionated eggs slightly; covering the eggs with a halocarbon oil; and injecting the posterior end of the egg near the primordial germ cells (pole cells) with a 14-16 μm glass needle (*1*). Injected embryos are then placed in a humid environment until hatching occurs. In several cases, removal of the egg chorion was not possible, but some survival was obtained by injection through an intact chorion. Approximately 80% of injected *Drosophila* embryos hatch and 40 to 50% of these embryos survive to adulthood under optimal conditions; however, most of these surviving adults (10 to 90%) may have injection-induced abnormalities or are sterile because of incompatibilities with their genetic backgrounds (*2*). Survival rates similar to that for *D. melanogaster* have been observed for other insect embryos following microinjection. For example, 3.2% of *Bombyx mori* (silkmoth) embryos and 6 to 15% of mosquito embryos developed to adults after an injection during the embryonic stage (*3-6*).

A related microinjection procedure, injection of the egg within a gravid female mite, *Metaseiulius occidentalis*, has provided a means of obtaining stable transformation of this organism (*7*). A 0.5% transformation frequency was obtained without using a vector system and survival of the injected progeny was reported to be an order of magnitude higher than that of injected mosquito embryos. Whether the stable transformation gained by this method is related to the unique reproductive biology of this phytoseiid mite or to introduction of exogenous DNA before fertilization of the egg, is not yet known. However, further study is warranted since this method may have application for transforming certain oviviparous insects such as the tsetse fly.

Electroporation. While microinjection has been used to successfully introduce plasmid DNA into insect embryos, there are some drawbacks to this procedure. Microinjection is a time-consuming process requiring considerable technical expertise. Additionally, the microinjection technique is not suitable when experimental procedures require large sample populations. An alternate method for introducing DNA into insect embryos is electroporation. Electroporation is a procedure where the insect embryos are suspended between two electrodes in a liquid medium containing the DNA construct and then are given a single or a series of high voltage pulses. Transient holes or pores are produced in the egg shell and/or embryonic membranes through which the DNA enters. The relative ease of the electroporation procedure allows large numbers of embryos to be treated within a short time. Also, since there is no physical insult to the developing embryo as with microinjection, electroporation provides a noninvasive means to introduce foreign DNA into target embryos.

This technique has been used to successfully introduce DNA into *Oryzias latipes* and *D. melanogaster* embryos (*8,9*) and cultured cells of *Spodoptera frugiperda* (*10*). In our laboratory, we have used electroporation to successfully introduce several reporter gene constructs into *Helicoverpa zea* embryos (unpublished data). This represents, to our knowledge, the first use of electroporation to deliver plasmid DNA into any lepidopteran embryo. We observed that 17 - 30% of the electroporated *H. zea* embryos harbored plasmid DNA sequences and survival rates of 20 -30% for the electroporated *H. zea* embryos. Whether our electroporation protocol can be modified for use with other lepidopterans will need to be addressed in future studies. Nevertheless, the rapidity and ease of the electroporation procedure makes this technique an attractive alternative to the microinjection process.

Alternate DNA Delivery Methods. Plasmid DNA has also been introduced into *D. melanogaster* and *Anopheles gambiae* embryos using the biolistic bombardment technique. Balderelli and Lengyel coated particles 1.2 μm in diameter with a DNA construct and blasted the microprojectiles shotgun-like, into samples of 10-20,000 dechorionated, preblastoderm *D. melanogaster* embryos using a commercially available bioparticle delivery system (*11*). Mialhe *et al.* used a similar procedure to treat samples containing 3-5,000 mosquito embryos (*12*). In the latter study, the best results were gained without removing the egg chorion, but before hardening had occurred (60-80 min postoviposition). Survival to adulthood was not reported in either study and frequency of transfection was variable but repeatable.

Viral infection is another method of DNA delivery that has not been exploited in insects. Recently, Hughes *et al.* demonstrated that *Mamestra brassicae* nuclear polyhedrosis virus (*Mb*NPV) sequences could be detected throughout the life cycle of *M. brassicae* (*13*). Whether *Mb*NPV integrates into the genomic DNA, such as a hepatitis B virus (*14*) or is maintained as a nuclear episome similar to herpes simplex virus (*15*) is unclear. Nonetheless, these results indicate that modified viral vectors may be used to deliver target genes to individuals throughout a population.

The use of viral vectors may also be an especially appropriate method for introducing gene constructs into hymenopteran parasitoids. Many of these parasitoid wasps transmit viruses of the polydnavirus family to their hosts (*16*). Interestingly, these polydnaviruses appear to integrate into the genomic DNA of the parasitoid (*17*). If the mechanism of polydnavirus integration into the host genome can be determined, these types of viruses might be useful as natural transformation vectors.

Enhancement of DNA Delivery. When stable transformation is the desired goal, strategies having the potential to increase the probability of gene integration into germline cells will enhance the success of any transfection scheme. Injection of gene constructs conjugated with yolk proteins into the hemocoel of female insects during egg development might be a means of obtaining selective uptake by oocytes prior to their fertilization (*18*). Other strategies might also consist of injecting the developing testes with the gene construct of interest (*7*) or transfecting sperm with the desired gene *in vitro* before artificial insemination (*19*). Furthermore, a cationic liposome DNA carrier was used successfully to transfect murine sperm (*20*), but this method did not yield transgenic mice when the transfected sperm were tested via *in vitro* fertilization. In insects, foreign DNA was shown to bind externally to sperm of *Lucilla cuprina* (Australian sheep blowfly) and *Apis mellifera* (honeybee) without benefit of a chemical mediator (*21*).

The transplantation of primordial germ cells (pole cells) between mutant lines of *Drosophila* has been used to create germline mosaics for genetic and developmental biology studies. Increasing the production of transgenic insects may be possible by the *in vitro* transfection of pole cells collected by the mass isolation technique of Allis *et al.* (*22*). Complexing the gene construct with a chemical mediator for DNA such as cationic liposomes, polybrene, DEAE-dextran or polylysine before injection, electroporation or bombardment of the embryos could increase the efficiency of transfection. The chemical mediator would hopefully extend the life of an introduced DNA construct through several cycles of cleavage divisions of the preblastoderm embryo and thus increase the chances for integration when using a gene vector or by random integration.

Transposons as Gene Transfer Vectors

The ultimate goal of any transgenesis procedure is the stable integration of the desired gene into the genome of the target organism. Thus far, transposon vectors have proven to be the most efficient vehicles to achieve genetic transformation. Transposons are mobile genetic elements that are widespread in the genomes of all living organisms (*23*). They can comprise upwards of 20% of the genome in the higher eukaryotes and are responsible for a variety of genetic phenomena, including DNA insertions, deletions, and chromosomal rearrangements (*23*). A genome will either harbor a few copies of the autonomous (transposase-competent) full-length element, along with numerous copies of internally deleted, defective elements or will be totally lacking the element (*24-26*).

The three best characterized insect transposons are P-element, *hobo* and *mariner* (*27*). Of these, P-element has been exploited with a great deal of success as a transformation vector for *Drosophila* and closely related species (*28*). However, we believe that *hobo* and *mariner* will eventually be exploited as transformation vectors for insect species. Unlike P-element, *mariner* and mariner-like sequences have been detected in a number of different insect species (*29*) whereas *hobo* belongs to a family of transposons (*h*AT family) whose members include Activator from *Zea mays* and Tam3 from *Antirrhinium majus* (*30*). DeVault and Narang have identified, cloned, and sequenced the open reading frames of two *hobo*-like transposons from *H. virescens* and *H. zea* (*31*). Additional studies have also identified *hobo*-like sequences in a number of other insect species (J. D. DeVault, unpublished observation).

The ubiquity of *mariner* and *hobo* elements across species suggest that mobilization of the *Drosophila hobo* element in a heterologous host can occur. Indeed, Atkinson *et al.* have demonstrated that the *D. melanogaster hobo* element can be mobilized in Australian sheep blowfly and in the housefly (*Musca domestica*) in the absence of vector-encoded transposase functions (*32*). We have observed mobilization of the *hobo* element in *H. zea* embryos as well as in cultured cells of *H. zea* and *Trichoplusia ni* (unpublished observations). As before, *hobo* excision was observed in these lepidopteran hosts in the absence of any vector-encoded transposase functions. These results, along with those of Atkinson *et al.*, illustrate the potential of *hobo* and other transposable elements as gene transfer vectors for a wide range of insects.

Use of Transgenic Technology in Insects

The ability to manipulate the genome of insects will have significant impact on the design and implementation of insect control programs. For example, transgenic insects could be constructed such that either male or female progeny is produced under certain conditions. This would allow the mass rearing of sex-specific insects that could be sterilized by conventional methods and released in areas of greatest infestation. Alternatively, transgenic insects could harbor an environmentally regulated promoter (i.e.: temperature inducible) linked to a gene that would induce sterility in the insect. This functional element could be a dominant mutation or as antisense mRNA directed against a sex determining gene. As before, these sterile insects could be mass released in areas of infestation to suppress the pest population. Finally, transgenic insects that carry a gene encoding a defective metabolic enzyme could be repeatedly mass released into the environment. This transgenic trait would be introduced into the progeny, and these progeny would have reduced fitness because of this disruption of a metabolic pathway.

Transgenic technology will also have a significant impact on the use of natural enemies to control pest species. Currently, the use of natural predators to control pests in agroecosystems is hampered by the application of pesticides to control pests for which there are no biological enemies. This often results in the elimination of the introduced insects as well as other beneficial insects. However, genes that encode resistance mechanisms to a number of pesticides including organophosphates, pyrethroids and cyclodienes have been cloned and characterized (*33*). The introduction of these genes into natural enemies of pest species will allow these agents to be released into areas where pesticides are still being used to control pests for which there are no biological enemies. Because of the engineered pesticide resistance, the transgenic natural enemy will be able to establish itself in the agroecosystem and control the targeted pest population.

Transgenic technology will also allow genetic manipulation of economically important insect species. For example, honeybees and silkmoths could be altered so that these insects are more resistant to bacterial and viral infections (*33*). Also, the production of silk by silkmoths could be increased by introducing additional fibroin genes into silkmoths. As these examples and those cited above demonstrate, genetic engineering of insects will revolutionize a number of disciplines in applied entomology.

Ecological Implications

There are several issues that immediately appear when considering the release of a genetically engineered organism. A primary concern is whether the introduced gene will stay within the intended species. In bacterial systems, the presence of a specific transposon at a critically high copy number prevents further mobilization by that element (*34*). This phenomenon is referred to as transposable immunity. If transposable immunity also occurs in eukaryotic species, the widespread occurrence of the *hobo* and *mariner* elements in nature should prevent the further spread of modified *hobo* or *mariner* elements containing the gene of interest into natural populations. Also, the potential dispersal of the these insects into other areas and the effect of the transgenic insect on other species present in the ecosystem will need to be determined prior to any mass release of transgenic insects. Finally, the fitness of the transgenic insect in the environment will need to be assessed before mass releases take place. Only after these concerns have been addressed will transgenic insects become incorporated into an integrated pest management program.

Summary

We have discussed a number of possible ways in which biotechnology could be used in developing new and innovative insect management programs. While these technologies are realizable, the base of knowledge surrounding the genetics, biochemistry, and developmental biology of agriculturally and medically important insects needs to be significantly increased. Only then, will the full power of molecular genetics be brought to bear on the control insect pest species.

Literature Cited

1. Ashburner, M. *Drosophila- A Laboratory Handbook*; Cold Spring Harbor Laboratory Press, Cold Spring Harbor, NY, 1989.
2. Pirrotta, V. In *Vectors- A Survey of Molecular Cloning Vectors and Their Uses;* Rodriguez, R. L.; Denhardt, D. T., Eds.; Butterworths: Boston, MA, 1988.
3. Miller, L. K.; Sakai, R. K.; Romans, P.; Gwadz, R. W.; Kautoff, P.; Coon, H. G. *Science* **1987**, *237*, 779-781.
4. McGrane, V.; Carlson, J. O.; Miller, B. R.; Beaty, B. J. *Am. J. Trop. Med. Hyg.* **1988**, *39*, 502-510.
5. Morris, A. C.; Eggleston, P.; Crampton, J. M. *Med. Vet. Entomol.* **1989**, *3*, 1-7.
6. Nikolaev, A. I.; Chkoniya, T. I.; Kafiani-Eristavi, K. A. *Molec. Biol.* **1991**, *25*, 896-904.
7. Presnail, J. K.; Hoy, M. A. *Proc. Natl. Acad. Sci. USA.* **1992**, *89*, 7732-7736.
8. Kamdar, P.; Von Allmen, G.; Finnerty, V. *Nucleic Acids Res.* **1993**, *20*: 3526.
9. Lu, J. K.; Chen , T. T.; Chrisman, C. L.; Andrisani, O. M.; Dixon, J. E. *Molecular Marine Biology and Biotechnology.* **1992**, *1*, 366-375.
10. Mann, S. G.; King, L. A. *J. Gen Virol.* **1989**, *70*, 3501-3506.
11. Baldarelli, R. M.; Lengyel, J. A. *Nucleic Acids Res.* **1992**, *18*, 5903.
12. Miahle, E.; Miller, L. H. *Nucleic Acids Res.* **1994**, *16*, 924-931.
13. Hughes, D. S.; Possee, R. D.; King, L. A. *Virology.* **1993**, *194*, 608-615.
14. Howard, C. R. *J. Gen. Virol.* **1986**, *67*, 1215-1235.
15. Mellerick, D. M.; Frazer, N. W. *Virology.* **1987**, *158*, 265-275.
16. Fleming, J. G. W. *Ann. Rev. Entomol.* **1992**, *37*, 401.
17. Fleming, J. G. W.; Summers, M. D. *Proc. Nat. Acad. Sci. USA.* **1991**, *88*, 9770.
18. Handler, A. M.; O'Brochta, D. A. *Ann. Rev. Entomol.* **1991**, *36*, 159.
19. Sawa, M. *Jpn. J. Genet.* **1991**, *66*, 297.
20. Bachiller, D.; Schellander, K.; Peli, J.; Ruther, U. *Molec. Reprod. Dev.* **1991**, *30*, 194.
21. Atkinson, P. W.; Hines, E. R.; Beaton, S.; Matthaei, K. I.; Reed, K. C. *Molec. Reprod. Dev.* **1991**, *29*, 1.
22. Allis, C. D.; Waring, G. L.; Mahowald, A. P. *Dev. Biol.* **1977**, *56*, 372.
23. Georgiev, G. P. *Eur. J. Biochem.* **1984**, *145*, 203-220.
24. Goldberg, M. L.; Sheen, J.; Gehring, W. J.; Green, M. M. *Proc. Natl. Acad. Sci. USA* **1983**, *80*, 5017-5021.
25. Caizzi, R.; Caggese, C.; Pimpinelli, S. *Genetics* **1993**, *133*, 335-345.
26. Engels, W. R. In *Mobile DNA*; Berg, D. E.; Howe, M., Eds. American Society for Microbiology: Washington, D. C., 1989;
27. Heilmann, L. J.; DeVault, J. D.; Leopold, R. A.; Narang, S. K. In *Genetic*

Applications to Arthropods of Biological Control Significance; Narang, S. K.; Bartlett, A. C.; Faust, R. M., Eds. CRC Press: Ann Arbor, MI, 1993. 167-189.
28. Rubin, G. M.; Spradling, A. C. *Science* **1982**, *218*, 348-353.
29. Robertson, H. M. *Nature* **1993**, *362*, 241-245.
30. Calvi, B. R.; Hong, T. J.; Findley, S. D.; Gelbart, W. M. *Cell* **1991**, *66*, 465-471.
31. DeVault, J. D.; Narang, S. K. *Biochem. Biophys.Res. Comm.* **1994**, *203*, 169-175.
32. Atkinson, P. W.; Warren, W. D.; O'Brochta, D. A. *Proc. Natl. Acad. Sci. USA* **1993**, *90*, 9693-9697.
33. Hoy, M. A. *Insect Molecular Genetics: An Introduction to Principles and Applications*; Academic Press: San Diego, CA, 1994; 431-475.
34. Kleckner, N. *Ann. Rev. Genetics* **1981**, *15*, 341.

Chapter 23

Monitoring Strategies for Early Detection of Lepidoptera Resistance to *Bacillus thuringiensis* Insecticidal Proteins

Steven B. Sims[1], John T. Greenplate[1], Terry B. Stone[1], Michael A. Caprio[2], and Fred L. Gould[3]

[1]Ceregen, 700 Chesterfield Parkway, North Chesterfield, MO 63198
[2]Department of Entomology, Mississippi State University, Mississippi State, MS 39762
[3]Department of Entomology, North Carolina State University, Raleigh, NC 27695

We discuss assay approaches for monitoring the sensitivity of Lepidoptera to *Bacillus thuringiensis* (Bt) insecticidal proteins and compare the relative sensitivity of larval feeding bioassays in which, respectively, mortality or growth inhibition were scored. *Heliothis virescens* (F.) and *Helicoverpa zea* (Boddie), major lepidopteran pests targeted for control by transgenic cotton, were used for assay comparison. Larval growth inhibition assays using sublethal CryIA(c) protein concentrations were considerably more sensitive than dose-response mortality assays. Growth inhibition assays were easy to set-up and read, and could readily deliver a diagnostic dose allowing for visual discrimination of resistant from susceptible phenotypes. The ability of a larval growth assay, combined with a diagnostic dose, to unambiguously separate resistant from susceptible insects was validated using a CryIA(c) protein resistant strain of *H. virescens* and F_1 hybrids derived by crossing the resistant strain to a susceptible *H. virescens* strain.

Threat of Insect Resistance to Transgenic Plants Producing Bt-Proteins

Transgenic plant technology has the potential to provide significant improvements in crop protection and benefit to growers. The "first generation" of transgenic plant products produce *Bacillus thuringiensis* (Bt)-derived insecticidal proteins that are pest-specific, environmentally safe, and extremely effective. Unfortunately, the long-term success and maintenance of transgenic plant effectiveness is threatened by the development of insect resistance (*1,2*). For example, many geographically isolated populations of the diamondback moth, *Plutella xylostella* (L.), have already developed field resistance to microbial preparations of Bt which has led to control failures (*2*). Several other species of Lepidoptera and Coleoptera that are now controlled by transgenic crops can develop high levels of resistance to Bt proteins under laboratory selection (*2*-4, Luttrell, R., Mississippi State University, unpublished data). Consequently, the most controversial issues accompanying the introduction of Bt-producing transgenic crops have centered around the potential for insect resistance and the

0097–6156/96/0645–0229$15.00/0

preemptive implementation of appropriate resistance management procedures (*1,2*).

Current resistance management theory suggests that a useful resistance management strategy for a transgenic crop involves combining an "optimal" dose of insecticidal protein with a refuge of non-transgenic plants (*5*). If resistance is a recessive or partially recessive trait, then most heterozygous individuals will be unable to survive an optimal dose (also referred to as a "high" dose) of the insecticidal protein. Genetic analysis of Bt-resistant Lepidoptera strains (*Plodia interpunctella* (Hübner), *P. xylostella, H. virescens*) generally supports the assumption that Bt-resistance is a partially (or incompletely) recessive character (*2,4*). The number of genetic elements contributing to resistance is less clear, even for a relatively well-studied species such as *H. virescens* (*4,6*). Genes for resistance will initially be rare in populations and homozygous resistant individuals, with the greatest potential for survival on transgenic plants, will initially be extremely rare compared to heterozygotes (*4,7*). Adequate refuges and successful production of susceptible insects will increase the probability that any resistant homozygote will mate with a susceptible individual to produce heterozygous progeny. This "assumed recessive trait + optimal dose + refuge" resistance management strategy is the cornerstone of several first-generation insect-control plant products.

Resistance Monitoring

An important, but sometimes neglected, component of all preemptive Bt resistance management strategies involves the simultaneous implementation of an efficient resistance monitoring program. Data from appropriate monitoring programs helps us to evaluate the effectiveness of resistance management strategies and permits early detection of resistant phenoypes. Under favorable circumstances, this would allow remedial measures to be implemented prior to control failures (*8*). Historically, the development and implementation of significant resistance monitoring procedures for chemical insecticides has followed, rather than preceded, the initial occurrence of control failures. The availability of field-derived resistant phenoypes permitted studies on level (magnitude) of resistance, genetics of resistance, and resistance mode of action. In addition, practical resistance monitoring assays were developed, validated using resistant insect strains, and used to study subsequent changes in insecticide susceptibility within resistant and non-resistant populations. Studies on *H. virescens* and *Helicoverpa armigera* (Hübner) populations resistant to pyrethroid insecticides represent good examples of this approach (*9,10*).

Diagnostic Doses

Resistance to conventional chemical and microbial insecticides has typically been monitored and compared using the log-dose probit mortality responses of insect strains. This approach allows calculation of a resistance ratio (the LD_{50} or LC_{50} of the field test strain divided by the LD_{50} or LC_{50} of a reference susceptible strain) and statistical comparison of the LD_{50}s and slopes of the probit regression lines (*11,12*). Similarly, "baseline" susceptibility studies on insects targeted for control by transgenic plants have generated LC_{50}s and slope estimates for different populations exposed to the specific Bt protein incorporated into a suitable diet matrix (*13-15*, Diveley, G., University of Maryland, unpublished data). However, LC_{50}s and slope estimates, although suitable for distinguishing resistant phenotypes at a high frequency, are not adequately sensitive for detecting resistance when the incidence of resistance is low, e.g. 10^{-3} - 10^{-4} (*7*). Diagnostic doses (i. e. doses that unambiguously discriminate between resistant and susceptible phenotypes) are a more efficient means of finding resistant phenotypes because all individuals tested provide useful data (*7,8*).

Assay Description

The dose-mortality response assays and larval growth inhibition assays that we compared were initiated in a similar manner. Approximately 24 mL of a liquid agar-based insect diet (*16,17*) with 20% of the water omitted was added to a 6 mL sample of test liquid (distilled water containing a dose of the CryIA(c) protein). Treated diet was blended using a Vortex mixer, poured into 96-well insect assay trays (Jarold Mfg. Co., St. Louis, MO), and allowed to cool and harden. Each well of the assay tray had a 2.0 mL capacity and contained 1.0 - 1.5 mL of treated diet. One 1st instar *H. virescens* or *H. zea* larva was added to each well. The wells were then covered with Mylar plastic and ventilated with a single insect pin hole. Assays were incubated at 28 ± 2°C and evaluated after 7 days by scoring the number of survivors (individuals showing movement when probed with a needle) per concentration or by weighing larvae in groups of 10 - 48 and calculating the mean larval weight. The dose-response function of treatments was fit using either probit analysis (mortality data) or non-linear regression analysis for larval weight data (*18*). The non-linear logistic model used was: weight = W_0 / [(1 + (concentration / $EC_{50})^B$]where W_0 is the expected control weight, concentration is the amount of CryIA(c) protein per mL of diet, EC_{50} is the effective concentration of CryIA(c) protein that is expected to reduce larval weight by 50%, and B is the logistic function slope parameter (*19*). For calculation of the EC_{99} values and 95% CIs, the modified equation used was: weight = W_0 / [(1 + (100-1) (concentration / $EC_{99})^B$].

Dose Mortality Response Evaluation

We re-evaluated the data of Stone and Sims (*14*) by examining the combined dose mortality responses of 12 strains of *H. virescens* and 15 strains of *H. zea* to purified 63 kDa (trypsin-activated) CryIA(c) protein. There was initial indication of significant differences in CryIA(c) protein susceptibility among population samples from distinct geographic locations. However, the present analysis combined all data to examine the potential of a single dose, using mortality as an endpoint, to discriminate between resistant and susceptible individuals over a significant proportion of each species' distribution. The results are presented in Figures 1 and 2. Each data point represents an assay determining the percent mortality response of 24 - 48 larvae exposed to the indicated dose. The total number of assays contributing to the analyses for *H. virescens* and *H. zea* were 234 and 456 respectively. The data sets were evaluated by probit analysis to estimate

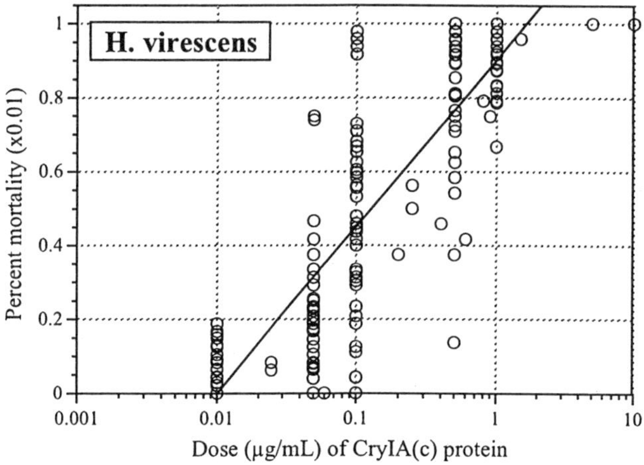

Figure 1. Mortality response of *Heliothis virescens* larvae to purified 63-kD CryIA(c) protein.

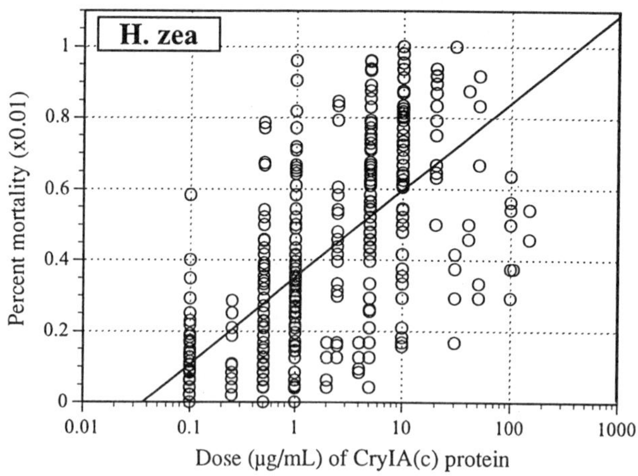

Figure 2. Mortality response of *Helicoverpa zea* larvae to purified 63-kD CryIA(c) protein.

LC_{99} values and 95% confidence limits. LC_{99} estimates for the 63 kDa protein were 3.3 μg/mL (95% CI = 2.3 - 5.3) for *H. virescens* and 6661 μg/mL (95% CI = 1003 - 2.12 x 10^5) for *H. zea.* Because transgenic cotton produces the non-activated, full-length CryIA(c) protein (~ 130 kD) that is approximately 2X the molecular weight of the trypsin-resistant core, the LC_{99} estimates for the full-length CryIA(c) protein are 6.6 μg/mL for *H. virescens* and 13322 μg/mL for *H. zea. H. zea* clearly is significantly less sensitive to the CryIA(c) protein than *H. virescens* and would require a very high concentration if the LC_{99} was used as a possible diagnostic dose.

Growth Inhibition Response Evaluation

Growth inhibition of larvae in response to purified full-length CryIA(c) protein was studied using *H. virescens* and *H. zea* from the USDA, Stoneville, MS laboratory colonies and two additional *H. zea* colonies initiated from Brooksville, MS. The results are presented in Figures 3 and 4. Each data point represents one assay determining the mean larval weight (mg) of a sample of 10 - 32 larvae in response to the indicated dose. The total number of assays contributing to the analyses for *H. virescens* and *H. zea* were 178 and 173 respectively. The data set for each species was fit by nonlinear regression to estimate EC_{99} values, i. e. the concentrations required to reduce larval weight to 1% that of the mean control weight, and 95% confidence intervals. EC_{99} values were 0.058 μg/mL (0.030 - 0.086) for *H. virescens* and 28.8 μg/mL (-7.4 - 65.1) for *H. zea*. These estimates are considerably lower (114-fold less for *H. virescens*, 463-fold less for *H. zea*) than the corresponding LC_{99} estimates for the full-length CryIA(c) protein.

Diagnostic Doses and Resistance Monitoring

The CryIA(c) EC_{99} diagnostic doses indicated for *H. virescens* (0.058 μg/mL) and *H. zea* (28.8 μg/mL) provide reasonable starting points for the dose-setting process. For *H. zea*, the EC_{98} (6.6 μg/mL, 0.1 - 13.0) might be more practical because it provides adequate discrimination (stunting) of susceptible larvae at a much lower concentration. In general, the lowest test concentration providing the requisite degree of larval growth inhibition should be selected. Final diagnostic doses are probably best achieved empirically by testing populations from across the geographic range of each species against 1 or 2 doses that bracket the doses proposed here. This multi-population dose-setting procedure was used to establish discriminating doses of microbial Bt products against Australian *Helicoverpa armigera* and *Helicoverpa punctigera* (*20*). A possible monitoring approach for obtaining initial information on the intensity of resistance would be to simultaneously use more than one diagnostic dose (see *21*) although a sequential testing procedure for *H. virescens and H. zea* would probably be more economical. In our studies, essentially all healthy larvae of both species tested on control diet were 3rd - 5th instars and weighed ≥ 10 mg (usually ≥ 100 mg) after 7 days. Therefore, it seems most practical to set the final diagnostic dose at a concentration preventing all, or most, susceptible larvae from reaching 3rd instar. Due to variability in larval growth rates, this criteria would involve concentrations producing a mean larval weight of 1.0 mg or less. Above this weight, a significant

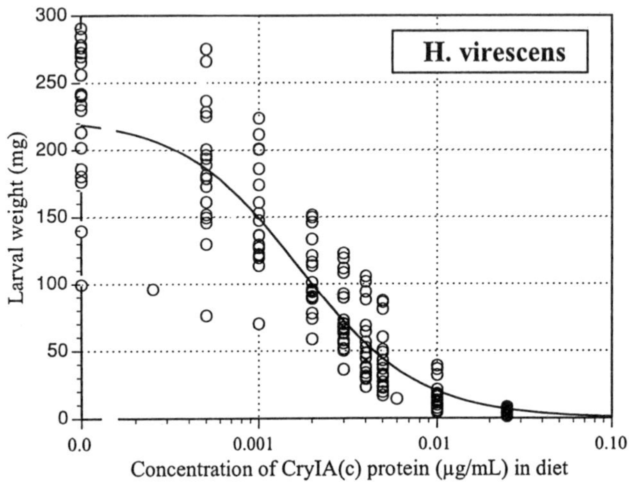

Figure 3. Growth inhibition of *Heliothis virescens* larvae in response to purified 130-kD CryIA(c) protein.

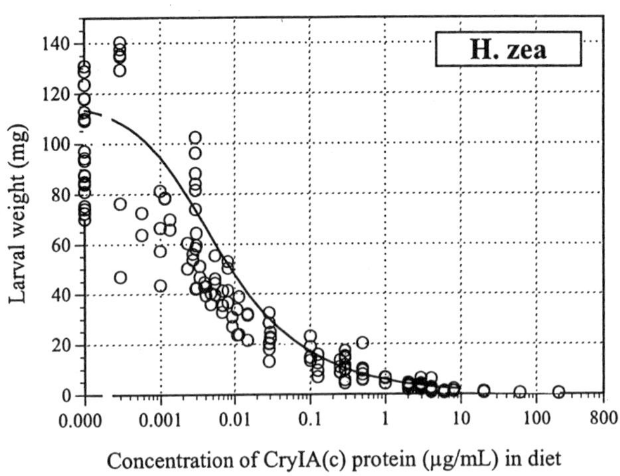

Figure 4. Growth inhibition of *Helicoverpa zea* larvae in response to purified 130 kD CryIA(c) protein.

percentage of susceptible larvae might still reach 3rd instar (Figure 5) and the incidence of false positives would be greater. We note that the proposed diagnostic EC_{99} concentrations for CryIA(c) protein will result in some larval mortality for both *H. virescens and H. zea* (*14,22*, Figs. 1 and 2). This would not reduce the efficiency of the growth assay because both dead and stunted larvae would be correctly classified as susceptible.

Additional Sampling Considerations

Species Identification. The two species involved, *H. virescens* and *H. zea*, are not equally susceptible to the CryIA(c) protein and require significantly different diagnostic doses. Larvae will therefore need to be identified before being placed on the appropriate test diet concentration. In contrast, *H. armigera* and *Helicoverpa punctigera* (Wallengren) in Australia have approximately equal susceptibility and can potentially be monitored using a single dose (*20*).

Subtle morphological differences between eggs of *H. virescens* and *H. zea* are not sufficiently consistent to provide reliable field identification (*23,24*). Species identification would require hatching and additional larval development. Larval characteristics do not allow reliable species discrimination between *H. zea* and *H. virescens* before the 3rd instar. The 3rd and later instars of *H. virescens* have a large retinaculum ("tooth") on the inner side of the mandible and short spines present on tubercles located on the dorsum of the 8th abdominal segment whereas *H. zea* larvae lack both the retinaculum and the spines (*25*). In addition to morphological characters for species identification, immunoassay test kits are currently being developed for differentiating between *H. zea* and *H. virescens*. Similar test kits (LepTon) based on monoclonal antibodies to species-specific lipophorins have been developed by Abbott Labs to reliably distinguish eggs and larvae of *H. armigera* from *H. punctigera*. Unfortunately, insects sampled and killed for species identification are unavailable for bioassay.

Sampling. Collection of eggs from host plants requires much effort but may fail to provide adequate sample material. Larvae could be collected on non-transgenic cotton but since both *H. zea* and *H. virescens* are polyphagous, other preferred host plants should not be overlooked. For example, in some locations sampling *H. zea* larvae from maize and *H. virescens* larvae from soybean or tobacco might be more efficient than collecting larvae from non-transgenic cotton. Collected larvae could complete development on artificial insect diet and the resulting adults mated. Individual pair, rather than mass, matings would be preferable to maximize the effective population sample size. Light trapping might be the best solution for ease of collecting sample material. Light trapping techniques are well-developed and could potentially supply all of the females needed to assess population sensitivity (*26*). Adult females would be collected in light traps, identified to species by wing scale pattern and color, and held for oviposition.

Sample Locations. Limited resources will obviously require that the number of locations sampled be restricted to a small subset of the possibilities. *H. zea* and *H. virescens* are major problems on 4 to 6 million acres of cotton in the United States and transgenic cottons could eventually be grown on a large percentage of this area.

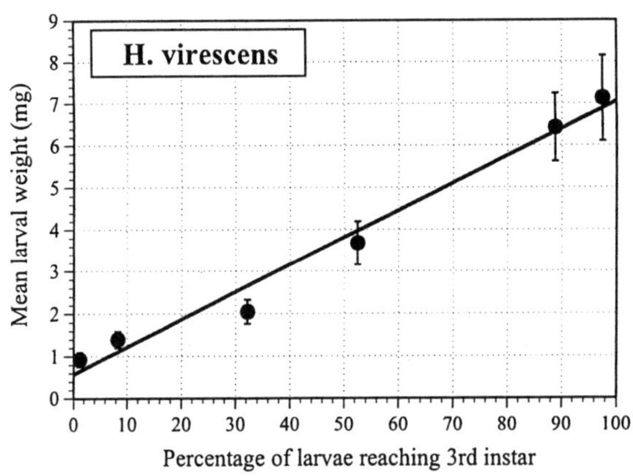

Figure 5. Relationship between *H. virescens* larval weight and percentage of individuals reaching 3rd instar on insect diet.

Sales data could help identify acreage with transgenic plants on both a state and county basis which could, in turn, be used to prioritize sampling locations. In addition to routine monitoring, intensive scouting of transgenic cotton fields might identify situations where numbers of larvae and feeding damage on transgenic plants is unusually high. After verifying that the transgenic plants involved are producing Bt protein, larvae should be collected, identified, and subsequently tested for Bt protein susceptibility.

Bioassay Techniques and Source(s) of Bt Protein Added to Diet. We found that lyophilized plant tissue containing a known concentration of CryIA(c) protein could readily be added to insect diet to provide a diagnostic dose for *H. virescens*. Because *H. zea* requires a significantly greater diagnostic dose concentration, plant powder cannot be used because of growth inhibition effects caused by gossypol and other cotton allelochemicals. Therefore, purified or partially purified protein would be required. This requirement might be met by using commercial CryIA(c)-containing microbial products such as MVP (*4*) or other transgenic microbes expressing only the CryIA(c) protein. We have used a low gelling point agar (Serva Feinbiochemica GmbH & Co. KG, Heidelberg), workable at temperatures of from 50 to 55°C, to avoid denaturing proteins. An inexpensive, soybean-based, pre-mixed diet (Southland Products, Lake Village, AK), has been convenient for testing *H. virescens, H. zea* and many other species of Lepidoptera.

Validation of the Diagnostic Dose Against CryIA(c) -Resistant *H. virescens*

Larvae from a North Carolina strain of *H. virescens* (YHD2) selected for > 1000-fold resistance to CryIA(c) protein (*4*) were used to validate the concept of a diagnostic dose in combination with a larval growth inhibition assay. CryIA(c) protein, within a lyophilized transgenic cotton leaf tissue matrix, was incorporated into insect diet at concentrations of 4, 20, 60, and 80 mg/mL. The concentrations of active CryIA(c) protein in these diets were determined to be approximately 0.24, 1.20, 3.6, and 4.8 ug/mL respectively by insect bioassay and ELISA (*19*). Diets containing appropriate concentrations of leaf tissue from non-transgenic C312 cotton were used as controls for weight comparisons. The results showed that resistant YHD2 larvae developed at a significantly faster rate on all CryIA(c) concentrations compared to larvae from a non-selected susceptible laboratory strain (YDK) (Fig. 6). The mean weight of presumptive heterozygotes for the resistance trait (i.e. YHD2 x YDK and YDK x YHD2) can be distinguished from the mean weight of YDK larvae reared on diet treated with 4 mg/mL of transgenic leaf powder (Fig. 7). However, a detailed analysis of individual growth rates (4) indicated that a significant proportion of susceptible YDK larvae grew at the same rate as presumptive heterozygotes. Nevertheless, the sensitivity of the growth assay and the potential for detecting any resistant heterozygotes significantly increases the probability of detecting resistance while it is still rare (*7*).

Discussion

Larval growth inhibition assays are considerably more sensitive than corresponding dose-mortality assays for detecting incipient changes in *H. virescens* and *H. zea*

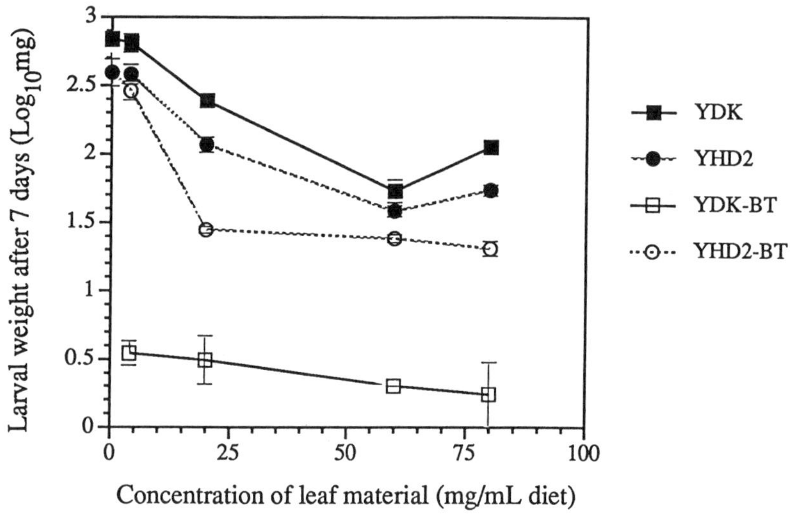

Figure 6. Effect of CryIA(c) protein, in transgenic cotton leaf tissue, on weight gain (± 1 SEM) of susceptible (YDK) and resistant (YHD2) *H. virescens* larvae.

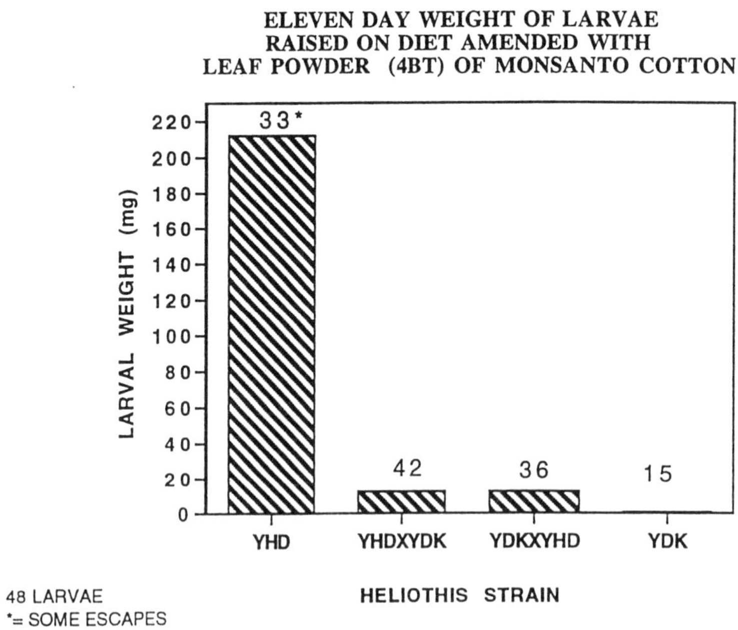

Figure 7. Growth response of CryIA(c) resistant, susceptible and reciprocal F1 hybrid *H. virescens* larvae to a discriminating dose (4 mg dry transgenic cotton tissue per mL diet) of CryIA(c) protein.

susceptibility to the Bt CryIA(c) protein. Since both assay types require ingestion of the insecticidal protein mixed into an insect diet matrix, little additional effort is required to set-up and score growth inhibition tests. Size differences between 1-2nd instar vs. 3rd instar larvae are usually obvious but chances for error in interpretation can be minimized by concurrently testing sample larvae on control diet to provide a direct size comparison. Diagnostic doses, used in combination with larval growth inhibition, are likely to be the most efficient means of tracking population susceptibility, especially when the assay can detect the decreased susceptibility present in resistant heterozygotes. Davidson (*27*) and Georghiou and Taylor (*28*) recognized the importance of diagnostic doses in testing for insecticide resistance and Roush and Miller (*7*) explored the genetic and logistical implications of using diagnostic doses. We estimated diagnostic doses for *H. virescens* and *H. zea* empirically, and suggest that the most practical approach for dose validation is to use individuals sampled from numerous populations within the geographic range of each species.

Dulmage and Martinez (29) were among the first to report that sublethal concentrations of Bt spore-crystal preparations in insect diet inhibit larval growth of *H. virescens*. More recently, Sims and Berberich (*19*) and J. Greenplate (Monsanto Co., unpublished data) demonstrated that extremely low concentrations (< 1 ng protein/mL of diet) of purified CryIA(b) and CryIA(c) proteins can be detected using a *H. virescens* larval growth inhibition assay. Gould et al. (*4*) described the use of a chronic exposure assay in which neonate *H. virescens* larvae were exposed to a sublethal concentration of CryIA(c) protein. This assay allowed differentiation of CryIA(c)-resistant and CryIA(c)-susceptible larvae based on larval weight. Growth inhibition assays are presently being evaluated for monitoring other cotton pest species such as the pink bollworm, *Pectinophora gossypiella* (Saunders) (Watson, T., University of Arizona, unpublished data), and might also be useful for monitoring Lepidoptera species with previously documented Bt resistance (*2*). For example, McGaughey and Beeman (*30*) suggested that the use of a mortality diagnostic dose for monitoring resistance in Indianmeal moth, *Plodia interpunctella* (Hübner), would not be effective until the resistance gene reached high levels. Accordingly, Halliday and Burnham (*31*) demonstrated that the greatest probability of identifying resistance in an Indianmeal moth population would occur with large sample sizes (~ 2000) and a high resistance gene frequency (0.05 to 0.10). Both the required sample size and minimum detectable gene frequency could be reduced by using a more sensitive assay technique coupled with a diagnostic dose. Another application of growth assays could be the analysis of allelic frequencies of resistance prior to field release of transgenic plants. One possible way to do this is to screen populations for individuals surviving to 3rd instar on an approximate EC_{99} concentration. Following transfer to fresh diet, completion of development, and adult mating, resulting progeny would be tested against an appropriate diagnostic dose for the presence of genetic factors having major effects on susceptibility. This approach, based on larval growth, might also be a more useful method for obtaining resistant insect strains compared to selection based on larval survival.

The highly vagile nature of adult *H. virescens* and *H. zea* makes it difficult to interpret estimates of interpopulation variation in Bt susceptibility (*32*). Interpopulation variation in susceptibility (*14*) may therefore reflect non-genetic variation or sampling error rather than genetically fixed geographic differences. For example, repeated bioassays determining LC_{50} and LC_{99} values for Bt proteins against larvae from single strains of the Colorado potato beetle and diamondback moth have demonstrated variability similar to that reported among geographic strains (*14,33*). The maximum LC_{50} and LC_{99} toxicity ratios (highest LC_{50} or LC_{99} divided by the lowest values) for Colorado potato beetle were 12.8 and >150 respectively. Comparable LC_{50} and LC_{99} toxicity ratios for diamondback moth were 3.7 and 10.2 respectively. More than 50% of the LC_{99} estimates differed significantly from the standard minimum value due to within-strain variability alone.

A practical, but often difficult, goal of resistance monitoring is to determine the relationship between laboratory-derived assay results and field control (*34-36*). Strains of insects selected, under laboratory conditions, for resistance to microbial or purified Bt protein preparations often remain susceptible to similar proteins when they are produced in transgenic plants. For example, a Colorado potato beetle (*Leptinotarsa decemlineata*) strain selected for > 60-fold resistance to microbial *B. thuringiensis tenebrionis* (*3*) could not survive as larvae nor reproduce as adults when fed on transgenic potato (*5*). Similarly, a strain of *H. virescens* highly resistant to purified CryIA(c) protein developed slowly on commercial transgenic cotton foliage (Gould, F., North Carolina State University, unpublished data). Conversely, field-selected, apparently homozygous Bt-resistant diamondback moths completed development on transgenic broccoli expressing CryIA(c) protein but F1 heterozygotes, produced by crossing the resistant line with a susceptible strain, did not complete development (*37,38*). Assays on existing laboratory-selected resistant strains might overestimate the potential field importance of these types of resistance. We conclude that it is critical to determine the relationship between resistance intensity and the ability of resistant phenotypes to develop on, and cause damage to, transgenic plant tissues. Without this information, the significance of various levels of resistance, as quantified on insect diet, will be unclear and it will be more difficult to suggest appropriate modifications, in response to resistance, to existing management programs.

Literature Cited

1. McGaughey, W. H.; Whalon, M. E. *Science* **1992**, *258*, 1451-1455.
2. Tabashnik, B. E. *Ann. Rev. Entomol.* **1994**, *39*, 47-79.
3. Whalon, M. E.; Miller, D. L.; Hollingworth, R. M.; Grafius, E. J.; Miller, J. R. *J. Econ. Entomol.* **1993**, *86*, 226-233.
4. Gould, F.; Anderson, A.; Reynolds, A.; Bumgarner, L.; Moar, W. *J. Econ. Entomol.* **1995**, *88*, 1545-1559.
5. Roush, R. T. In *Proceedings of Biotechnology for Integrated Pest Management*; Persley, G., Ed.; CAB International, Wallingford, UK, 1996, In press.
6. Heckel, D. G. *Biocont. Sci. Tech.* **1994**, *4*, 405-417.
7. Roush, R. T.; Miller, G. L. *J. Econ. Entomol.* **1986**, *79*, 293-298.
8. ffrench-Constant, R. H.; Roush, R. T. In *Pesticide Resistance in Arthropods*; Roush, R. T.; Tabashnik, B. E., Eds.; Chapman and Hall, New York, N.Y., 1990; pp. 4 -38.

9. Plapp, F. W. Jr. In *Pesticide Resistance in Arthropods*; Roush, R. T.; Tabashnik, B. E., Eds.; Chapman and Hall, New York, NY, 1990; pp. 237-260
10. Forrester, N. W.; Cahill, M.; Bird, L. J.; Layland, J. K. *Bull. Entomol. Res. Suppl. Ser.*, Suppl. No. 1, CAB International; **1993**, 132 p.
11. Staetz, C. A. *J. Econ. Entomol.* **1985**, *78*, 505-510.
12. Robertson, J. L.; Preisler, H. K. *Pesticide Bioassays With Arthropods.* CRC Press, Boca Raton, FL, 1992; 127 p.
13. Siegfried, B. D.; Marcon, P. C.; Witkowski, J. F.; Wright, R. J.; Warren, G. W. *J. Agric. Entomol.* **1995**, *12*, 257-263.
14. Stone, T. B.; Sims, S. R. *J. Econ. Entomol.* **1993**, 86, 989-994.
15. Wan, L. Ms. Thesis, Mississippi State Univ., Mississippi State, MS, **1995**, 137 p.
16. Reese, J. C.; English, L. M.; Yonke, T. R.; Fairchild, M. L. *J. Econ.Entomol.* **1972**, *65*, 1047-1050.
17. King, E. G. Jr.; Hartley, G. G. In *Advances in Insect Rearing for Research and Pest Management;* Anderson, T. E.; Leppla, N. C., Eds.; Westview Press, Boulder, CO, 1992; pp. 159-172.
18. *SAS/STAT User's Guide: Statistics*; SAS Institute, Cary, NC, 1992.
19. Sims, S. R.; Berberich, S. A. *J. Econ. Entomol.* **1996**, *89*, 247-251.
20. Forrester, N. W.; Forsell, L. *Pacific Rim Conf. on Biotechnology of Bacillus thuringiensis and its Impact on the Environment.* Taipei, Taiwan, **1994**.
21. Scott, J. G.; Roush, R. T.; Rutz, D. A. *J. Agric. Entomol.* **1989**, *6*, 53-64.
22. MacIntosh, S. C.; Stone, T. B.; Sims, S. R.; Hunst, P. L.; Greenplate, J. T.; Marrone, P. G.; Perlak, F. J.; Fischhoff, D. A.; Fuchs, R. L. *J. Invert. Pathol.* **1990**, *56*, 258-266.
23. Bernhardt, J. L.; Phillips, J. R. *Southwest. Entomol.* **1985**, *10*, 236-238.
24. Greenstone, M. H. *J. Econ. Entomol.* **1995**, *88*, 213-218.
25. Neunzig, H. H. *North Carolina Agr. Exp. Stn. Tech. Bull.* 196. **1969**; 76 p.
26. Hartstack, A. W.; Hendricks, D. E.; Lopez, J. D.; Stadelbacher, E. A.; Phillips, J. R.; Witz, J. A. In *Economic Thresholds and Sampling of Heliothis Species on Cotton, corn, Soybeans and Other Host Plants.* Southern Coop. Ser. Bull. 213. **1979**; pp. 105-131.
27. Davidson, G. *Misc. Publ. of the Entomol. Soc. Amer.* **1960**, *2*, 93-94.
28. Georghiou, G. P.; Taylor, C. E. *Proc. 15th Inter. Cong. Entomol., Washington, D. C.*, **1976**, 759-785.
29. Dulmage, H. T.; Martinez, E. *J. Invert. Biol.* **1973**, *22*, 14-22.
30. McGaughey, W. H.; Beeman, R. W. *J. Econ. Entomol.* **1988**, *81*, 28-33.
31. Halliday, W. R.; Burnham, K. P. *J. Econ. Entomol.* **1990**, *83*, 1151-1159.
32. Fitt, G. P. *Ann. Rev. Entomol.* **1989**, 34, 17-52.
33. Robertson, J. L.; Preisler, H. K.; Ng, S. S.; Hickle, L. A.; Gelernter, W. D. *J. Econ. Entomol.* **1995**, *88*, 1-10.
34. Ball, H. J. *Bull. Entomol. Soc. Amer.* **1981**, *27*, 261-262.
35. Denholm, I.; Sawicki, R. M.; Farnham, A. W. In *Proceedings, British Crop Protection Conference, Pests and Diseases, Brighton;* British Crop Protection Council, Croydon, England, 1984; pp. 527-534.
36. Sawicki, R. M. In *Combating Resistance to Xenobiotics. Biological and Chemical Approaches*; Ford, M. G.; Holloman, D. W.; Khambay, B. P. S.;

Sawicki, R. M. Eds.; Ellis Horwood, Chinchester, England, 1987; pp. 105-117.
37. Roush, R. T. *Biocont. Sci. Tech.* **1994**, *4*, 501-516.
38. Metz, T. D.; Roush, R. T.; Tang, J. D.; Shelton, A. M.; Earle, E. D. *Molec. Breed.* **1995**, *1*, 309-317.

Chapter 24

The U.S. Environmental Protection Agency's Role in Pesticide Resistance Management

Sharlene R. Matten[1], Paul I. Lewis[2], Gail Tomimatsu[3], Douglas W. S. Sutherland[4], Neil Anderson[4], and Tobi L. Colvin-Snyder[4]

[1]Environmental Fate and Effects Division (7507C), [2]Special Review and Reregistration Division (7508W), [3]Biopesticides and Pollution Prevention Division (7501W), [4]Biological and Economic Analysis Division (7503W), Office of Pesticide Programs, U.S. Environmental Protection Agency, 401 M Street Southwest, Washington, DC 20460

The U.S. Environmental Protection Agency (EPA) has historically considered pesticide resistance management as an important component of environmentally sound pest management practices. However, EPA does not have an official policy or standard data requirements in place. This paper will consider: (1) how the Agency has considered pesticide resistance management under the Federal Insecticide Fungicide Rodenticide Act (FIFRA) when making emergency exemption decisions (e.g., oxytetracycline), special review decisions (e.g., EBDCs), and registration decisions (e.g., synthetic pyrethroid insecticides, and plant-pesticides producing *Bacillus thuringiensis* endotoxins); and (2) how the Agency is continuing to evaluate and refine the role pesticide resistance management has in the Agency's regulatory decisions.

The problem of pest resistance to pesticides is a worldwide concern. The U.S. Environmental Protection Agency has considered the development of pesticide resistance and pesticide resistance management in its regulatory decisions. With a greater focus on use reduction of the higher risk pesticides, the EPA believes that it is very important to implement effective resistance management strategies.

The development and spread of resistance to pesticides is generally associated with increases in frequency and rate of application of pesticides. To combat resistance, growers often resort to pesticides which may have increased toxicities or, may apply mixtures of pesticides. The pesticide to which resistance has developed can no longer be used effectively against one or more pests, and may be replaced by the use of a pesticide which may pose higher risks to humans or the environment. Preventing or managing resistance is in the user's and public's interests, because it might result in decreased risk. By averting the undesirable effects associated with pesticide resistance, successful resistance management strategies can result in significant pollution prevention and cost savings.

Practical, sound technical strategies must be developed and implemented to manage pesticide resistance. Successful implementation requires cooperation by all those involved in pesticide use: industry, public sector research and extension, user groups, and government regulatory agencies.

How EPA Regulates Pesticides

The EPA regulates pesticides under two major statutory authorities: the Federal Insecticide, Fungicide, and Rodenticide Act (FIFRA) (*1*) and the Federal Food Drug and Cosmetic Act (FFDCA) (*2*). Under FIFRA, EPA has the authority to regulate the development, sale, distribution, use, storage, and disposal of pesticides. To be registered, FIFRA requires that a pesticide will not cause "unreasonable adverse effects" to human health or the environment. EPA determines if a pesticide would cause an unreasonable adverse effect by considering "the economic, social, and environmental costs [risks] and benefits" of the use of the pesticide.

Under FFDCA, EPA is responsible for determining the amount of pesticide residue that is allowable in raw and processed agricultural commodities when they enter commerce. The statute gives broad authority to protect against human dietary risks that might be posed by the use of any pesticide in food for humans, or as feed for animals.

The Role of Pesticide Resistance in EPA Regulatory Decisions

The Historical Role of Pesticide Resistance in EPA Regulatory Decisions. EPA has considered pesticide resistance when making certain regulatory decisions. The Agency has addressed pesticide resistance issues under a number of sections of FIFRA including: Sections 3, 6, and 18.

Pesticide resistance has been a factor in many decisions to grant "emergency exemptions" under Section 18 that allowed use of an unregistered pesticide in an emergency situation where significant economic loss would occur under a non-routine situation. For example, in 1995, greater than 30% of the more than 400 requests for emergency or crisis exemptions under Section 18 were requested, in part, because resistant pest populations have rendered the registered alternatives ineffective.

EPA has also considered pesticide resistance when making determinations of whether unreasonable adverse effects would occur if registered uses of a pesticide are maintained. This determination is a component of the Agency's Special Review process (formerly known as the Rebuttable Presumption Against Registration process).

Historically, pesticide resistance has not been a consideration upon determining whether a new pesticide should be registered. However, beginning in the late 1980s, in specific cases in which pesticide resistance development has been a concern, EPA has worked with some pesticide registrants to develop appropriate pesticide label language to advise pesticide users on ways to avoid or delay the onset of pesticide resistance. Registration labels have included statements related to resistance management that include recommending the use of alternative pesticides if resistance were already a factor. Specific instructions were developed for the use of synthetic pyrethroid insecticides on cotton to reduce the potential for resistance.

The November 23, 1994 Federal Register notice of the Plant-Pesticide Proposed Policy also indicated that the Agency was considering how to best encourage development of agricultural practices that will minimize resistance development to plant-pesticides, i.e., pesticidal substances produced in plants (*3*). For example, plant-pesticides include insecticidal toxins genetically engineered into plants.

Refining the Role EPA Plays in Pesticide Resistance Management. In August 1992, the Assistant Administrator requested that an Office of Pesticide Programs (OPP) workgroup be formed following discussions at OPP's FIFRA Science Advisory Panel meetings and letters from Public Interest Groups regarding potential for development of pesticide resistance to *Bacillus thuringiensis* (Bt) foliar insecticide sprays because of the pending introduction of Bt plant-pesticides. At this time, the Pesticide Resistance Management Workgroup (PRMW) was formed. The PRMW includes scientists from several scientific disciplines. The workgroup considers EPA's role concerning the resistance management of conventional, biological, and genetically-engineered pesticides.

The following list summarizes the PRMW's accomplishments on regulation and policy for pesticide resistance management:

(1) Established a list of appropriate factors to be considered in developing a pesticide resistance management plan. This list was approved by the March 1, 1995 Subpanel on Plant-Pesticides of the FIFRA Science Advisory Panel.

(2) Recommended reporting requirements for incidents of pesticide resistance development that are included in the revision of the adverse effects reporting rule (FIFRA Section 6(a)2 Rule, in draft at the time this chapter was written).

(3) Recommended EPA policy to allow emergency exemptions to be granted under certain conditions for two or more unregistered pesticides for the purpose of avoiding or delaying the buildup of pest resistance to these two compounds, i.e., resistance management. State pesticide regulatory bodies have requested these changes.

(4) Recommended revising EPA policy to include resistance management criteria for issuing special local needs (FIFRA section 24(c)) registrations. EPA proposed a change in policy in the draft guidance for special local needs registrations in which EPA would allow a special local needs registration to avoid or delay the buildup of pest resistance under certain conditions (*4*). State pesticide regulatory agencies have requested pesticide resistance management be part of the guidance document.

(5) Developed criteria for determining when pesticide resistance management plans should be implemented for granting experimental use permits (FIFRA Section 5) and registration of a new active ingredient (FIFRA Section 3). At the time this chapter was written, these criteria were under internal EPA review. The PRMW believes that resistance management should be considered for all pesticides, but the Workgroup is not recommending that data for resistance management or specific labeling be required for all pesticides. A screening process is being discussed to identify classes of pesticides, including new modes of action, with concerns for resistance management.

How EPA Considers Pesticide Resistance In Specific Types of Regulatory Actions

Emergency Exemptions. Section 18 of FIFRA authorizes EPA to allow use of a pesticide that is not registered in an emergency situation. These emergency exemptions may only be requested by state authorities or other federal agencies. The Agency considers an emergency situation as a result of one or more non-routine events: e.g., loss of effective pest controls (registered pesticides or cultural practices), lack of feasible alternative practices, introduction of a new pest, endangerment of public health, and significant economic losses. The granting of an emergency exemption is a temporary privilege; the Agency is not inclined to grant repeated exemptions unless there is evidence of reasonable progress toward registration of the pesticide use, absence of effective pest management alternatives, or where there are other extenuating circumstances. For repeat requests to be authorized, EPA may require applicants to demonstrate that they are attempting to find innovative alternative solutions.

Claims that a pest has developed resistance to a pesticide should have documentation of a consistent, or seasonal loss of efficacy and laboratory verification of resistant pest populations. Adequate documentation may also consist of results from susceptibility tests in which samples from a suspected resistant pest population are collected from an actual use site and a sample from a known susceptible pest population are both treated with the pesticide under the same conditions, and there is a comparative loss of pesticidal susceptibility in the suspected resistant population. Demonstration that laboratory pest populations have developed resistance generally are not considered to provide adequate evidence of pesticide resistance problems in the field, because differences in laboratory and field conditions may result in differences in pest susceptibility to a pesticide. Resistance may also be shown by a series of field tests over a number of years which show an upward trend in the pesticide dosage required for control or a reduction in efficacy (% controlled). EPA recognizes that, in some cases, when the suspected pesticide resistance problem first develops, there may not be sufficient time to gather documentation in the field. In these cases, EPA may grant an emergency exemption based on available information; however, the Agency requires the applicant to provide documentation of resistance if any repeat exemptions are requested. Expert opinion concerning historical perspectives or substantive evidence of loss of the registered pesticides' efficacy may be used in some circumstances.

Approximately 30% of all emergency exemption requests in the last 3 years have been made, at least in part, due to suspected or proven resistance to registered alternatives. Typically, an emergency exemption is granted for use of one pesticide to use as a substitute for the pesticide to which pests have developed resistance. Some examples of recently granted emergency exemptions are: (1) cryolite insecticide to control Colorado potato beetle resistant to chlorinated hydrocarbon, organophosphate, and synthetic pyrethroid insecticides on potatoes in several states; (2) myclobutanil fungicide to control benomyl-resistant *Sphaerotheca macularis* on strawberries in California; (3) oxytetracycline bactericide to control streptomycin-resistant *Erwinia amylovora* which causes fire blight on apples in Michigan, Washington, and Oregon; (4) lactofen herbicide to control paraquat- and diquat-resistant nightshade weeds in tomatoes and peppers in Florida; and (5) quinclorac herbicide to control propanil-resistant barnyard grass in rice in Arkansas.

More recently, EPA has granted emergency exemptions for use of two or more pesticides to manage resistance. Prior to 1985, it was EPA's policy not to grant such requests. EPA, in response to concerns expressed by state regulatory officials, announced a change in policy to allow emergency exemptions for use of two or more pesticides from different chemical classes to manage resistance in 1985 (*5*). In 1992, the Agency further refined this policy (*6*). Emergency exemptions may only be authorized for resistance management in cases where documented pest resistance to the registered alternative(s) has already developed, a pest control emergency exists, currently registered pesticides are ineffective, and a significant economic loss is expected to result. An example of an emergency exemption that has been granted for resistance management is avermectin and cyromazine insecticides to control *Liriomyza* leafminers resistant to organophosphate, carbamate, and synthetic pyrethroid insecticides on tomatoes in Florida.

Special Review Decisions. A second area in which EPA considers pesticide resistance and pesticide resistance management is in the benefits assessments for Special Review decisions (Section 6). A registered pesticide is placed in Special Review when EPA has determined there may be an unreasonable adverse effect associated with use of a pesticide. For example, the Agency may receive new data indicating that a pesticide is a carcinogen. In Special Review, EPA determines if use of a pesticide causes unreasonable adverse effects by weighing the risks to human health and the environment against the benefits.

One example where pesticide resistance played an active role in assessing the benefits during the special review process was for the ethylene bisdithiocarbamate (EBDCs) fungicides. These fungicides include mancozeb, maneb, metiram, and nabam. EBDCs are major agricultural fungicides controlling several important fungal pathogens on over 40 fruit and vegetable crops. There are no reports of pest resistance under field conditions after more than 40 years of use. Upon review of the benefits for EBDCs, the Agency concluded that EBDCs are an important tool in fungicide resistance management. For example, EBDCs in combination with benomyl function in resistance management by controlling apple scab (*Venturia inaequalis*), sooty blotch (*Gloeodes pomigena*) and fly speck (*Shizothyrium pomi*) on apples. EBDCs in combination with copper function in resistance management by controlling bacterial spot (*Xanthomonas vesicatoria*) resistance on peppers and tomatoes. The importance of EBDCs for pesticide resistance management was considered both qualitatively (decrease in fruit quality) and quantitatively (decrease in fruit yields) by EPA in estimating the fungicide's benefits. The uses of EBDCs were maintained on numerous commodities, in part, because of the benefits of EBDCs in fungicide resistance management (*7*).

Registration Decisions. EPA has no formal policy or guidelines on how pesticide resistance management should be considered in making registration decisions. The Agency is currently determining how to refine the role of pesticide resistance and pesticide resistance management in its regulatory decisions for all pesticides. EPA, in cooperation with registrants, has addressed the development of pest resistance and pesticide resistance management through the development of specific use statements on some pesticide labels. In addition, EPA has reviewed several pesticide resistance management strategies that were voluntarily submitted to the Agency by pesticide registrants.

Registration of Conventional Pesticides. One example of industry and EPA voluntary cooperation was the development of risk mitigation measures and use instructions to mitigate the development of resistance for synthetic pyrethroids. The industry (i.e., the registrants) formed a Pyrethroid Working Group, which developed programs that were reviewed by an OPP liaison group (G.D. Thompson and P.K. Leonard, this volume). The immediate issues were: aquatic organism risk mitigation and tobacco budworm resistance management. The synthetic pyrethroids include permethrin, bifenthrin, esfenvalerate, lambda-cyhalothrin, cyfluthrin, cypermethrin, fenproprathrin, zeta-cypermethrin, and tralomethrin. As a result of the Pyrethroid Working Group's efforts, the labels for synthetic pyrethroids include appropriate spray drift mitigation measures, a section on the development of resistance, and language indicating that the use of the product should conform to resistance management strategies established for the local use areas. If resistance is suspected, the label states that products with a similar mode of action, e.g., other synthetic pyrethroids, may not provide adequate control and that the user should consult with the local company representative or agricultural advisor for the best alternative method of control. As a result of these efforts, a tri-state (Arkansas, Louisiana, Mississippi) resistance management plan for cotton insect control has been developed by research and extension entomologists to control tobacco budworm populations.

Other examples of how the PRMW has evaluated the potential for resistance to develop include the following:

(1) The Workgroup has had several discussions with registrants, representatives of the Insecticide Resistance Action Committee, and other groups on resistance management strategies.

(2) The Workgroup has considered the potential for resistance to develop for herbicides being registered for use on herbicide-tolerant crops, specifically, bromoxynil-tolerant cotton.

(3) The Workgroup has reviewed voluntary resistance management plans for conventional chemical pesticides, e.g., metalaxyl fungicide.

(4) The Workgroup has discussed with the registrant the development of appropriate resistance management strategies and label instructions for imidacloprid, a novel insecticide registered in 1994 for the control of Colorado potato beetle on potatoes and for the control of whiteflies on cotton.

Registration of Plant-Pesticides. The PRMW has reviewed plant-pesticide resistance management strategies which have been voluntarily submitted by the registrants. EPA uses the PRMW's reviews of the management plans to make suggestions to registrants to help them improve their management plans, and, when necessary, establish conditions for registration of plant pesticides. The EPA believes that resistance management is critical to the long-term viability of plant-pesticides. For example, if no resistance management plan is implemented for Bt plant-pesticides, it is expected that widespread pest resistance would develop in less than 5 years after transgenic crops have been grown uniformly over large areas following registration. Because the pesticidal proteins in Bt plant-pesticides, Cry delta endotoxins, are also widely used in a variety of Bt foliar spray products on many crops, resistance development to Bt plant-pesticides would also affect efficacy of foliar Bt products.

The PRMW has identified seven elements that need to be addressed to develop an adequate resistance management plan. A subpanel of the FIFRA Science Advisory Panel approved of these seven factors on March 1, 1995. These elements are: (1) knowledge of pest biology and ecology, (2) appropriate gene deployment strategy, (3) appropriate refugia (primarily for insecticides), (4) monitoring and reporting of incidents of pesticide resistance development, (5) employment of IPM, (6) communication and educational strategies on use of the product and (7) development of alternative modes of action. These elements are discussed briefly below:

(1) Knowledge of the pest biology and ecology of the pests involved, e.g., gene flow, mating behavior, flight range, larval movement is needed. Information should be gathered on the primary target pests as well as additional susceptible non-target pests. The obtained information should include a determination of the probability that the pest(s) will develop cross and multiple resistance.

(2) Use of appropriate dose deployment strategies can lower the likelihood of resistance development. The purpose of a particular dose deployment strategy is to reduce the chance selecting resistance alleles to fixation in a breeding population of the target pest.

High dosage expression of genes encoding pesticidal proteins will theoretically eliminate all but rare homozygous resistant individuals. The expectation is that 100 percent of heterozygous individuals will be killed by the high dose as well as homozygous susceptible individuals and that homozygous resistant individuals will be so rare as to be insignificant.

Low dosage expression of a gene encoding a pesticidal protein is a strategy that would allow a percentage of susceptible genotypes to survive and yet limit crop damage. This strategy has several limitations. Resistant and partially resistant genotypes will reproduce at a higher rate than the susceptible pest, leading to rapid loss of efficacy. The amount of allowable, non-economic plant damage is dependent on the crop agronomic practices, and the environment which would restrict the predictability of the Bt control protein. The effective use will require additional pest management strategies that could result in the use of higher risk pesticides.

In tissue-specific expression, the pesticidal protein is produced only in economically important tissues of the plant, such as only in the leaves, but not in pollen. Current technology usually only allows tissue-preferential expression rather than tissue-specific expression. Low expression in some tissues could allow sub-lethal dose selection for resistant pests. In temporal expression, the pesticidal protein is produced only at specific times during the growing season. Both types of expression can limit the total amount of pest exposure to the pesticide.

Constitutive expression provides that the pesticidal protein will be expressed throughout all tissues of the plant. Given adequate expression, no part of the plant should allow survival of the target pest.

Inducible expression could be activated by feeding damage or chemical application. However, wound-inducible expression could allow economic damage before a pesticidal protein reached a controlling level. Chemically inducible expression would require additional grower resources for scouting, purchase and application of the chemical used for induction.

(3) Appropriate refugia may assist in slowing the development of resistance to insecticides by preserving a population of susceptible individuals. Structured refugia coupled to a high dose expression strategy are currently considered key resistance management factors for plant-pesticides producing Bt delta endotoxins. Effective refuges allow survival of susceptible pests and subsequent intermating with individuals that may have been selected for resistance. Therefore, the major contribution of a refuge is in diluting the impact of insecticide-resistant individuals, should they occur. Viable refugia may include: (a) existing non-transgenic crops, (b) weeds present in the cropping system, (c) refuge crops specifically planted by growers, (d) naturally occurring weed hosts, (e) crop rotation, (f) seed mixes of transgenic seed and non-transgenic seed, or (g) block plantings of transgenic seed and non-transgenic seed.

(4) Monitoring of pest populations in treated fields is important to determine the frequency of pest-resistant target insects. Continued resistance monitoring and reporting of pesticide resistance incidents allows for observation of trends in pest susceptibility to pesticides and provides information on whether or not currently followed resistance management practices are mitigating the development of resistance.

(5) Selection of appropriate IPM practices could delay the development of pesticide resistance. Examples could include removal of overwintering habitat and crop rotation with crops not susceptible to the target pest.

(6) Communication and educational strategies on use of the product are critical to successful resistance management. In order for any resistance management strategy to be accepted and successful, the user must be educated on the unique features of using crops expressing pesticidal proteins to help manage resistance development. A multi-level approach to educating the growers should include the development, dissemination, and explanation of technical bulletins, presentations at local grower meetings and state organization meetings, and training seminars at the Regional Level. Pertinent research results should be made available to the users. Material should also describe the expected pest behavior on the plants so that the grower can have adequate knowledge to use the product effectively. The grower must be made aware that additional pesticide use may be unnecessary to control the target pest.

(7) Development of alternative modes of action, when available, is another useful tool for resistance management. Multiple gene and alternate gene strategies are two approaches to use of alternative modes of action. For example, the plant may be engineered to contain two or more genes that each code for a different pesticidal protein. Because a pest is less likely to survive exposure to two different toxins, particularly if the mode of action of these toxins is different, development of resistant populations is less likely than if only one gene is present.

Reviews of resistance management plans that have been completed by the PRMW include: (1) the *Bacillus thuringiensis* (Bt) CryIIIA delta endotoxin produced in potato to control Colorado potato beetle (registered May 1995), (2) the Bt CryIA(b) delta endotoxin produced in field corn to control European corn borer (registered in August 1995), and 3) the Bt CryIA(c) delta endotoxin produced in cotton to control pink bollworm, cotton bollworm, and tobacco budworm (registered October 1995).

When reviewing resistance management plans for Bt plant-pesticides, the PRMW has primarily focused on the ecological factors that might influence the likelihood of resistance occurring which impinge on resistance management options for deploying Bt plants. These factors include the following: impacts cross-resistance may have on microbial Bt sprays, impacts on other non-target lepidopteran species, impacts on beneficial insects, and impacts on minor pests. In addition, because lepidopteran insects will be exposed throughout the growing season to Bt delta endotoxins produced in plants, different resistance management approaches are needed.

Case Study - Summary of EPA Review of the CryIIIA Pesticide Resistance Management Strategy

The Agency granted a conditional unlimited registration of the plant-pesticide *Bacillus thuringiensis* subspecies *tenebrionis* Colorado potato beetle (*Leptinotarsa decemlineata*, CPB) Control Protein in May 1995, the first unlimited registration of a plant-pesticide. A review of the proposed pesticide resistance management strategy for the CryIIIA CPB control protein produced in potatoes was part of the decision process for registering this plant-pesticide. Below, is a summary of the Agency's review of the registrant's resistance management plan for the CryIIIA delta endotoxin produced in potatoes.

The Agency concluded that registrant's plan is a workable pesticide resistance management strategy for CryIIIA delta endotoxin in potatoes because it includes all of the necessary elements to reduce the selection pressure on the target pest, CPB, and therefore the likelihood for the rapid development of resistance. Specifically, the Agency and the subpanel of the FIFRA Science Advisory Panel (meeting held March 1, 1995) indicated that the most important elements of a resistance management plan are selection of an appropriate expression system (e.g., constitutive, high dose expression strategy), refugia, monitoring, and education (including IPM). The components of registrant's plan are: integration of agronomic and other pest management strategies (IPM), monitoring of CPB populations for resistance, high dosage expression of the protein, refugia as hosts for susceptible insects, development of novel CPB control proteins, and user education.

The Agency and the Science Advisory Panel agreed that registrant's resistance management strategy for the potato variety expressing the CryIIIA delta endotoxin, although adequate for the present, should be further refined in the future as additional data is made available. Many of the specific questions with respect to monitoring for resistance development and strategies to retard resistance development can best be evaluated when the potatoes expressing the CryIIIA delta endotoxin have been in commercial production for several years. This situation is no different from any resistance management strategy developed for any pesticide.

The Agency recommended that the registrant continue to voluntarily work with EPA on further development and refinement of the resistance management strategy for the CryIIIA delta-endotoxin expressed in potatoes to control the development of CPB resistance as additional data becomes available. The Agency recommended the following specific areas for further development:

(1) Information concerning reproductive strategies of CPB with respect to gene flow, particularly regarding adult movement, larval movement, behavioral responses including mating studies.

(2) Refugia strategies.

(3) A specific monitoring plan which should include educating the growers in regular monitoring for resistant individuals, eradication procedures if resistant individuals are detected, appropriate sampling procedures, and development of a discriminating dose assay.

(4) Education strategies for the users on appropriate use of this product.

(5) IPM recommendations at the local level. The Agency recommends that crop rotation and other cultural practices be employed to prevent replanting transgenic potatoes in or adjacent to the same fields year after year.

(6) Novel CPB control mechanisms with different modes of action.

Conclusions

The U.S. Environmental Protection Agency considers pesticide resistance management important in determining environmentally sound pest management practices. The Agency has considered the development of pest resistance and pesticide resistance management under FIFRA when making emergency exemption decisions, special review decisions, and registration decisions. The Agency is continuing to evaluate and refine the role pesticide resistance management has in the Agency's regulatory decisions.

The Workgroup believes that pesticide resistance management strategies should be developed in cooperation with a number of partners including industry, USDA, user groups, and academic authorities. These strategies must be flexible enough to adapt to additional data as it is gathered, and tailored to specific geographic regions, the specific pest-crop-pesticide complex, or other circumstances.

Acknowledgements

We are grateful to the other members of the PRMW for their valuable discussions on the topics discussed here: Laurel Celeste, Pat Cimino, Leonard Cole, Frank Ellis, Mary Beth Gleaves, Phil Hutton, Robert Rose, Dennis Szuhay, and Steven Tomasino. In addition, we appreciate the review that Janet Andersen (Acting Director, Biopesticides and Pollution Prevention Division) has given this manuscript.

The views expressed in this article are those of the authors and do not necessarily represent those of the United States government.

Literature Cited

1. U.S. Congress. 1988. Federal, Insecticide, Fungicide, and Rodenticide Act, as amended. 7 U.S.C 136 *et seq.*
2. U.S. Congress. 1988. Federal Food Drug and Cosmetic Act, as amended. 21 U.S.C. 321 *et seq.*
3. U. S. Environmental Protection Agency (EPA). 1994. Plant-Pesticides Subject to the Federal Insecticide, Fungicide, and Rodenticide Act (FIFRA) and the Federal Food, Drug, and Cosmetic Act (FFDCA); Proposed Policy, *Federal Register* **59**: 60496, November 23, 1994.
4. U.S. EPA. 1994. Draft Guidance on 24(c) Registrations. December 7, 1994.
5. U. S.EPA. 1985. Exemption of Federal and State Agencies for Use of Pesticides Under Emergency Conditions; Proposed Rules, *Federal Register* **50:** 13948, April 8, 1985.
6. U. S. EPA. 1992. Emergency Exemptions Under Section 18 of the Federal Insecticide Fungicide and Rodenticide Act: I. Guidance for State and Federal Agencies, June 22, 1992.
7. U.S. EPA, 1992. Ethylene Bisdithiocarbamates (EBDCs); Notice of Intent to Cancel and Conclusion of Special Review. *Federal Register* **57:** 7484, March 2, 1992.

INDEXES

Author Index

Affiliation Index

Subject Index

V

W

X